COAL PETROLOGY
ITS PRINCIPLES, METHODS, AND APPLICATIONS

by

R.M. Bustin[1], A.R. Cameron[2], D.A. Grieve[3], and W.D. Kalkreuth[2]

Geological Association of Canada
Short Course Notes
Volume 3

Victoria, British Columbia
May 8-10, 1983

Sponsored by the Geological Association of Canada

[1]Department of Geological Sciences, University of British Columbia, Vancouver, British Columbia, V6T 2B4
[2]Institute of Sedimentary and Petroleum Geology, Calgary, Alberta, T2L 2A7
[3]B.C. Ministry of Energy, Mines and Petroleum Resources, Fernie, British Columbia, V0B 1M0

First printing - May, 1983
Second revised edition - Spring, 1985

These notes were designed primarily for the use of short course participants at the 1983 Victoria GAC Annual Meeting and for others who desire an introduction to the subject. The volume has not been openly reviewed, and only minimal editing has been done. The authors retain responsibility for the contents. Use of trade and brand names does not imply endorsement.

Canadian Cataloguing in Publication Data

Main entry under title:
Coal petrology

(Geological Association of, Canada short course notes ; v. 3)
Bibliography: p.
ISBN 0-919216-28-5

1. Petrology. 2. Coal - Geology. 3. Geology, Economic. I. Bustin, R. Marc. II. Geological Association of Canada. III. Series.

QE431.5.C62 1985 552 C85-098682-6

International Standard Book Number: 0-919216-28-5

The GEOLOGICAL ASSOCIATION OF CANADA is Canada's national society for the geosciences. It was established in 1947 to advance geology and its understanding among both professionals and the general public. The GAC membership of 3,000 includes representative of all geological disciplines from across Canada and many parts of the world employed in government, industry and academia. There are specialist dvisions for environmental earth sciences, geophysics, mineral deposits, paleontology, Precambrian, sedimentology, tecontics, and volcanology. Regional sections of GAC have been set up in Victoria, Vancouver, Edmonton, Winnipeg, and St. John's, and there are affiliated groups in Toronto and the Maritimes.

GAC activities include the organization and sponsorship of conferences, seminars, short courses, field trips, lecture tours, and student and professional awards and grants. The Association publishes the quarterly journal *Geoscience Canada* and the quarterly newsletter *Geolog*, a *Special Paper* series, Short Course Notes, and several other continuing series. GAC also maintains liaison with other earth science societies and provides advice to government and the public on geological issues. The Association was incorporated under the Canada Corporations Act in January, 1984. For information contact: Geological Association of Canada, Department of Earth Sciences, Memorial University of Newfoundland, St. John's, Newfoundland, A1B 3X5, Canada.

Geological Association of Canada
Department of Earth Sciences
Memorial University of Newfoundland
St. John's, Newfoundland
Canada A1B 3X5

Printed by: Jasper Printing, Edmonton

TABLE OF CONTENTS

1	INTRODUCTION	1
1.1	Historic Overview	1
2	ORIGIN OF COAL	7
2.1	Evolutionary Development of the Flora	7
2.2	Climate	8
2.3	Depositional Environment	9
2.3.1	Sedimentology of Coal Measures	9
2.3.1.1	Paralic Coal Deposits	9
2.3.1.1.1	Back-Barrier Coals	9
2.3.1.1.2	Deltaic Coals	10
2.3.1.1.3	Interdeltaic and Coastal Plain Coals	14
2.3.1.2	Limnic Coal Measures	14
2.3.2	Sedimentation of Coal	15
2.3.2.1	Type of Deposition	15
2.3.2.2	Plant Communities	15
2.3.2.3	Depositional Setting: pH, Sulphur, Bacterial Activity and Nutrient Supply	18
2.3.2.4	Redox Potential	19
2.4	Origin of Sapropelic Coals	20
2.5	Coalification	21
2.5.1	Biochemical Stage of Coalification	21
2.5.2	Geochemical Stage of Coalification	23
2.5.3	Causes of Coalification	30
3	SYSTEMS OF COAL RANK CLASSIFICATION	31
4	PETROGRAPHIC COMPOSITION OF COAL	35
4.1	Nature of Macerals	35
4.1.1	Petrographic Properties	36
4.1.1.1	Huminite/Vitrinite Group	39
4.1.1.2	Liptinite Group	39
4.1.1.3	Inertinite Group	48
4.2	Lithotypes and Microlithotypes	49
4.2.1	Lithotypes	49
4.2.1.1	Definition	49
4.2.1.2	Stopes-ICCP System	49
4.2.1.3	Other Lithotype Approaches	54
4.2.1.4	Lithotypes in Low Rank (Lignites-Subbituminous)	59
4.2.2	Microlithotypes	61
4.3	Mineral Matter in Coal	63
4.3.1	Clays	67
4.3.2	Carbonates	67
4.3.3	Iron Sulphides	67
4.3.4	Silicas	71
4.3.5	Other Mineral Groups	71
4.3.6	Trace Elements in Coal	71
5	OXIDATION	73
5.1	Petrography	73
5.2	Depth of Oxidation	74
5.3	Predicting Fresh Coal Characteristics	74
5.4	Spontaneous Combustion of Coal	75

6	SAMPLING AND SAMPLE PREPARATION	81
6.1	Sampling Problems and Methods	81
6.2	Sample Preparation	82
6.2.1	Blocks	82
6.2.2	Pellets	84
6.2.3	Thin Sections	85
6.2.4	Kerogen	86
6.2.5	General Remarks	86
7	PETROGRAPHIC METHODS	89
7.1	Reflectance of Coal Macerals	89
7.1.1	Optical Characteristics of Vitrinite	89
7.1.2	Instrumentation	91
7.1.3	Preparation and Setting Up	92
7.1.4	Measurement Procedure	93
7.1.5	Choosing Vitrinite for Measurement	93
7.1.6	Calculating and Presenting Data	94
7.2	Random Point Count Analysis — Macerals and Microlithotypes	94
7.2.1	Methods	94
7.2.2	Point Counting	96
7.2.3	Statistical Evaluation	97
7.3	Fluorescence Microscopy	98
7.3.1	Quantitative Fluorescence Photometry	102
7.3.1.1	The Fluorescence Microscope	102
7.3.1.2	Sample Preparation	104
7.3.1.3	Monochromatic Fluorescence Photometry	104
7.3.1.4	Spectral Fluorescence Photometry	104
7.3.1.5	Alteration	107
7.3.1.6	Computerization	107
7.3.2	Applications of Fluorescence Microscopy	109
7.3.2.1	Application in Coal Petrology	109
7.3.2.2	Application in Kerogen Studies	109
7.3.2.3	Spectral Fluorometric Analysis of Soluble Organic Matter (Extract)	109
7.3.2.4	Fluorescence Properties of Vitrinites and Their Relation to Technological Properites of Coals	111
7.3.2.5	Characterization of Solid Bitumens	114
8	RELATIONSHIP OF PETROGRAPHIC PROPERTIES TO OTHER COAL PROPERTIES	117
8.1	Vitrinite Reflectance and Coal Rank	117
8.2	Relationship of Coal Petrography to Chemical Parameters	118
8.3	Relationship of Coal Petrography to Rheological Properties	121
8.4	Application of Petrography to Classification	123
9	GEOLOGICAL APPLICATIONS	127
9.1	Seam Correlation	127
9.1.1	Macroscopic Criteria	127
9.1.2	Microscopic Criteria	129
9.1.3	Formation Correlation	133
9.2	Interpretation of Geological Structure	135
9.2.1	Timing of Coalification and Structural Events	135
9.2.2	Structural Interpretation Based on Timing of Coalification	141
9.2.3	Other Applications	142
9.3	Applied Coal Petrography — Environments of Coal Accumulation	142
9.3.1	Microlithotype Environmental Interpretations	145
9.3.2	Lithotype Environmental Interpretations	147
9.4	Application of Coal Petrology to Geothermal and Basinal Studies	148
9.4.1	Time-temperature Coalification Models	149
9.4.2	Geothermal Gradient Determinations	152
9.4.3	Paleoisotherms and Isoreflectance Lines	155
9.4.4	Magmatic Studies	156

10	TECHNOLOGICAL APPLICATIONS	159
10.1	Coal Preparation	159
10.1.1	Principals of Coal Preparation	159
10.1.2	Application of Coal Petrology	160
10.2	Combustion	162
10.3	Carbonization	167
10.3.1	Coke Quality	167
10.3.1.1	Chemical Properties	168
10.3.1.2	Physical Properties	169
10.3.2	Qualities of Coking Coal	171
10.3.3	Petrographic Methods for Calculation of Coking Properties	174
10.3.3.1	Strength	174
10.3.3.1.1	Method of Ammosov *et al*. and Schapiro *et al*.	174
10.3.3.1.2	Method of Benedict *et al*.	178
10.3.3.1.3	Method of Mackowsky and Simonis	185
10.3.3.1.4	Method of Brown *et al*.	187
10.3.3.1.5	Method of Steyn and Smith	187
10.3.3.2	Relationship of Petrography to Other Coking Parameters	189
10.4	Liquefaction	196
10.4.1	Principles and Methods	196
10.4.2	Influence of Rank and Petrographic Composition of Coals on Conversion	199
10.4.3	Characterization of Residual Material	200
10.5	Oil and Gas Exploration (Kerogen Analysis)	204
10.5.1	Determination of Vitrinite Reflectance in Kerogen	207
10.5.2	Examples of Application	207
10.5.3	Petrographic Classification of Organic Matter in Sedimentary Rocks	212
REFERENCES		217

LIST OF TABLES

Table 1. List of Canadian coal petrology laboratories. 5

Table 2. Comparison between Carboniferous, Tertiary, and modern peat-forming floras and type of resulting coal. 8

Table 3. Effects of oxygen availability on transformation of peat. 20

Table 4. ASTM classification of coals by rank. 31

Table 5. NCB (Britain) coal classification system. 33

Table 6. Correlation of the huminite macerals of brown coals and lignites with the vitrinite macerals of bituminous coals. 35

Table 7. Macerals of liptinite and inertinite groups. 35

Table 8. Coal macerals: classification, origin, and properties. 36

Table 9. Boundaries for thickness of vitrain bands and frequency of occurrence in Schopf classification. 55

Table 10. The correlation between microlithotypes after Stopes and those distinguished in N.S.W. (Australian) coals. 56

Table 11. CSIRO (Australia) terms for the macroscopic description of coal. 57

Table 12. Distribution of lithotypes in some coking coals of the Illawarra Coal Measures, Australia. ... 58

Table 13. Nominal maceral contents of various lithotypes distinguished in some coals of Illawarra Coal Measures, Australia. ... 59

Table 14. Calculated and analysed maceral contents in samples of Wongawilli and Bulli seams, Illawarra Coal Measures, Australia. ... 59

Table 15. Macropetrographic classification of lignite. ... 60

Table 16. Microlithotype classification. ... 62

Table 17. Minerals in coal. ... 66

Table 18. Results of experiments showing change in setting time relative to different Paraplex mixtures. ... 83

Table 19. Variation in reflectance results ($\bar{R}_o max$) obtained by 11 Canadian labs. ... 94

Table 20. Range of reflectance results ($\bar{R}_o max$) obtained by 11 British labs. ... 94

Table 21. Microlithotype analysis using a 20-point graticule. ... 97

Table 22. Maceral analysis from point counting of a low-volatile bituminous coal; Canadian Coal Petrography Group. ... 98

Table 23. Fluorescence colours and intensities for the different maceral groups at different coal ranks using blue light excitation. ... 99

Table 24. Suggested limits of ASTM rank classes in terms of vitrinite reflectance. ... 117

Table 25. Classification of coal types used on Figure 56. ... 127

Table 26. Calculation of the interval maturation and total maturation of horizon A. ... 152

Table 27. Petrographic components and mass percentages in the different size grades. ... 162

Table 28. Calculated petrographic component recovery and efficiency of separation by sizes. ... 163

Table 29. Calorific values for maceral groups. ... 164

Table 30. Combustible content of fly-ash samples. ... 165

Table 31. Analyses of coals. ... 166

Table 32. Petrographic analyses. ... 166

Table 33. Proposed limits for various chemical components of blast-furnace burdens including ore, stone, and coke. ... 168

Table 34. Rating of coking coals for blending. ... 172

Table 35. Summary of coke prediction methods. ... 173

Table 36. Procedure of Bethlehem Steel for grouping reactives and inerts. ... 179

Table 37. Fluidity (Gieseler) and FSI data on whole coal and vitrain samples. ... 193

Table 38. Qualitative behaviour of coal constituents during the liquefaction process. ... 201

Table 39. Relationships between catagenesis of coal and dispersed solid organic matter and occurrence of oil and gas deposits. 208/209

Table 40. Petrographic classifications of organic constituents in rocks including oil-source rocks. 216

LIST OF PLATES

Plate 1. Changes in the petrographic appearance of coal with increasing rank. 27

Plate 2. Petrographic appearance of cannel and boghead coals. 29

Plate 3. Macerals of the huminite/vitrinite group. 41/43

Plate 4. Macerals of the liptinite group. 45

Plate 5. Macerals of the inertinite group. 47

Plate 6. Macroscopic appearance of coal lithotypes. 51

Plate 7. Representative photomicrographs of microlithotypes. 65

Plate 8. Mineral matter in coal. 69

Plate 9. Mineral matter in coal. 70

Plate 10. Photomicrographs of oxidized coal. 77

Plate 11. Coal stained with safranin "O." 79

Plate 12. High reflecting organic material (OM) in whole rock (A) and in concentrate (B) from a Devonian shale, western Germany. 87

Plate 13. Macerals of the liptinite group from various rank levels after blue light excitation. 101

Plate 14. Macroscopic and microscopic views of folded coal from the Groundhog Coalfield, British Columbia. 143

Plate 15. Photomicrographs of coke. 181

Plate 16. Photomicrographs showing influence of coal rank on structure of coke walls. 195

Plate 17. Examples of residual materials obtained from liquefaction of peat and subbituminous coal. 203

Plate 18. Micropetrological features of kerogen. 215

LIST OF FIGURES

Figure 1. Back-barrier depositional models for coal deposition. 10

Figure 2. Development of a delta sequence showing sites of peat accumulation. 11

Figure 3. Distribution of peat deposits on the Fraser River Delta. 12

Figure 4. Peat deposit of the distal lower delta plain at Boundary Bay. 13

Figure 5. Peat deposits of the lower delta plain at Lulu Island. 13

Figure 6. Successional sequence of plant biofacies, Fraser River Delta. 16

Figure 7. Open sedge grass marshes, Fraser River Delta at Lulu Island (from Styan, 1981). 17

Figure 8. Salt marsh peat, Fraser River Delta at Boundary Bay. 17

Figure 9. Sulphur content of brackish marine peats at Boundary Bay (A) and freshwater peats at Pitt Meadows (B), Fraser River Delta. 19

Figure 10. Stages of coalification based on physical and chemical properties of the coal. 22

Figure 11. Changes in the chemical and physical properties of coal with increasing rank. 24

Figure 12. Van Krevelen diagram showing the coalification tracks of liptinite, inertinite, and huminite-vitrinite. 25

Figure 13. Variation of various chemical and physical parameters with rank. 32

Figure 14. Grey levels of liptinite, vitrinite, and inertinite macerals in a bituminous coal; l = liptinite, v = vitrinite, i = inertinite. 37

Figure 15. Changes in reflectances of liptinite, vitrinite, and inertinite macerals with increasing rank (modified from Alpern and Lemos de Sousa, 1970). 37

Figure 16. Comparison of etched (right) and unetched (left) portion of a vitrinite (from Stach, 1982). 38

Figure 17. Macroscopic profile of Dekoven seam (Pennsylvanian age) from southern part of Illinois Basin (from Cameron, 1978). 54

Figure 18. Macroscopic log of coal seam prepared according to method of Goscinski *et al*. 56

Figure 19. Macroscopic profile of "D" coal zone, Hat Creek Coalfield, British Columbia, described according to Australian system. 57

Figure 20. Relationship between vitrain and vitrinite contents of petrographic intervals in No. 10 seam, Sparwood, British Columbia (from Cameron, 1978). 58

Figure 21. Macroscopic profile of lignite bed (Souris seam), Estevan area, Saskatchewan. 61

Figure 22. Diagrammatic representation of microlithotype classification shown in Table 16. 62

Figure 23. No. 6 seam adit. 75

Figure 24. C-H-O diagram showing the shifts in elemental composition of coal with progressive oxidation. 80

Figure 25. Anisotropy of vitrinite. 90

Figure 26. Schematic diagram of the components of the reflectance microscope. 92

Figure 27. Typical example of reflectance frequency histogram. 95

Figure 28. 20-point graticule used for microlithotype analysis. 96

Figure 29. Schematic diagram illustrating apparatus for spectral fluorescence photometry in reflected light, modified after International Handbook of Coal Petrography (1975). 102

Figure 30. Function of beam-splitting mirror (modified after Cook, 1980). 103

Figure 31. Example for procedure of correction of fluorescence spectra (Table) and illustration of sample curve, background curve, and final corrected curve for the maceral resinite in a lignite from Hat Creek, British Columbia. 105

Figure 32. Changes of fluorescence spectra of sporinite with increase of rank (after Ottenjann *et al.*, 1974). 106

Figure 33. Relation between volatile matter (maf) and red/green quotient of sporinites at different rank levels (modified after Ottenjann *et al.*, 1974). 107

Figure 34. Alteration of bituminous groundmass at different rank levels (modified after Teichmüller and Ottenjann, 1977). 108

Figure 35. Fluorescence microscope, computer (Apple II), monitor, printer, and plotter used at ISPG for spectral fluorescence photometry. 108

Figure 36. Relationship between vitrinite reflectance and fluorescence properties of alginite, sporinite, and cutinite (modified after Ottenjann, 1982). 110

Figure 37. Variations of rank parameter with increasing depth (after Hagemann and Hollerbach, 1983). ... 111

Figure 38. Variation of fluorescence parameters of crude oils (after Hagemann and Hollerbach, 1983). 112

Figure 39. Proportion of fluorescent vitrinite (% of total vitrinite) versus mean vitrinite reflectance for coal samples of two wells in western Germany (well Kohlhaus and well Uedem 1, circles and triangles, respectively). 113

Figure 40. Relation between rank (% Rm vitrinite reflectance) and fluorescence intensities of vitrinites at 546 nm (after Ottenjann *et al.*, 1981). 113

Figure 41. Relation between softening point and vitrinite reflectance (A) and fluorescence intensities of vitrinite (B). After Ottenjann *et al.* (1981). 114

Figure 42. Relation between Swelling Index and the red/green quotient of Carboniferous coals (after Ottenjann *et al.*, 1981). 114

Figure 43. Genetic classification of solid bitumens (after Jacob, 1976). 115

Figure 44. Bitumen fluorescence intensities versus random reflectance (modified from Jacob, 1976). 116

Figure 45. Relationship of volatile matter and reflectance of vitrinite from European coals. 118

Figure 46. Relationship between vitrinite reflectance ($\bar{R}_o m$) and volatile matter, hydrogen and carbon contents, and the ratio hydrogen/carbon (from Teichmüller and Teichmüller, 1982). 119

Figure 47. Volatile matter contents of Australian coals as a function of vitrinite reflectance and maceral composition (from Strauss *et al.*, 1976). 120

Figure 48. Hydrogen content of Australian coals as a function of vitrinite reflectance and maceral composition (from Strauss *et al.*, 1976). 120

Figure 49. Free swelling index of western Canadian coals as a function of vitrinite reflectance and inert content (from Pearson, 1980). 122

Figure 50. Maximum dilatation of western Canadian coals as a function of vitrinite reflectance and inert content (from Pearson, 1980). 122

Figure 51. Maximum Gieseler fluidity of western Canadian coals as a function of vitrinite reflectance and inert content (from Pearson, 1980). 123

Figure 52. Coking coal groups and international coals plotted as a function of vitrinite reflectance and inert contents (from Pearson, 1980). 124

Figure 53. Predicted maximum Gieseler fluidity of coals of the Fernie Basin, southeast British Columbia, based on samples collected in the field. 125

Figure 54. Coking coal classification based on vitrinite reflectance and dilatation (Marshall, 1976). 126

Figure 55. Coking coal classification based on vitrinite reflectance and Gieseler fluidity (Okuyama *et al.*, 1970). 126

Figure 56. Distribution of lithotypes in intervals of column samples from the southern Illinois Basin (from Cameron, 1978). 128

Figure 57. Stratigraphic cross-section of lithotypes of the Harbour seam in the Princess and Florence collieries, Sydney Coalfield, Nova Scotia (from Cameron, 1971). 129

Figure 58. Lateral extent of tonstein bands in western Europe (ICCP, 1963). 130

Figure 59. Maceral composition of Kootenay Group coals from southeast British Columbia, based on relative position within the group. 131

Figure 60. Panel diagram showing vertical distribution of Kootenay Group seams with high and low vitrinite content (from Cameron, 1972). 132

Figure 61. Microlithotype composition of several Australian coal seams from various locations (from Taylor and Warne, 1960). 132

Figure 62. Vitrinite reflectance and microlithotype composition of several Australian coal seams from various locations (from Taylor and Warne, 1960). 133

Figure 63. Vitrinite content profiles of the Bulli seam, Sydney Basin, New South Wales (from Shibaoki and Smyth, 1975). 134

Figure 64. Maceral content profiles of the Balmer seam, Fernie Basin, southeast British Columbia (from Cameron and Babu, 1968). 134

Figure 65. Composition profiles of the Tracy seam, Sydney Coalfield, Nova Scotia (from Hacquebard, 1952). 135

Figure 66. Idealized relationships between folded coal seams and isorank lines in cross-section in cases of: (a) pre-folding; (b) post-folding; and (c) pre- and post-folding (or syntectonic coalification) (from Pearson and Grieve, in press). 136

Figure 67. Cross-section through the Bochum megasyncline (Ruhr Basin) showing the pre-tectonic 22% isovol approximately parallel to the Sonnenschein seam (from Teichmüller, 1982). 137

Figure 68. Cross-section of inclined coal seams in Sydney Coalfield, Nova Scotia, with horizontal (post-tectonic) isoreflectance lines (from Hacquebard and Donaldson, 1970). 137

Figure 69. Cross-section of Coal Creek Mountain, Fernie Basin, southeast British Columbia, with reflectance values on four coal seams in the Mist Mountain Formation (Kootenay Group). 138

Figure 70. Comparison of pre- and post-folding coalification gradients and total coalification gradient for 10-seam (Balmer) at Michel and B-seam at Coal Creek in the Fernie Basin, southeast British Columbia (from Pearson and Grieve, in press). 138

Figure 71. Idealized rank relationships in hangingwall (h.w.) and footwall (f.w.) of major faults: a - normal fault, coalification pre-faulting; b - thrust faulting, coalification pre-faulting; c - normal fault, coalification post-faulting; and, d - thrust fault, coalification post-faulting. 139

Figure 72. Isoreflectance lines in the Paleozoic in the region of the Ostsauerland anticline, Germany. A major break in the pattern of the isoreflectance lines coincides with the Altenburener lineament. 140

Figure 73. Decrease in reflectance with increasing depth in overturned strata in the Peace River Coalfield, British Columbia (Kalkreuth, unpublished). 141

Figure 74. Thrust fault in an area of pre-faulting coalification detected in borehole by a repetition of rank values (calorific values in this case). 141

Figure 75. Coal and moor types and resulting coal microlithotypes (modified from Teichmüller, 1962 in Teichmüller, 1982). 144

Figure 76. Variation in microlithotypes and macerals in a typical seam in the Sydney Coalfield (modified from Hacquebard *et al.*, 1967). 146

Figure 77. Coal facies in the Lloyd Cove seam, Nova Scotia (modified from Hacquebard and Donaldson, 1969). 147

Figure 78. Development of miospore phases in relation to formation of a split in a coal seam. 148

Figure 79. Seam formation curve after the methods of Tasch (1960). 149

Figure 80. Relationship between coal rank (as determined by reflectance) and time of coalification and temperature (modification of Karweil's nanogram (1956) using Teichmüller's 1971 reflectance values). 150

Figure 81. Application of Lopatin's (1971) method to evaluating the maturation (coalification) history of a basin. 151

Figure 82. Rank gradient ($R_o$max/km) and isopach map showing original thickness (in metres) of Narabeen Group (N), Hawkesbury sandstone (H), and Wianomatt Group (W), all of which overlie the coal measures, southern Sydney Basin. 153

Figure 83. The integral of the burial history versus the square of the median vitrinite reflectance from borehole of known temperature in the Rhine Graben (modified from Buntebarth, 1978). 155

Figure 84. Isoreflectance map (in V types) showing regional variations in coalification of Upper Paleozoic rocks of the Atlantic Provinces (modified from Hacquebard and Donaldson, 1970). 156

Figure 85. Vertical variation in vitrinite reflectance adjacent the Whin sill. 157

Figure 86. Vitrinite on an ash-free basis versus specific gravity for two Fording River coals (modified from Bustin, 1982). 160

Figure 87. Recovery of petrographic components vitrinite plus liptinite (V + E) and inertinite plus ash (I + A) in the undersized product (modified from Tompos, 1982). 161

Figure 88. Variation in separation efficiency of vitrinite plus liptinite (V + E) and inertinite plus ash (I + A). See text for explanation. 161

Figure 89. Relationship between volatile matter and calorific value for coals of different ash content (from Mackowsky, 1982). 164

Figure 90. Diagram illustrating interaction of coal properties and coking conditions and influence upon coking capacity (from Mackowsky, 1982c). 167

Figure 91. Relation between Micum 40 and JIS 15 indices with ASTM stability factor (from Zimmerman, 1979). 170

Figure 92. Graph of optimum ratios; reactive to inert components (modified from Zimmerman, 1979, after Schapiro *et al.*, 1961). 175

Figure 93. Strength index graph; coke strengths for different V types with variable quantities of inerts (from Zimmerman, 1979, after Schapiro *et al.*, 1961). 175

Figure 94. Relationship of coke stability factor to Strength Index and Composition Balance Index (modified from Zimmerman, 1979, after Schapiro *et al*, 1961). 176

Figure 95. Correlation of predicted and determined coke stability factors (from Schapiro and Gray, 1964). 177

Figure 96. Relationship of coke size, coal rank, and utilization characteristics to stability diagram (modified from Zimmerman, 1979). 178

Figure 97. Curves for determination of inert character of pseudovitrinite (from Benedict, Thompson, and Wenger, 1968). ... 182

Figure 98. Relationship of effective inert content, reflectance, and ASTM Stability Factor, (a) high-volatile A and medium-volatile bituminous coal, (b) low-volatile bituminous (modified from Benedict, Thompson, and Wenger, 1968). ... 183

Figure 99. Curves for predicting ASTM stabilities on blends (modified from Thompson and Benedict, 1978). ... 184

Figure 100. Relationship of petrographic composition and rank to ¼ inch hardness factor of coke (from Brown *et al.*, 1964b). ... 186

Figure 101. Relationship of petrographic composition and rank to 1½ inch shatter index of coke (from Brown *et al.*, 1964b). ... 186

Figure 102. Graph showing relationship of reactive semifusinite to vitrinite content (from Smith *et al.*, in press). ... 188

Figure 103. Relationship of reactives/vitrinite reflectance to volatile matter content (from Smith *et al.*, in press). ... 188

Figure 104. Graph for determining M40y and M10y indices from reflectance and Composition Balance Index (from Steyn and Smith, 1977). ... 188

Figure 105. Relationship of coke porosity (A) and coke wall thickness (B) to CO_2 weight loss (reactivity) (from Benedict and Thompson, 1980). ... 189

Figure 106. Relationship of effective inert content (Bethlehem Steel method) to coke reactivity (from Thompson and Benedict, 1978). ... 189

Figure 107. Relationship of rank (by reflectance), and content of organic inerts to volume change in coke oven charges (from Gray *et al.*, 1978). ... 190

Figure 108. Relationship of rank (by reflectance), and inert content to pressures developed in coke ovens (from Gray *et al.*, after Thompson *et al.*, 1965). ... 191

Figure 109. Relationship of vitrinite reflectance to Gieseler fluidity (from Paulencu and Readyhough, 1974). ... 192

Figure 110. Relationship of rank (by reflectance) and inert content to Free Swelling Index (FSI) (modified from Pearson, 1980). ... 192

Figure 111. Relationship of rank (by reflectance) and inert content to Gieseler fluidities (after Pearson, 1980). ... 192

Figure 112. Typical hydrogen-to-carbon ratios by weight in Canadian fossil fuels and refined products (after Taylor, 1979). ... 196

Figure 113. Proposed structural elements of coal (after Given and Peover, 1960). ... 196

Figure 114. Schematic diagram illustrating formation of hydrogen-rich products and coke from coal (modified after Whitehurst *et al.*, 1980). ... 197

Figure 115. Principal routes to liquid fuels from coal (modified after Taylor, 1979). ... 197

Figure 116. Typical liquid yields and thermal efficiencies in liquefaction (modified after Taylor, 1979). ... 198

Figure 117. Dependence of conversion on reactive maceral content (after Given, 1975). ... 199

Figure 118. Relation of vitrinite reflectance on conversion of coal of similar composition (after Whitehurst *et al.*, 1980). ... 199

Figure 119. Petrographic control of coal utilization (after Davis *et al.*, 1976). 200

Figure 120. Generalized scheme of distribution of organic material (OM) in a sedimentary rock and methods used in coal petrology and organic geochemistry to characterize maturation level (rank) and type of organic material. 204

Figure 121. General scheme of hydrocarbon generation. 205

Figure 122. Examples of kerogen evolution paths. 205

Figure 123. Correlation of the coal rank scale with various maturation indices and the zones of petroleum generation and destruction (modified after Dow, 1974). 206

Figure 124. Rank to depth relationships for different vitrinite populations observed in borehole Hibernia P-15. 210

Figure 125. Relation of mean maximum reflectance of vitrinite ($\bar{R}_o max$) and the abundance of alginite (after Hutton and Cook, 1980). 211

Figure 126. Kerogen maturation profile for a Louisiana Gulf Coast well (modified after Dow, 1974). 211

Figure 127. Coalification maps of northwest Germany for the Lower Cretaceous (Wealden) and the Carboniferous (boundary Westphalian C/D) based on vitrinite reflectances, measured mainly on finely dispersed organic material. Economic oil and gas deposits are related to particular coalification stages (from Stach, 1982). 213

1 INTRODUCTION

Coal petrology, as its name suggests, is the study of the organic and inorganic component make-up of coal, its origin and subsequent geological history, and its properties as they relate to its composition and history. The key approach to coal petrology is microscopic, and thus the greater share of this manual is devoted to coal petrography. Relative to the more conventional uses of the microscope in the earth sciences, coal petrography is a young field. Nonetheless, it has proven its value in applications to many geological and technological problems, including: interpretation of depositional, structural and geothermal histories of coal basins; prediction of utilization potential of coal, most notably in the area of coke-making; and petroleum exploration.

This manual will be of use to coal petrologists who need a further grounding in the geological principles of their science, and who are interested in exploring more of its possible applications. Perhaps more importantly, this manual will provide a valuable background to those geologists, chemists, mine and process engineers, and others who presently apply or are considering application of coal petrology to their own field.

The manual is roughly divided into three major topics suggested by its title. Topic one (Chapters 2 through 5) includes the basic principles of coal petrology, including the formation of coal, the origin and nature of coal macerals, lithotypes, microlithotypes and mineral matter, and oxidation of coal. Topic two (Chapters 6 and 7) covers the most commonly used methods involved in sampling, sample preparation, measurement of coal reflectance, random point-counting and fluorescence. Topic three (Chapters 9 and 10) covers many of the geological and technological applications of coal petrology.

The authors wish to thank Dr. A. Sutherland-Brown, formerly of the B.C. Ministry of Energy, Mines and Petroleum Resources, General Chairman of the 1983 GAC-MAC-CGU Annual Meeting, for initially suggesting the course. We are grateful to Dr. W. J. McMillan of the B.C. Ministry of Energy, Mines and Petroleum Resources for coordinating production of the initial manuscript and to Dr. J.W. Kramers of the Alberta Geological Survey for production of the second printing. We thank Mrs. D. Bulinckx for typing the final manuscript and to Elisabeth Elsby for typing the revisions for the second printing. We also acknowledge our employers for permitting time to prepare the manual and for making necessary technical services available to us. The authors accept responsibility for all errors and omissions.

1.1 **Historic Overview**

The birth of coal petrology is often associated with two names and the dates of two important publications. These are Marie Stopes (1919) and Reinhardt Thiessen (1920). Both were paleobotanists; Stopes was English, Thiessen American. Two contemporaries of Stopes and Thiessen should also be considered as pioneers whose ideas contributed much to the science of coal petrology. These are the German scientists, H. and R. Potonié (H. Potonié, 1920; R. Potonié, 1924).

Stopes and Thiessen are important because their ideas on classification and terminology have persisted into more recent time. Both used the microscope to study coal, although in Stopes' early work she used microscopic observations to supplement macroscopic descriptions. Thiessen relied much more heavily on the microscope. Both used the thin-section technique, although Stopes later made observations in reflected light. Thiessen's ideas on classification formed the basis of the U.S. Bureau of Mines system (Parks and O'Donnell, 1956). The Stopes (later Stopes-Heerlen) system has been adapted to reflected light studies, much modified and enlarged, and now is the method of coal petrographic description accepted by all countries including the United States.

Prior to Thiessen and Stopes coal petrological investigations had been carried out in rather sporadic and isolated fashion by various researchers in England, France, Germany and the United States. One of the important results of these early workers was confirmation of the land-plant origin of coal. Another result of more limited importance involved the distinction between cannel and boghead coals, with microscopy demonstrating that the latter was of algal origin. This result played a part in the famous 'Torbanite' litigation in Britain in the mid-19th century when microscopic evidence was used to determine whether or not the 'mineral' at Torbane Hill should be regarded as coal.

The years 1919 to 1920 mark a turning point for coal petrology. Subsequent research became more systematic, more people became involved, and a group of specialists appeared whose main interest was coal petrology rather than paleobotany or mineralogy with coal as a sideline. The decade of the 1920's saw the application of reflected light techniques to coal petrology, and in this connection the pioneer work of Stach in Germany is noteworthy (Stach, 1927). In the late 1920's and early 1930's Thiessen, Stopes, and Stach were joined by others, including Duparque in France, Seyler in Britain, and Jurasky and Hoffmann in Germany, in the investigation of coal composition by petrographic means (Thiessen, 1925; Thiessen and Sprunk, 1934, 1936; Stopes, 1935; Seyler, 1929; Hickling and Marshall, 1932; Duparque, 1934; Jurasky, 1936, 1939; Hoffmann, 1933; Hoffmann and Jenkner, 1932). A good review of major contributions made by some of the early British coal petrologists is given by Murchison (1978). In 1935 Stach published the first edition of his textbook on coal petrology (Stach, 1935) which 40 years later was updated, and just recently has appeared in a revised third edition (Stach *et al.*, 1975, 1982). This book must be regarded as the most important current reference book on coal petrology.

Nearly all of the early studies were concerned with what might be described as the characterization of coal by petrological methods. Such studies attempted to identify and classify various components, both macroscopic and microscopic, to trace their origin back to original plant parts and products, and to speculate on the processes which converted them to coal. A considerable volume of such background information had accumulated by the mid-1930's, and because of varied research approaches and ideas on classification, confusion had arisen in concepts and terminology. To resolve some of these difficulties a conference of coal scientists met at Heerlen in the Netherlands in 1935. One of the important people at this conference was Marie Stopes, and an important result was the formation of the Stopes-Heerlen system of nomenclature which forms the basis of our presently used scheme of petrographic terminology. Another important result of this meeting was the introduction of the concept and term 'maceral' by Stopes to designate the fundamental microscopic constituents of coal, analogous to the minerals of inorganic rocks. The maceral concept is with us today as an essential part of coal petrological terminology. A relatively complicated system of maceral classification was proposed at Heerlen (Stopes, 1935; Jongmans *et al.*, 1936), some of which is not used today. Nevertheless the outline of the three main maceral groups, namely, vitrinite, liptinite (or exinite), and inertinite, was clearly identified and is still in use today. The Heerlen meeting was a milestone for coal petrology but it left a number of problems unresolved. One of these concerned changes in the nature of macerals with changes in coal rank (degree of metamorphism), and how such changes could be reconciled with petrographic nomenclature. Spackman in 1958 re-examined the maceral concept and proposed revisions which would integrate maceral and rank changes.

The 1930's saw the introduction of a new technique which has become an essential part of coal petrology today, namely the measurement of reflectance and its use as an indicator of coal rank. Hoffmann and Jenkner in Germany were the first to explore this important petrological tool (Hoffmann and Jenkner, 1932). Somewhat later Seyler applied the technique to the study of British coals (Seyler, 1943). Early reflectance measurements were made using a Berek microphotometer; later this was supplanted by photomultiplier systems which yield more reliable results (Davis, 1978a; see also Section 7.1.2).

In the decade of the 30's some scientists were beginning to explore the relationships between petrographic composition and the technological behaviour of coal. It was noted that coals rich in the macerals vitrinite and exinite behaved quite differently in fluidity, gasification and extraction experiments than did coals with high inertinite (for example, high fusain) contents. Papers during this period, reporting on the results of technological testing of coal, also began to report petrographic data. For example, a number of publications during this period by the U.S. Bureau of Mines, reporting results on carbonization and gasification of American coals, also recorded the petrographic compositions of the tested coals, although little attempt was made to directly correlate petrography and technical test data. It was not until the late 1940's and 1950's that the industrial possibilities of coal petrology came to be investigated seriously at laboratories such as the Bergbau-Forschung in Germany, Cherchar in France, the Illinois Geological Survey, and the Pennsylvania State University in the United States, and at several industrially-oriented research institutes in the Soviet Union.

In North America the 10 years between 1955 and 1965 were not overly prosperous for the coal industry; nevertheless, in 1956 the first industrial coal petrology laboratory was established at the U.S. Steel Corporation. A few years later a second laboratory was set up at Bethlehem Steel. The emphasis in these laboratories was on the use of coal petrography to predict coke quality, thus reducing the need for the more costly and lengthy tests which traditionally had been used. In this activity these laboratories were largely successful. The U.S. Steel method was a modification of a method described earlier by Ammosov *et al.* (1957) in the Soviet Union. The paper by Schapiro *et al.* (1961) of U.S. Steel, describing this method as applied to American coals, has been one of the most often quoted coal quality papers in North America. Other methods of coke quality prediction utilizing coal petrography which appeared in the period from 1960 to 1969 include those of Brown *et al.* (1964b), Benedict *et al.* (1968), and Mackowsky and Simonis (1969). All these methods have two common premises: one being that the petrographic constituents of coal can be divided into 'reactive' and 'inert' categories based on their behaviour during carbonization; and the second being that rank, usually expressed by reflectance, is a critical factor that must be included in any predictive formula (see Section 10.3.3).

An important milestone in the 1950's was the establishment of the International Committee for Coal Petrology with the aim of advancing the science through standardization of methodology and terminology. This committee has established a practice of meeting every year and one of the first fruits of their efforts was the publication in 1957 of a copiously illustrated Glossary or Handbook of coal petrological terms. A second edition of this appeared in 1963 followed by supplements in 1971 and 1975. The supplements contain not only descriptions of new terms, particularly with relation to brown coal petrography, but also concise descriptions of coal petrographic techniques. The committee is organized into subcomissions and working groups which carry out cooperative studies testing new techniques or investigating problem areas. The results are reported at the annual meeting and recommendations as to acceptance or modifications in technique and terminology are discussed and voted upon.

Since the 1960's there has been a great increase in the number of people participating wholly in coal petrology or making use of petrology in their investigations. Coal petrology is now a many-faceted discipline, including studies on the petrology of peat (the precursor of coal), studies on the evaluation and causes of rank or coalification, fundamental research on the chemical and physical properties of macerals, studies relating petrology to the sedimentology of coal seams, studies on the source material (kerogen) of oil and gas, investigations into weathering and oxidation, and a host of studies relating directly to technological applications such as coking, liquefaction, and combustion. Indeed a new organization in North America is called "Society for Organic Petrology". The name itself and the variety of backgrounds and interests of people participating in this society and indicative of the broadening spectrum of studies formerly incorporated in the term "coal petrology". The total literature from such varied effort has become quite voluminous and is impossible to cite properly within this context. The third edition of Stach's Textbook of Coal Petrology contains an exhaustive list of important references, while Bostick and Nicksic (1975) published an extensive bibliography relating mainly to dispersed carbonaceous matter in sedimentary rocks. Representative publications pertaining to the varied spectrum of coal studies indicated above should be mentioned:

(1) Peat Studies (Cohen and Spackman, 1977; Styan and Bustin, 1983a, 1983b).

(2) Metamorphism of Coal/Rank (Karweil, 1956; Lopatin, 1971; Bostick, 1979; Teichmüller, 1979)

(3) Chemical and Physical Properties of Macerals (Kroger *et al.*, 1957a, 1957b, 1957c, 1957d; Van Krevelan, 1961; Given *et al.*, 1966; Goodarzi and Murchison, 1972; Allan and Douglas, 1977)

(4) Sedimentology of Coal Seams (Hacquebard *et al.*, 1967; Smith, 1968; Mackowsky, 1968; Shibaoka and Smyth, 1975)

(5) Oil and Gas Exploration (Teichmüller, 1971, 1974a; Tissot *et al.*, 1974; Hacquebard, 1975; Alpern, 1980; Cook *et al.*, 1981)

(6) Weathering and Oxidation (Chandra, 1958, 1962; Benedict and Berry, 1964; Gray *et al.*, 1976; Bustin, 1980; Marchioni, 1983)

(7) Coking (Cook and Edwards, 1971; Steyn and Smith, 1977; Koensler, 1980; Pearson, 1980)

(8) Liquefaction (Given *et al.*, 1975a, 1975b; Kalkreuth and Chornet, 1982)

(9) Combustion (Shibaoka, 1969; Nandi *et al.*, 1977)

Two new techniques to come on the scene in the last two decades are the use of automated microscopy and the application of fluorescence in the identification of certain macerals. Automated microscopy has a distinct appeal for coal petrologists, who, following the conventional microscopic procedures normally used, must spend long hours gathering the raw data of their studies. Automation relieves this tedium and also promises much more rapid and objective collection of data. These characteristics of rapidity and objectivity are especially attractive to industrial laboratories which often need to have data quickly on a large number of samples.

Automated coal petrography, at least in North America, got its start at Bituminous Coal Research in the early 1960's when Berry and his associates helped develop the AMEDA system (Berry in Berry *et al.*, 1967). The AMEDA, and other systems which have followed it, produce data in the form of a reflectance profile or reflectogram of the whole coal. One of the problems has been to equate such a profile with a conventional maceral analysis. However, for many purposes a reflectogram has been found to be a quite satisfactory way of expressing coal composition. Additional discussion on automated microscopy may be found in Davis (1978a). At present in Canada there are several laboratories, which routinely examine coal by automated techniques (Hampson *et al.*, 1980).

Fluorescence microscopy is based on the fact that certain constituents in coal fluoresce when irradiated with ultra-violet or blue light. It was applied to coal as early as 1936 by Schochardt, but it was not until the 1970's that equipment and methodology had evolved sufficiently to make this technique a reliable tool in coal petrology. Fluorescence provides two kinds of information. First, it greatly assists in the identification of hydrogen-rich constituents, particularly of the liptinite group, and second, it provides another method of measuring maturity or rank, because fluorescence properties of macerals change with rank. Fluorescence has been particularly useful in examining low-rank coals such as lignites, because fluorescence phenomena are especially strong at this end of the rank scale. The most important application of fluorescence is in the petrological examination of potential hydrocarbon source rocks, or, more specifically, the kerogen contained within these rocks. The varying fluorescence properties of the constituents of kerogen provide an important diagnostic tool. Representative papers relative to the theory, methodology, and use of fluorescence are those by Jacob (1974), Teichmüller, (1974b), Ottenjann *et al.* (1975), Van Gijzel (1975), Teichmüller and Wolf (1977), and Creaney (1980).

In Canada coal petrological studies started with the establishment of a laboratory by the Geological Survey of Canada and the Nova Scotia Department of Mines in Sydney, Nova Scotia in 1948. Petrological studies were initiated there by P. A. Hacquebard and his colleagues. In the late 1950's a petrology laboratory was set up at the Mines Branch (now CANMET in Ottawa, and the first studies by L. H. King utilized petrological data along with chemical information to characterize naturally occurring bitumens. A few years later this laboratory, with B. N. Nandi and J. Jorgensen, began studies on the petrology of coking coals. Later in the 1960's the Steel Company of Canada (STELCO) and Dominion Foundry and Steel (DOFASCO), both in Hamilton, Ontario, became the first industrial organizations in the country to initiate petrological studies, in this case related to basic research as well as quality control.

In the decade of the 1970's a number of institutions in government, industry, and universities began coal petrological investigations including some related to hydrocarbon exploration. Laboratories currently carrying out petrological studies in Canada are listed in Table 1. From the aforementioned it should be clear that coal petrology has grown considerably in stature since the days of Stopes and Thiessen, although the number of practitioners in the art is still relatively small. Even so, it is almost certain that important contributions have inadvertently not been mentioned in this review. With this in mind the reader's attention is directed once again to the reference list in Stach's Textbook (1982), as well as to an excellent recent review of the science by Neavel (1981).

TABLE 1. LIST OF CANADIAN COAL PETROLOGY LABORATORIES

Laboratories	Principal Investigator(s)	Main Activities
Memorial University St. John's, Nfld.	R. F. Hyde	Rank Studies
Atlantic Coal Institute Sydney, Nova Scotia	J. Pilgram	Petrology studies (contract basis)
Geological Survey of Canada Atlantic Geoscience Center Dartmouth, N.S.	P.A. Hacquebard	Petrology of eastern Canadian coals; east coast hydrocarbon exploration
INRS Quebec City, P.Q.	Y. Heroux	Regional rank studies
Fuels Research Center CANMET, Ottawa	B. N. Nandi J. Jorgensen	Coking coals, combustion, conversion*
STELCO Hamilton, Ontario	P. J. Readyhough G. Patterson	Coking coals
DOFASCO Hamilton, Ontario	A. Hampson	Coking coals
Algoma Steel Sault St. Marie, Ontario	M. Khan	Coking coals
University of Western Ontario, London, Ontario	G. Winder	Presently studying Ontario lignites**
University of Regina Regina, Saskatchewan	J. Potter	Conversion, petrology of Saskatchewan coals, teaching
Alberta Research Council Edmonton, Alberta	S. Parkash	Conversion
Geological Survey of Canada ISPG, Calgary, Alberta	A. Cameron F. Goodarzi W. Kalkreuth	Regional petrological studies of western Canadian coals, hydrocarbon exploration, conversion, mineral matter
Petro-Canada Calgary, Alberta	P. Gunther	Hydrocarbon exploration, petrology of coals on company properties
Esso Resources Ltd. Calgary, Alberta	S. Creaney J. Allan	Hydrocarbon exploration, petrology of coals on company properties, geochemistry
Westar Mining Sparwood, B.C.	G. Mayor	Analysis of exploration and mine samples
University of British Columbia, Vancouver, B.C.	R. M. Bustin	Petrological studies related to a number of geological and technological projects, peat studies, teaching
B.C. Ministry of Energy, Mines & Pet. Res., Fernie and Victoria, B.C.	D. Grieve (Fernie) W. Kilby (Victoria)	Regional petrological studies on B.C. coals and related technological studies
D. E. Pearson & Associates Victoria, B.C.	D. E. Pearson J. S. Pearson	Consultant: Petrological analyses of coal; coal geology and utilization

*Conversion refers to liquefaction or gasification studies, mainly the former at present.
**This laboratory is currently being established.

2 ORIGIN OF COAL

Any deposit in which organic matter comprises 50 per cent by weight or 70 per cent by volume can be considered coal (Schopf, 1956). In modern sediments, and in rocks, organic matter tends to occur in three natural associations (Tissot and Welte, 1978): (1) remnants of algae, micro-organisms and phyto- and zooplankton; (2) highly degraded remnants of higher plants; and (3) less degraded remnants of higher plants. Depending on the type and degree of maturation of the organic material, either petroleum source rocks, oil shales or coal will result (see Figure 12).

The first association of organic material, generally referred to as 'Type I' organic matter, is the principal organic component of oil source rocks. If the Type I organic matter comprises a significant portion of the rock, the rock may be referred to as an oil shale, or, if the organic component is greater than 50 per cent by weight of the rock it may be referred to as coal. Type II organic material, depending on the abundance of organic matter, may also be referred to as a source rock, oil shale (if immature) or coal. Type III organic material, when present in adequate amounts, comprises coal, but, because this type of organic matter has little oil generating potential, rocks containing lesser amounts of Type III organics are not considered source rocks or oil shales but are simply referred to as carbonaceous. Coal originating from Types I and II organic material are generally referred to as sapropelic coals. Sapropelic coals are further subdivided into cannel coals, which may be rich in spores, and boghead coals, which are rich in algal remains. Coals originating from Type III organic material comprise the humic coals. Humic coals are by far the most abundant and constitute the major coal resources of the world. Accumulation and preservation of organic material in sufficient quantities to form humic coals is generally restricted to swamp, marsh, or bog environments, and most coal seams have been interpreted to have originated from peat deposited in these environments. Factors affecting the accumulation and preservation of peat and, ultimately, the transformation of peat into coal, are at best poorly understood.

As a result of detailed petrographic and sedimentological studies of coal measures and recent peats, it is apparent that the evolutionary development of the flora, the climate and the depositional environment are all factors contributing to the formation of coal (Teichmüller and Teichmüller, 1982).

2.1 **Evolutionary Development of the Flora**

The flora in any marsh, swamp, or bog environment is a product of the climate and the factors described in section 2.3 on depositional environments. The evolution of the flora, however, ultimately determines what plant forms were incorporated into the peat. Although coals as old as Middle Huronian have been described (Teichmüller and Teichmüller, 1982), it was not until the Late Precambrian that thalloid plants evolved, and not until the Early Silurian that the first vascular plants appeared. Only in the Early Devonian did herbaceous plants such as *Rhynia, Hornea*, and *Asteroxylon* become established. These early plants, and somewhat later herbaceous vascular cryptogams, were adapted and restricted to swampy soils and were characterized by a widespread shallow root system (Francis, 1961). In the Early Devonian articulates and pre-ferns developed. Progymnosperms with secondary wood and arborescent habitat evolved in the Middle to Late Devonian, leading to large gymnosperm trees of the Carboniferous that were completely developed for life on dry land. During the Late Carboniferous, in the northern hemisphere, luxuriant and diverse swamps existed which were characterized by gymnosperms such as *Cordaitales* and *Pteridospermae* and by lycopod trees (club mosses) such as *Sigillaria* and *Lepidodendron*, which attained heights up to 30 metres, and by tree-size Equisetales such as *Calamites* and a variety of *Filicales* (ferns) and *Pteriodosperms* (Francis, 1961). In the southern hemisphere, during the Late Carboniferous and Permian periods on the other hand, the climate was cool and the flora was characterized by tree fern-like forms such as *Glossopteris, Gangamopteris,* and *Pecopteris*. In the Mesozoic gymnosperms, such as *Ginkophyta, Cycadophyta*, and conifers, became the main coal-producing plants. In the Late Mesozoic ferns and gymnosperms (except conifers) became less prominent, partly due to competition from the newly evolved flowering plants (angiosperms). In the Late Cretaceous more specialized and diverse angiosperms became progressively more abundant and the prominence of conifers declined.

Although the evolutionary development of the plants has changed the composition of peat-forming plant communities through time, the effects of plant evolution on coal formation has been more subtle. By the major coal-forming periods of the Carboniferous, plants were diversified and adapted to the extent that all ecological niches in the swamps were occupied by species not unlike those of the Tertiary or those found in modern peat-forming areas. Teichmüller (1968), for example, was able to draw a direct comparison between plants in coal-forming swamps of the Carboniferous and Miocene in Germany, and such a comparison can be readily extended to include modern peat swamps (Table 2). Carboniferous floras, however, were less diverse than younger floras, and their chemical composition was different and therefore the resulting coals may differ. Specialized tissues of herbaceous angiosperms, for example, are more susceptible to decay so that Tertiary coals may be enriched in the resistant wax and resinous components. Tertiary conifers, which are more resistant to decay, may thus be over-represented in the coal relative to their prominence within the peat swamp.

2.2 **Climate**

Climate controls the rate of plant growth, type of plant growth and, to a large extent, decompositional rates, all of which ultimately determine whether or not peat will form. Warm wet tropical and subtropical climates are most favourable for the development of forest swamps. Such climates are characterized by both high rates of plant accumulation and degradation. In Florida and the Mississippi Delta, peats have been reported accumulating at rates of one-half to 1 millimetre per year (Fisk, 1960) and in Borneo rates of accumulation of up to 4 millimetres per year occur (Anderson, 1964). In temperate and cool climates, rates of plant growth are much slower but the rate of decomposition is also slower. Temperate peats of the Fraser Delta have accumulated at rates of about 1 millimetre per year. In humid climates where rainfall exceeds evaporation, raised bogs occur. In temperate and cool climates the main vegetation of raised bogs is *Sphagnum* moss whereas in tropical climates the flora is dominated by trees of low diversity.

Coal seams and modern peats can accumulate wherever the rates of accumulation are greater than decomposition. Peat accumulates in high Arctic climates as well as in tropical climates. Most ancient coals probably originated in temperate to tropical climates. The Gondwanaland coals however, accumulated in cool climates and some Gondwanaland coal seams are interbedded with glacial tills. Coal seams that are deposited in tropical or subtropical climates are generally characterized by the presence of bright bands originating from woody material. Coal seams originating in cool climates are more commonly thinly banded and 'finely detrital' as a result of the paucity of large trees within the peat swamps (Teichmüller and Teichmüller, 1982).

TABLE 2. COMPARISON BETWEEN CARBONIFEROUS, TERTIARY, AND MODERN PEAT-FORMING FLORAS AND TYPE OF RESULTING COAL (MODIFIED FROM TEICHMÜLLER, 1982)

Origin of Coal

	Type of Plant Life			
Condition	Carboniferous	Miocene	Modern Fraser Delta	Type of Resultant Coal
Open Water	Algae, Plankton	Algae, Plankton	Algae, plankton	Cannel or Boghead - rich in spores or pollen grains
Shallow Water	(Possibly Calamites or Lycopods)	Reeds and weeds	Reeds and weeds	Wax-rich mixed coals, rich in spores or pollen grains
Wet Soil or Peat	*Pteridophyta, e.g. Equisitineae* or *Lycopodineae*	*Myrica* *Taxodium* Fungi	*Spria* *Myrica* *Sphagnum*	Mixed Coal - rich in spores
Dry land	Gymnosperms	*Sequoia*	*Pinus contorta* *Betula L.*	Woody coal-low in spores

2.3 **Depositional Environment**

To a great extent the lateral distribution, thickness, composition and quality of coal are determined by the depositional environment. For the formation and preservation of significant peat deposits, a depositional setting is required in which there is high organic productivity and slow continuous subsidence, such that the groundwater table is at or close to the peat surface, and in which the peat swamp is protected for protracted lengths of time from invasion by marine waters or clastic sediment influx. These conditions apparently occurred in the past and presently occur in paralic or limnic environments within areas that are relatively stable tectonically, such as foredeeps or intracratonic basins.

The deposition of coal can be considered at two levels: (1) a macroscopic level in which the sedimentology of the coal (peat) is placed in the perspective of surrounding sediment, for example the sedimentology of coal measures; and (2) a microscopic level in which an attempt is made to interpret the sedimentology within the peat swamp, bog, or marsh and which generally involves detailed studies of the coal or peat. These levels are discussed in sections 2.3.1. and 2.3.2., respectively.

2.3.1 *Sedimentology of Coal Measures*

From detailed and extensive studies of coal measures in the Appalachian coalfields, Horne *et al*. (1978) found that the thickness, lateral extent, and quality of the coal is affected directly by the depositional environments that preceded, coexisted with, and followed deposition of the peat. The sedimentary environment that preceded the coal swamp shapes the topography on which the coal develops which, in turn, affects the thickness and lateral extent of the coal. Environments that coexist laterally with the swamp directly affect the lateral continuity of the seams which, together with the environment within the swamps, directly determine the composition of the coal. Post-depositional environments further affect the peat as a result of channelling, which may remove part or all of the peat and by later, post-depositional, introduction of brackish or marine waters which may erode the peat and affect the peat chemistry.

2.3.1.1 **Paralic Coal Deposits**

In paralic (coastal marine) settings significant peat accumulation can occur in several distinct environments, all of which result in coal seams with characteristic geometry. These are: (1) back-barrier; (2) deltas; and (3) coastal and interdelta plains.

2.3.1.1.1 Back-Barrier Coals

Back-barrier coals accumulate landward of pre-existing or contemporaneous barrier islands (Figure 1). The coal swamps presumably evolve as a result of infilling of the lagoons and result in pod-shaped coals that parallel the adjacent barrier and thus parallel depositional strike (Horne *et al*., 1978). Recent peat deposits have been described from back barriers from the Guadalupe Delta by Donaldson *et al*. (1970), and more recently from the Snuggedy Swamp on the South Carolina coast by Staub and Cohen (1979). The Snuggedy Swamp has apparently evolved in response to the recent rise in sea level. Initially, salt marshes developed between Pleistocene barrier islands. The salt marsh surface was then colonized by a freshwater vegetation which eventually formed freshwater peat islands which later coalesced to form an extensive swamp surface. Staub and Cohen (1979) suggest that coals produced from a similar setting to that of the Snuggedy Swamp would be of two types: thin and discontinuous coals which form from the salt marsh peats, or thicker coals with numerous splits will form in freshwater environments. Splays in the Snuggedy Swamp originate from crevassing of tidal channels and from fires. Burning of the peat surface results in formation of charcoal and depressions in the peat that may then promote flooding. In the Snuggedy Swamp, Staub and Cohen (1979) found the thickest occurrences of peat to be proximal to the barrier and to run parallel to the barrier island or at right angles to it in areas coincident with former tidal inlets. The distribution and thickness of the peat is considered to be related to the size of the barrier, width of the lagoon and the rate of sea level rise (or alternatively subsidence). The latter controls the rate of expansion of the swamp and determines whether or not freshwater peat is inundated by saltwater or buried by sediment.

Coal seams interpreted as back-barrier deposits have generally been found to be thin, laterally discontinuous, and high in sulphur, nitrogen, and ash (Weimer, 1977; Horne *et al*., 1978; Balsley, 1980). In the Cretaceous Blackhawk Formation in Utah, for example, Balsley (1980) found that coals of lagoonal origin

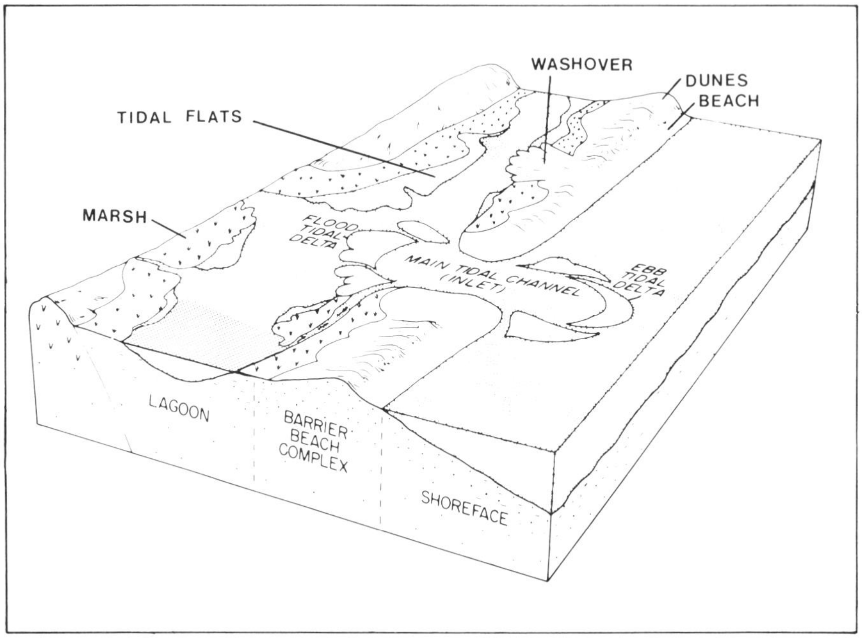

Figure 1. Back-barrier depositional models for coal deposition. Modified from Reinson (1979).

were invariably less than a few centimetres thick, whereas the overlying delta plain coals were up to 6.3 metres thick. Relatively thick (1 to 3 metres) and laterally extensive coals of back-barrier origin have recently been described by Cotter (1982) from the Sabina Basin of northern Mexico. These coals are interbedded with bioturbated mudstone which, in turn, overlie shoreface and foreshore deposits. Cotter (1982) has suggested the thicker mineable seams developed during periods of long-term environmental stability, whereas more numerous and thinner coals developed in areas of rapid progradation and periodic marine transgression.

2.3.1.1.2 Deltaic Coals

Most ancient coal deposits are considered to have been formed in deltaic environments and many modern deltas are the site of significant peat deposition. Depending on the delta plain configuration, major coal deposits may occur in a variety of subenvironments.

On the modern river-dominated Mississippi Delta, Fisk (1960) and Frazier and Osanik (1969) have described the major peat-forming environments. During progradation of the delta, peats initially form from sedge and brackish marshes adjacent to levees on the lower delta plain (Figure 2). As the levees enlarge the marshes may prograde over larger areas of the interdistributary bays, while in drier, elevated, areas cypress-gum forest swamps develop. Following abandonment of the distributary channels, and with continued subsidence, the swamp-marsh complex is progressively inundated by marine waters. Alternatively if peat accumulation can keep pace with subsidence, the swamp-marsh complex may expand and build over the abandoned distributary. Within the transition between the lower and upper delta plains on the Mississippi Delta, the interdistributary bays are shallow and thus support more widespread peat development.

On modern wave-dominated deltas, such as the Nile and Niger deltas, shallow and widespread delta front sands produce a broad platform that enables laterally continuous peats to accumulate. The lower coastal plain and delta plain of the Niger is characterized by the occurrence of coastal barrier islands and tidal inlets. Allen (1970) has described broad belts of mangrove swamps in the axial part of the Niger Delta and behind sandy barrier islands. The swamps are vegetated intertidal flats and the mangroves act as sediment binders. On the lower floodplain alluvial sediments are presently encroaching across the swamp deposits.

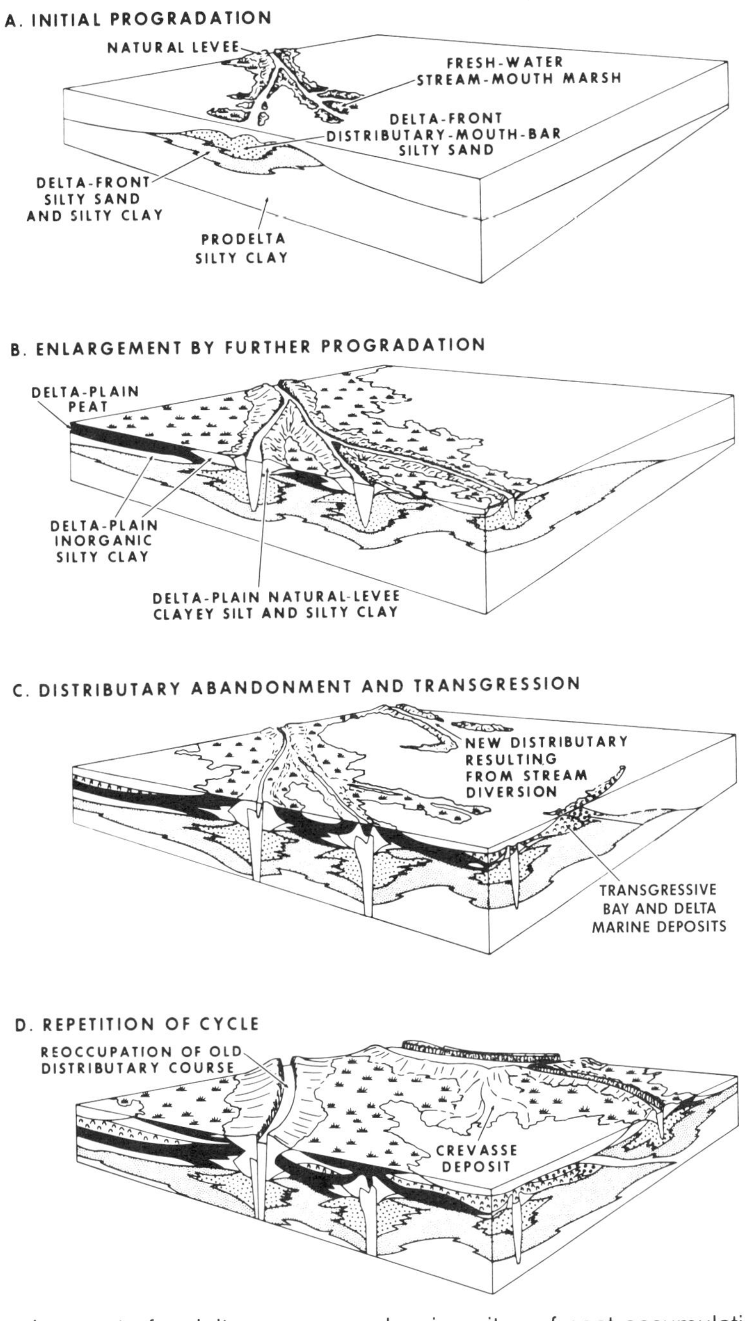

Figure 2. Development of a delta sequence showing sites of peat accumulation. Modified from Frazier and Osanik (1969).

On the western part of the Niger Delta, on the other hand, dendritic swamp channels and high level swamp deposits are encroaching on abandoned alluvial ridges. On the eastern part of the delta, where little fluvial sediment is being introduced, the floodplain is being progressively overgrown by tidal swamps.

Tide-dominated deltas, such as the Klang-Langet River Delta of Malaysia, are characterized by a large number of tidal channels, islands, and offshore tidal current ridges (Leeder, 1982, p. 188). During delta progradation the tidal current ridges emerge and are coated with intertidal and supratidal sediments (Leeder, 1982, p. 188). On the Klang-Langat delta peat accumulates from extensive mangrove swamps that ring the delta and occupy islands within the delta (Coleman *et al.*, 1970). Here the mangrove swamps extend inland from neap high water mark, and woody vegetation is highly diversified extending through a wide range of salinities and tidal conditions. Once freshwater swamps are established, rates of accumulation are high (up to 10 centimetres per 100 years). In such areas the accumulation of organic material may result in peat deposits higher than adjacent channel levees. The peat may then control the distribution of channels and channels that do transect the peat may have no natural levee deposits.

The Fraser River Delta is an example of peat accumulation in a humid temperate climate of moderate river, tidal, and wave energy. On the Recent lobe of the Fraser River Delta peat accumulation has actively occurred on the distal lower delta plain, the transition zone between upper and lower delta plains, and the alluvial plain (Styan and Bustin, 1983b; Figure 3). Distal lower delta plain peats developed from widespread salt and brackish marshes and were not influenced appreciably by fluvial activity (Figure 4). Lateral development of the marsh facies were controlled by compaction and eustatic sea level rise. The resulting thin, discontinuous peat network contains numerous silty clay partings.

Lower delta plain-upper delta plain transitional peats initially developed from interdistributary brackish marshes and were later fluvially influenced as the delta prograded (Figure 5). The thickest peats occur in areas where distributary channels were abandoned earliest. The peats are underlain by a relatively thin sequence of fluvial deposits which in turn is underlain by a major coarsening upward delta front and prodelta sequence.

Alluvial plain peats accumulated in backswamp environments of the floodplain. Earliest sedge-clay and gyttjae peats developed over thin fining upward fluvial cycles or are interlaminated with fine-grained flood deposits. Thickest accumulations occur where peat fills small avulsed flood channels. The overlying sedge-grass and Sphagnum biofacies are horizontally stratified and commonly have sharp boundaries with fine-grained flood sediments. At active channel margins sedge-grass peats are intercalated with natural levee deposits consisting of silty clay. These levees reduce both the number and size of crevasse splay deposits.

Coal originating from peats of the different environments of the Fraser Delta would vary markedly in character. Peats of the lower delta plain are thin and lenticular with numerous splits, whereas peats from the lower to upper delta plain are laterally extensive and of variable thickness and quality. Basal portions of the peat contain numerous splits whereas upper portions are of higher quality. Peats from the upper delta plain-alluvial plain are thick, isolated and laterally restricted and characterized by a low ash content.

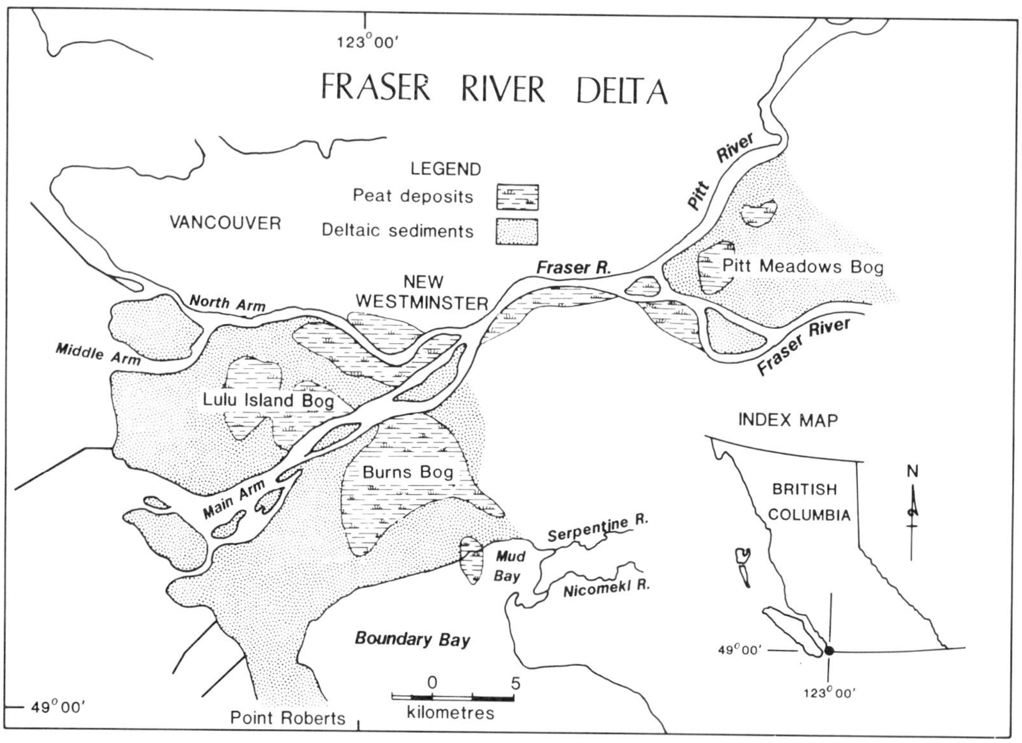

Figure 3. Distribution of peat deposits on the Fraser River Delta. Modified from Styan and Bustin (1983a).

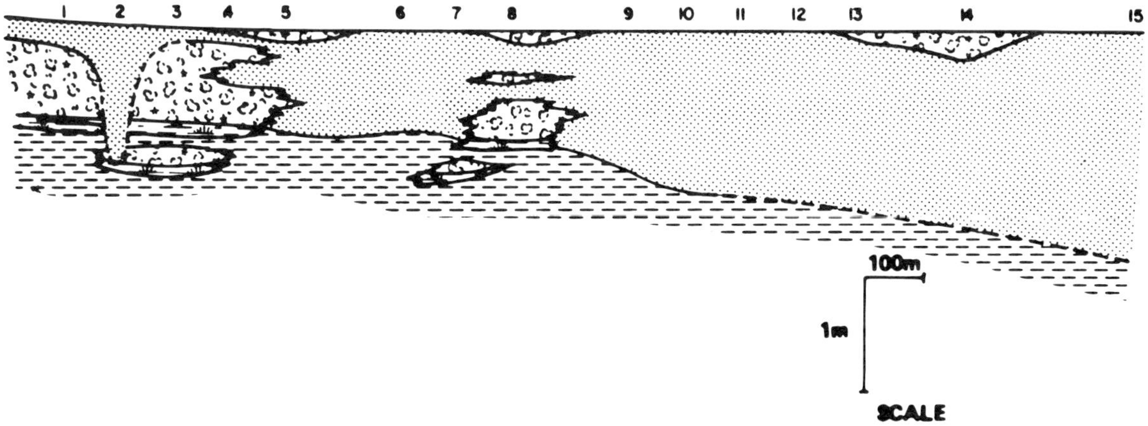

Figure 4. Peat deposit of the distal lower delta plain at Boundary Bay. Refer to the legend on Figure 5.

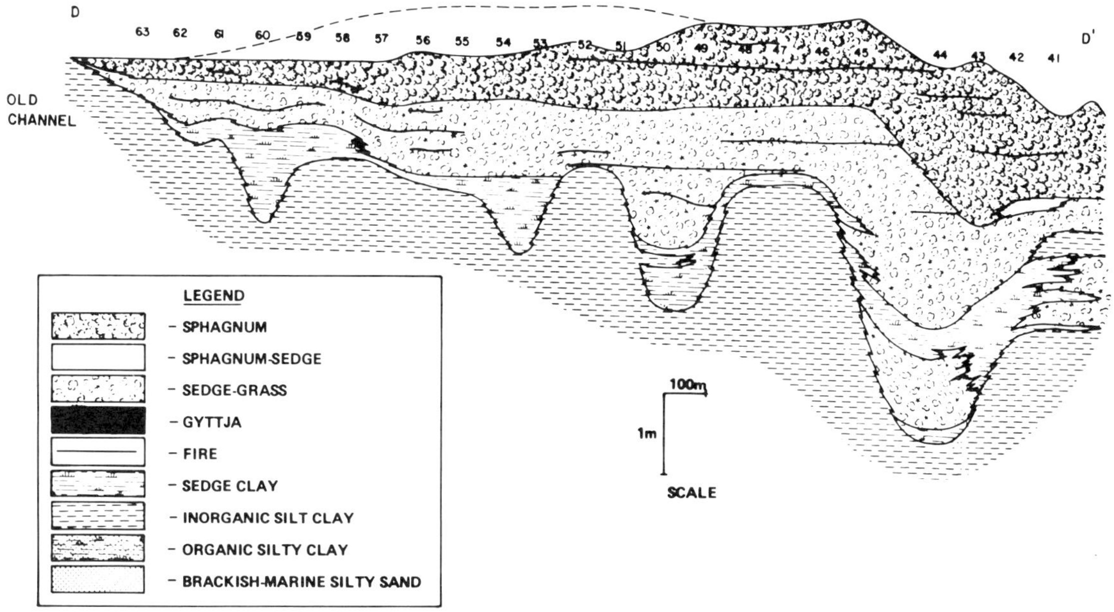

Figure 5. Peat deposits of the lower delta plain at Lulu Island.

Many detailed studies of ancient deltaic deposits have demonstrated the relationship between delta configuration, subsidence rates, and coal distribution. On the lower delta plain of river-dominated deltas, such as those described by Ferm (1974) in the Pocahontas and Dunkard Basins, the coal seams are thickest parallel to depositional dip, pass laterally into interdistributary bay sediments, and, proximal to coeval channels, contain numerous splits originating from flood events. On wave-dominated deltas like the Late Cretaceous Mesaverde Group of Wyoming (Horne *et al.*, 1982), the Blackhawk Formation of Utah (Balsley, 1980), or the Gates Formation of northeastern British Columbia (Carmichael, 1982), the delta plain platform is broad and thick and laterally continuous coals are present. Rahmani (1981) has interpreted coals of the Late Cretaceous Horseshoe Canyon Formation at Drumheller, Alberta as deposits of a tide-dominated delta. Here laterally extensive coals of uniform thickness cap the delta sequences. Rahmani (1981) suggests the peat-forming swamps prograded over the tidal areas during delta growth. Discontinuity of some seams is attributed to breaks in the peat swamp behind bar complexes.

On the upper delta plain-alluvial plain of most deltas freshwater swamps exist in the floodplain-backswamp environment adjacent to channels. Coal seams originating in this environment are typically elongate parallel to depositional dip and may be locally thickened. In the Blackhawk Formation alluvial plain coals are thickest closest to the coast and decrease in abundance and thickness landward (Balsley, 1980). In the Appalachian coalfields upper delta plain-alluvial plain coals vary markedly in thickness, contain numerous partings, and post-depositional channels and wash-outs are common (Horne *et al.*, 1978).

The rate of delta subsidence has a major effect on the lateral distribution and thickness of coal (Donaldson, 1974; Ferm, 1974). Slow subsidence during delta growth results in greater lateral continuity of facies, including coal; more gradual lateral facies changes; a larger and more elongate delta lobe; and a greater spatial separation between facies. On slowly subsiding deltas (or coastlines) during abandonment of the delta lobe, peat accumulation may keep pace with subsidence enabling the swamp to coalesce over the abandoned lobe, forming laterally continuous seams. Termination of peat accumulation in these areas may result from reoccupation of the delta lobe by distributaries, increased subsidence, and transgression or edaphic controls. Coals that develop in rapidly subsiding deltas, such as the Tertiary Rockdale Delta of Texas (Fisher and McGowen, 1969), are not as laterally continuous, contain more splays as a result of flooding and change facies over short distances. Slowly subsiding deltas prograde great distances over thin delta front-prodelta deposits, whereas rapidly subsiding deltas are characterized by little lateral shifting of facies with time, and vertical stacking of facies is more frequent.

2.3.1.1.3 Interdeltaic and Coastal Plain Coals

Many of the great peat swamps of the world are best described as coastal plains. These swamps have developed on subsiding coasts of low profile where a high water table exists and there is little clastic sediment influx. Major coastal plain swamps include the Everglades of Florida, Okefenokee Swamp of Georgia, and coastal plains of Surinam and Guiana along the northern coast of South America. These swamps are large and include a varied freshwater and brackish water flora.

In the past, few ancient coals have been considered interdeltaic or coastal plain in origin. Classification of coals as interdeltaic or coastal rather than simply 'deltaic' is a result of recent refinements in sedimentological studies. Balsley (1980) has recently described coals which he considers of interdeltaic origin from the Blackhawk Formation in Utah. These coals are generally thin (locally up to 2 metres) and can be traced laterally for no more than 1 kilometre.

The Balmer seam of the Late Jurassic-Early Cretaceous Mist Mountain Formation in southeastern British Columbia may have been deposited on a broad coastal plain. The Balmer seam is up to 15 metres thick and is part of a carbonaceous unit of great lateral distribution that includes most of the Fernie and adjacent coal basins, and overlies shallow marine sandstones of the Morrissey Formation (Gibson, 1979). Some of the laterally continuous coals of Pennsylvanian age in the eastern United States were likely deposited on a coastal plain. The Colchester No. 2 seam, which extends for a distance of 1 100 kilometres and possibly for 1 900 kilometres, and which required a depositional surface 'whose flatness is possibly not matched on earth today' (Wanless *et al.*, 1969), was likely deposited on a broad coastal plain.

2.3.1.2 **Limnic Coal Measures**

Limnic coals are those coals that are considered to have been deposited in intermontane regions where the water level was controlled by local conditions rather than regional (sea level) (Falini, 1965). Limnic coal measures commonly result from local subsidence in fault-bounded basins and are generally characterized by thick seams of limited lateral extent.

The organic material of limnic coal deposits may be allochthonous or autochthonous. In the Valdarno area of Italy, Falini (1965) described stumps in growth position indicating the organic material grew in situ. Kim (1979), on the other hand, has argued that the Tertiary Hat Creek deposit of British Columbia originated from hypautochthonous peat. Kim has suggested that the Hat Creek deposit originated in a broad north-south-trending marsh with little or no water circulation. The basin slowly and intermittently subsided such that hypautochthonous peat and sediment were deposited. All coarse sediment was deposited near the basin margin and only finely dispersed sediment and peat accumulated in the basin centre. In the basin, which is

0.5 kilometres wide and 25 kilometres long, a total thickness of 550 metres of coal accumulated (Campbell *et al.*, 1977). Sedimentation in the Hat Creek Basin was terminated by rapid subsidence that resulted in deposition of lacustrine deposits (Kim, 1979).

In their study of the Tertiary Lower Wasatch and Upper Fort Union Formation in the Powder River Basin, Wyoming, Ethridge *et al.* (1981) have described intermontane (limnic) coals associated with fluviatile sediments. They discriminate between two types of coal: relatively thick (24 metres), clean (average ash 6.5 per cent), and laterally extensive coal that occupies belts peripheral to the major channels; and thin coals (0.3 to 2.1 metres) which commonly contain shale interbeds and are of limited areal extent. Ethridge *et al.* (1981) argue that the thick coals originate adjacent to the main trunk streams, downflow from goundwater recharge and in a groundwater discharge area. The discharge would provide a high water table and nutrients to support a luxuriant flora. They further argue that the swamps may have accreted faster than surrounding fluvial deposits, which would account for the low ash content of the coal. The discontinuous coal seams, on the other hand, are interpreted as inter-channel deposits in areas of better drainage and thus lower groundwater levels.

2.3.2 *Sedimentation of Coal*

The sedimentology of coal cannot be considered separately from that of the coal measures; however, the accumulation and the preservation of peat is also, in part, a response to variations that existed within the peat swamp and which are not necessarily manifest in the enclosing strata. Teichmüller and Teichmüller (1982) have outlined the various factors that determine the coal facies and much of the following discussion is based on their summation. 'Coal facies' refer to the genetic types of coal which, together with diagenesis, determine the mineral and organic composition of the coal.

The coal facies is determined mainly by:

(1) type of deposition (autochthonous, allochthonous);

(2) plant communities;

(3) depositional milieux, including pH, bacterial activity, and sulphur availability; and

(4) redox potential.

2.3.2.1 **Type of Deposition**

Most coal seams are considered to have been derived from plant material that accumulated in situ (autochthonous) or material transported short distances (hypautochthonous). Evidence for an autochthonous origin of most coals includes traces of rooting, stumps in growth position, and occurrence of seat earths.

Allochthonous coals are generally rich in mineral matter and are thus of poorer quality. Isolated coal fragments or carbonaceous partings in clastic sediments are probably derived from drifted material.

2.3.2.2 **Plant Communities**

In modern peat swamps plant communities are adapted to the particular ecological conditions present, particularly water table, salinity, and climate. Fraser (in Francis, 1961) and Teichmüller and Teichmüller (1982) recognize four basic swamp types:

(1) open water with aquatic plants;

(2) open water with sedges and reeds;

(3) forest swamps; and

(4) moss swamps.

In modern swamps such a succession of plant communities (1 through 4) represent, the natural succession during infilling of a lake or pond starting with gytjja to reed, forest, and moss peats. What plant communities exist within the swamp will vary, however, with time (vertically) and laterally, depending on local conditions.

The successional sequence of plant biofacies on the Fraser River Delta, British Columbia, are shown on Figure 6 (Styan and Bustin, 1983b). Even though peat is accumulating in a variety of depositional settings, there is a natural succession of similar peat types. Open sedge-grass marshes have produced low moor peats which have accumulated quickly as a result of high nutrient supply (Figure 7). These marshes modify the environment sufficiently during their development to restrict sediment supply except during flood or storm events. Gramineae and Cyperaceae, the principal peat-forming families, have extensive root systems which quickly stabilize any sediment trapped by their vertical stems. Variability within these two plant groups has allowed several niches to be colonized. A salt marsh community is actively forming peat at Boundary Bay (Figure 8), while freshwater species are producing similar organic accumulations at Pitt Poulder on the upper delta plain. The sedge-grass marshes are the initial stage in a complex succession. Accompanying silt and clay deposition, marshes eventually raise the substrate sufficiently to minimize the effects of any marine influence. In so doing they establish quiet freshwater niches into which woody shrubs and *Sphagnum* spp. can invade. Colonization occurs soon after the cessation of inorganic sedimentation. However, where salinity is initially high, several metres of sedge-grass peat is required to accumulate before *Sphagnum* can develop. Once established, *Sphagnum* restricts water flow and lowers both nutrient supply and pH. Ericaceous genera, which are able to tolerate low nutrient supply, quickly replace freshwater marsh species. Water-filled depressions of various depths occur between these raised areas. Deeper pools, averaging 0.5 metres in depth, are floored with a dense bottom layer of *Sphagnum apiculatum*, or liverworts, and are covered by a luxuriant stand of *Nuphar lutea*. These depressions are gradually filled by detrital plant material.

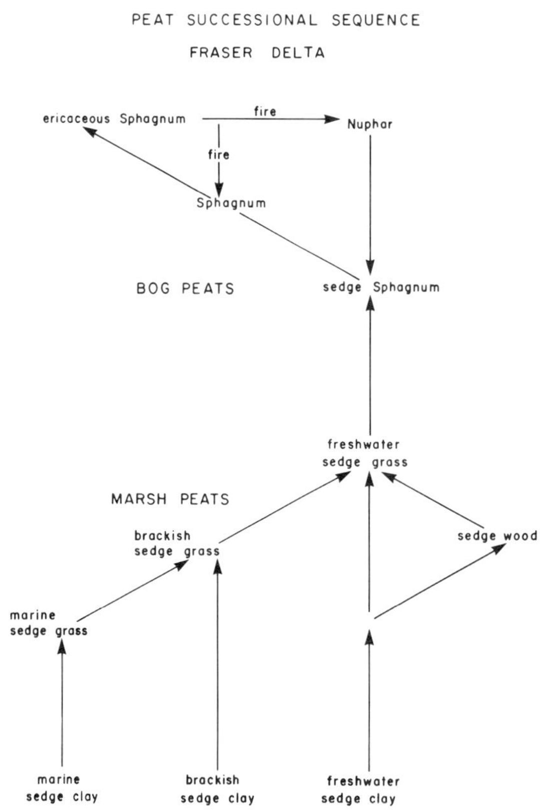

Figure 6. Successional sequence of plant biofacies, Fraser River Delta. Modified from Styan and Bustin (1983d).

In the ericaceous *Sphagnum* biofacies, the final stage of the successional sequence, fire controls both species composition and bog evolution. As a result of changing substrate elevation, pH, and nutrient supply, components of the successional sequence may be absent or repeated. Hebda (1977) utilized the influence of a fire to explain the characteristic hummocky topography in Burns Bog on the lower delta plain. The stratigraphic sequence predicted by his model: charcoal, followed by a thin layer of gyttja, sedge, *Sphagnum* peat, and finally ericaceous *Sphagnum* peat was observed throughout the bog. Pure *Sphagnum* peat also directly overlies charcoal horizons. In these situations prolonged burning must have formed depressions well below the water table. The pools of water which eventually formed were conducive only to

Figure 7. Open sedge grass marshes, Fraser River Delta at Lulu Island (from Styan, 1981).

Figure 8. Salt marsh peat, Fraser River Delta at Boundary Bay. The peat is presently being eroded by transgressing marine waters.

Sphagnum growth. Stagnant ditches observed in the bogs at the present time have a luxuriant growth of *Sphagnum* covering the bottom. The extent to which any fire burns is dependent upon the level of the water table at that time.

In warm humid climates the proportion of arborescent vegetation is much greater than in the temperate areas such as the Fraser River Delta. In the Okefenokee Swamp, the Everglades and on the Mississippi Delta the pioneer vegetation are specialized trees such as *Taxodium distichum* (swamp cyprus) and *Rhizophora mangle* (red mangrove, Fisk, 1960; Cohen and Spackman, 1976). In open water areas gytjja accumulates from mud and organic detritus that is blown in and from water plants including algae.

The different plant communities contribute different amounts and types of organic matter to peat. Mosses and herbaceous plants contribute higher quantities of fine detrital material because of their high rates of decomposition. Peat composed of arborescent vegetation, which contains abundant lignin and celluose, is more resistant to degradation and thus forms a woody peat. If *Sphagnum* is present, the resulting low pH (3 to 5) may retard bacterial decomposition and the associated peat will undergo less decomposition and thus be enriched in cellulose and hemicellulose which would be decomposed under more alkaline conditions.

2.3.2.3 **Depositional Setting: pH, Sulphur, Bacterial Activity and Nutrient Supply**

The depositional setting of coal (peat) has a decided effect on composition. Peats deposited in a brackish-marine environment are commonly rich in sulphur, hydrogen, nitrogen, and ash. Because of the higher pH, resulting from the neutralizing effects of seawater, bacterial degradation is much greater in brackish-marine environments. Most bacteria operate best under neutral to weakly alkaline conditions, whereas fungi can tolerate a pH as low as 4. With moderation of pH by seawater, anaerobic bacteria reduce the sulphate available from seawater facilitating the formation of sulphide (mainly pyrite) resulting in high sulphur peats (and coals). Casagrande *et al*., (1977) compared sulphur contents of marine peats from the Florida Everglades and freshwater peats of the Okefenokee Swamp of Georgia. They found in the marine peats the sulphur content increased with depth in the peat up to 10 per cent, most of which is pyritic sulphur. In the nonmarine peats the total sulphur content is lower with mean sulphur values of less than 0.2 per cent. Casagrande *et al*. (1977) attribute the high sulphur contents in the marine peats to the availability of sulphate, and a near neutral pH which enables production of hydrogen sulphide by *Desulfovibrio desulfuricans*. Studies of total sulphur in peats of the Fraser Delta (Styan, 1981) show similar results: in brackish marine peats at Boundary Bay total sulphur averages 6 per cent, whereas freshwater peats from Lulu Island have sulphur contents of less than 1 per cent (Figure 9).

The high hydrogen and nitrogen content of brackish-marine peats results from the incorporation into the peat of bacterial proteins and metabolic products. Casagrande (1971) found, for example, that in some Florida peats micro-organisms contributed at least 5 to 10 per cent of the total organic matter. Coals originating from brackish-marine peats thus have higher sulphur, nitrogen, and hydrogen contents resulting in anomalously high volatile yields (high bitumen content) and lower softening temperatures during carbonization than nonmarine coals of similar rank. The high ash content of brackish-marine peats (and coals) may result from intense degradation of the organic matter resulting in concentration of the inorganic fraction.

Coal seams overlain by marine strata and coals deposited in calcium-rich environments have similar compositions as brackish-marine coals (Teichmüller and Teichmüller, 1982). The high sulphur content of coals overlain by marine strata is attributed to the availability of sulphate ions and moderating pH of the marine waters. In calcium-rich swamps there is a similar moderate pH which promotes bacterial degradation of the peat, resulting in higher than normal hydrogen, nitrogen, and sulphur contents and in some cases the formation of dopplerite (calcium humate). Cecil *et al*. (1979 in Teichmüller and Teichmüller, 1982) found in their study of Pennsylvanian coals that the low ash and sulphur coals were mainly nonmarine in origin; where limestone occurred in adjacent strata they found moderate ash and sulphur contents; and where the coals were adjacent to marine strata, the coals were found to have high sulphur and ash contents.

Freshwater swamps, particularly in temperate to cool climates, may have a low pH. *Sphagnum* peats on the Fraser Delta have pH values between 3 and 4, and associated plant remains are consequently very well preserved. Waksman (in Teichmüller and Teichmüller, 1982) found in raised bogs (which have low pH) higher concentrations of cellulose, hemicellulose, fats, and waxes, but little protein. Under more neutral pH conditions, the cellulose and hemicellulose would be degraded and greater quantities of bacterial protein would be incorporated.

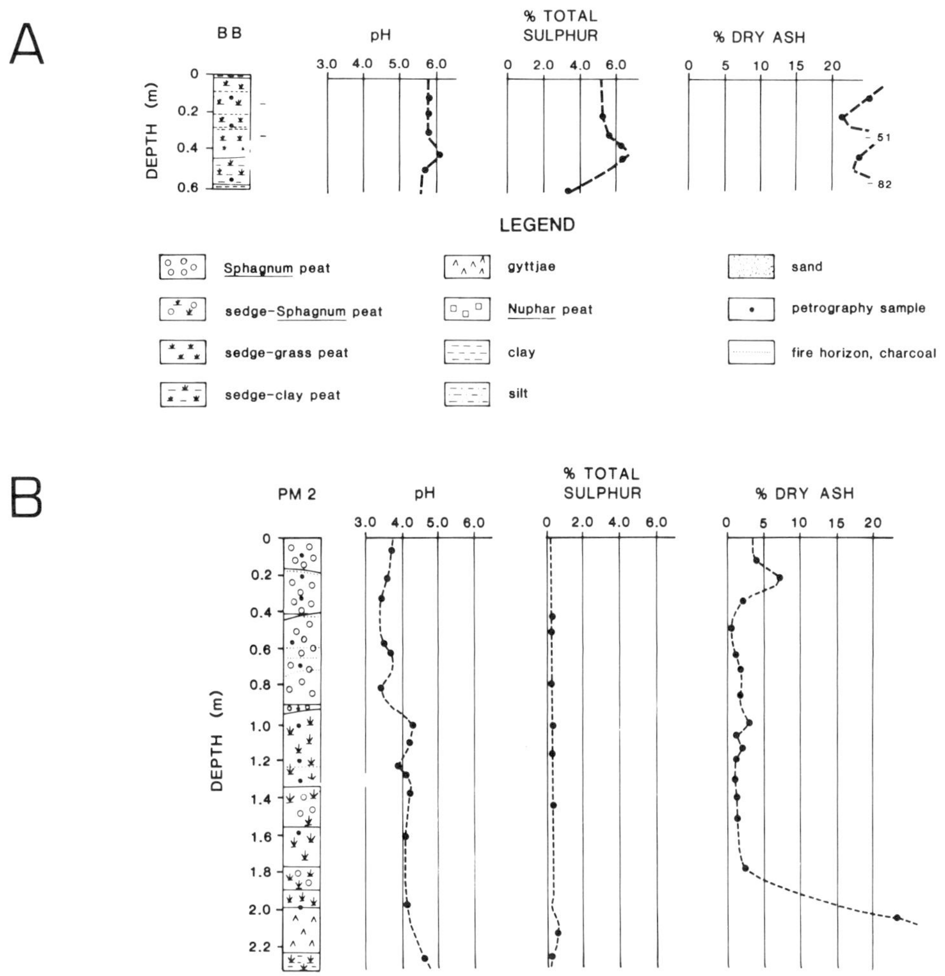

Figure 9. Sulphur content of brackish marine peats at Boundary Bay (A) and freshwater peats at Pitt Meadows (B), Fraser River Delta.

The nutrient supply to the swamp affects the type and diversity of species. Low moors are generally rich in nutrients as a result of groundwater inflow, whereas high moors are nutrient poor because the moisture is derived mainly from rainfall (Teichmüller and Teichmüller, 1982). The flora of low moors is thus more diverse and luxuriant, while high moors are generally characterized by low diversity of species. Most coals are considered to have been deposited in nutrient-rich (eutrophic) swamps.

2.3.2.4 **Redox Potential**

The oxidation-reduction potential (Eh) of the peat is of paramount importance in determining whether or not peat, if formed, will be preserved (Table 3). If an abundant supply of oxygen is available from flowing groundwater or, if there is access to the atmosphere, the peat may oxidize completely to gaseous products. With lower oxygen availability, mouldering occurs as a result of aerobic bacteria and fungal activity. Peatification occurs with restricted oxygen supply but some oxygen is required to form humic acid from lignin. During putrefaction anaerobic bacteria consume oxygen and organics, resulting in the generation of methane, carbon dioxide, and hydrogen.

TABLE 3. EFFECTS OF OXYGEN AVAILABILITY ON TRANSFORMATION OF PEAT (MODIFIED FROM POTONIE, 1920 in TEICHMÜLLER AND TEICHMÜLLER, 1982)

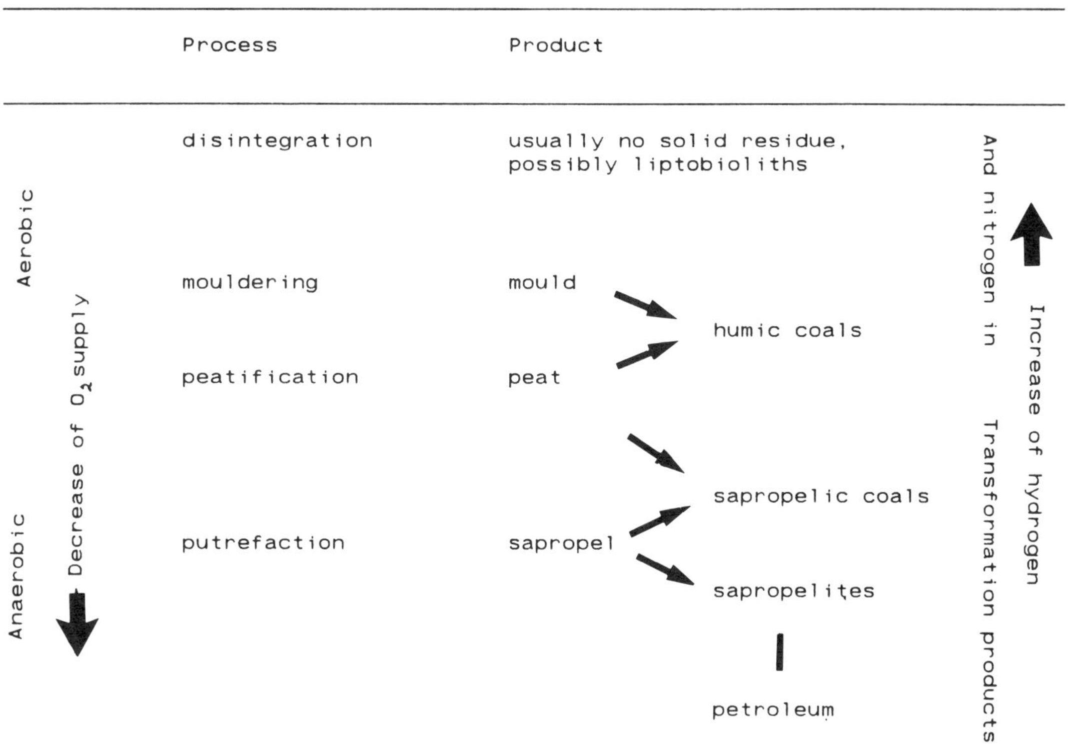

2.4 **Origin of Sapropelic Coals**

Sapropelic coals are deposited in subaquatic environments under anaerobic conditions (Teichmüller, 1982). They are composed of allochthonous material such as spores, pollen, and degraded peat and autochthonous algal material and water plants. Sapropelic coals are generally thin (less than 50 centimetres), lenticular, and generally occur at the base or top of coal seams (Francis, 1961; Moore, 1968). They are dull brown in colour, have a uniform texture, and conchoidal fracture.

Cannel coals are composed of degradation products of peat that accumulate in shallow basins. During degradation of peat the resistant plant parts which are waxy and resinous are concentrated at the expense of the more easily degraded peat components. Most cannel coals contain high contents of vitrinite though always in a finely comminuted form. Boghead coals are derived mainly from algal material that grow insitu. They form in basins in which transport of allochthonous organic material is restricted but where the water is sufficiently aerated to enable algae to flourish. Boghead coals may also be deposited in small, shallow, basins where drainage was received from restricted seepage through peat layers or aerated water free of humic material (Moore, 1968). When high proportions of inorganic material are present the boghead coals may grade to oil shales. In some basins boghead coals may be deposited in the centre where sufficient oxygen exists (because of the paucity of detrital organics), whereas cannel coals may accumulate near the margin.

Recent examples of sapropelic 'muds' are accumulating in glacial lakes in Russia where sapropels up to 40 metres thick have been reported (Teichmüller, 1982). On the modern Fraser River Delta and other peat swamps, peats rich in sapropelic constituents are presently accumulating in small ponds on the peat surface. On the Fraser Delta the ponds result from periodic fires that burn into the peat surface forming depressions.

2.5 **Coalification**

Coalification refers to the progressive transformation of peat through lignite, sub-bituminous, bituminous, semi-anthracite, anthracite to meta-anthracite coal (Figure 10). The degree of transformation of the coal within the series is referred to as the coal rank. White (1908, in Francis, 1961) suggested that two stages of coalification could be recognized in the progressive transformation of peat to coal: an early stage of biochemical coalification; and, a later stage of metamorphism or dynametamorphism. White's metamorphic stage is now generally referred to as the geochemical stage of coalification (Teichmüller and Teichmüller, 1982).

2.5.1 *Biochemical Stage of Coalification*

The biochemical stage of coalification encompasses those changes that occur within the peat swamp during deposition and in coals up to bituminous rank. The most intense alteration of the peat occurs at the surface to a depth of about 0.5 metres (Teichmüller and Teichmüller, 1982). In this region aerobic bacteria, fungi, and actinomyces degrade the organic material. The components most susceptible to degradation are in order: (1) protoplasm, (2) chlorophyll, (3) oils, (4) carbohydrates (starch, cellulose, lignin), (5) epidermis, seed coats, (6) pigments, (7) cuticles, (8) spore and pollen exines, (9) waxes, and (10) resins (Waksman and Stevens in Francis, 1961). With increasing depth of burial, aerobic bacteria are replaced by anaerobic bacteria and microbial life progressively diminishes and is essentially absent at depths greater than 10 metres (Teichmüller and Teichmüller, 1982).

Depending on the degree of degradation (which is a function of water table level, rate of burial, pH, etc.), the easily hydrolizable compounds such as protoplasm, chlorphyll, etc., are converted to gaseous products such as carbon dioxide, ammonia, methane, and water, which are lost. The other metabolites of micro-organisms and the remaining plant constituents are incorporated into the peat. The lignin, cellulose, and amino acids plus carbohydrates, together with some tannins, resins, and waxes, are progressively oxidized forming humic acids.

The transformation of these components to humic acids is poorly understood (Wollrab and Streibl, 1969). Oxygen is considered to be necessary for the transformation of lignin to humic acids (Teichmüller and Teichmüller, 1982). Such a transformation is apparently facilitated by fungi and aerobic bacteria. Cellulose is more easily degraded by hydolysis than lignin and thus is lost more rapidly. Flaig and Schulze (*in* Wollrab and Streibl, 1969) suggest that degradation of cellulose produces reactive ketoacids which show a strong tendency toward aromatization. With increasing depth and advent of reducing conditions, the humification of lignin, cellulose, and other compounds continues by intramolecular-reduction processes during which oxidized humic acids form on one hand, and strongly reduced compounds form on the other hand (Teichmüller and Teichmüller, 1982). The humic acids that form are the most characteristic products of peatification. They are dark brown to red-brown and by definition are soluble in alkalies. Humic acids can be extracted with alkalies up to the subbituminous coal stage.

Tissot and Welte (1978, p. 200) describe the formation of humic substances as occurring in two separate stages: firstly, plant material is mechanically disintegrated and depolymerized into aromatic, phenolic, and carboxylic groups with the aid of micro-organisms, and secondly, the humic polymers undergo 'random repolymerisation and polycondensation' of the molecular types. With progressive diagenesis the humic acids lose the acid groups and turn into humins which are insoluble in alkalies (Teichmüller and Teichmüller, 1982). The process of humification is followed by gelification (Teichmüller and Teichmüller, 1982), 'a physico-colloidal process by which the humic matter passes through a soft plastic gel stage and takes on a swollen appearance.' The resulting organic material is referred to as huminite.

The lipid (hydrogen-rich) fraction of the peat, which includes spore and pollen exines, fats, oils, balsams, etc., are very resistant to aerobic and anaerobic degradation. As a result these components do not undergo humification or gelification.

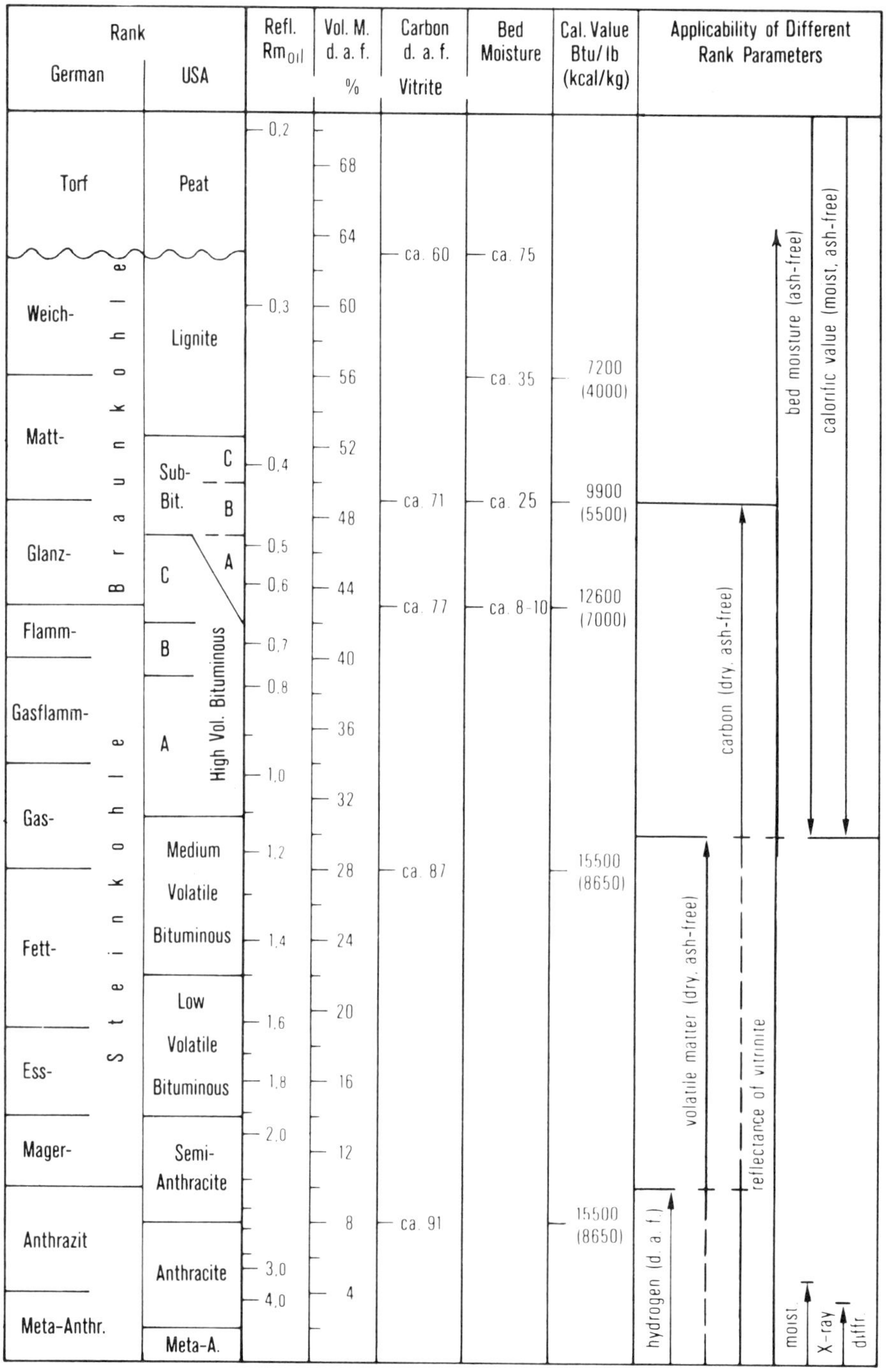

Figure 10. Stages of coalification based on physical and chemical properties of the coal. Modified from Teichmüller and Teichmüller (1982).

Carbon-rich components of the coal, which originate from peat fires, oxidation, moulding, and fungal attack, undergo little or no alteration during the biochemical stage of coalification. Such components, even at the peat surface, are high in carbon, low in hydrogen, and have a high degree of aromatization and condensation (Teichmüller, 1982b).

With progressive burial of the peat, chemical and physical changes occur. As a result of compaction, moisture content decreases and the calorific value increases. Teichmüller (1968) found that moisture decreased about 10 per cent to a depth of 100 metres. In the lignite coal stage moisture content decreases at a rate of 4 per cent per 100 metres and at the subbituminous stage moisture loss is about 1 per cent per 100 metres. With increasing depth of burial there is an associated increase in total carbon, and the amount of free cellulose progressively decreases. Differentiating lignite from peat is based on the presence of free cellulose in peat and the higher moisture content (greater than 75 per cent), low carbon content (daf. less than 6 per cent), and its ability to be cut. In the transformation from peat to lignite Ting (1977) found a four-fold reduction in volume of the organic matter. As the rank further increases, the porosity of the coal decreases and the functional groups -OH, -COOH, -OCH_3, C = O groups and ring oxygen diminish, and the total carbon content increases. In the transformation of lignite to subbituminous coal the remaining lignin and cellulose are humified, and the humic acids lose their functional groups becoming insoluble in alkalies (KOH). The total volatile content of the coal shows little change. At the subbituminous C stage the coal becomes bright and lustrous as a result of gelification of the humins (Teichmüller and Teichmüller, 1982).

Compaction and chemical alteration of the peat and progressive biochemical coalification result in notable petrographic changes. The huminite becomes swollen as it is gelified and its reflectivity in incident light increases and in transmitted light it becomes slightly darker coloured and increasingly opaque. The lipid-rich fraction (liptinite) of the coal shows little change in normal reflected light. In fluorescent light (UV irradiation), however, the peak intensity shifts from 500 nm wavelength in peats to 560 to 580 nm wavelength in lignite (Teichmüller and Teichmüller, 1982). The inertinite shows no visible change. As a result of compaction, woody and other tissues become compacted and the cells flattened.

2.5.2 *Geochemical Stage of Coalification*

The progressive changes of the whole coal with progressive burial are summarized on Figure 11. With increasing depth of burial the carbon content increases and the oxygen content decreases. Hydrogen continues to comprise about 5 to 6 per cent of the coal up to a rank of semi-anthracite after which it decreases (Cooper and Murchison, 1969). The volatile content of the coal decreases with increased coalification, as a result of removal of aliphatic and alicyclic groups accompanied by increased aromatization of the remaining carbon. At the low-volatile bituminous stage (87 to 89 per cent carbon) many coal properties pass through a minimum or maximum (Teichmüller, 1982). Most of the oxygen has been lost, the hydrogen content is high, the density and heat of wetting are at a minimum, and the fluidity during carbonization is at a maximum.

The main by products of coalification are methane, carbon dioxide, and water. Water is lost in the early stages of coalification and the ratio of methane to carbon dioxide progressively increases with rank, particularly in the range of semi-anthracite to anthracite (Cooper and Murchison, 1969). Large amounts of methane may be generated during coalification. Juntgen and Karweil (in Teichmüller, 1982) report that 200 litres of methane per kilogram of coal are generated between the bituminous and anthracite coal stages.

With progressive coalification there is an associated loss in coal volume as a result of compaction and expulsion of volatiles and water. The amount of volume reduction depends on the initial composition of the peat. Teichmüller and Teichmüller (1982) report a 7 to 1 compression of Pteridosperm peat up to low-volatile bituminous stage, whereas Carboniferous gyttjae over the same range had undergone a 20 to 1 loss of volume.

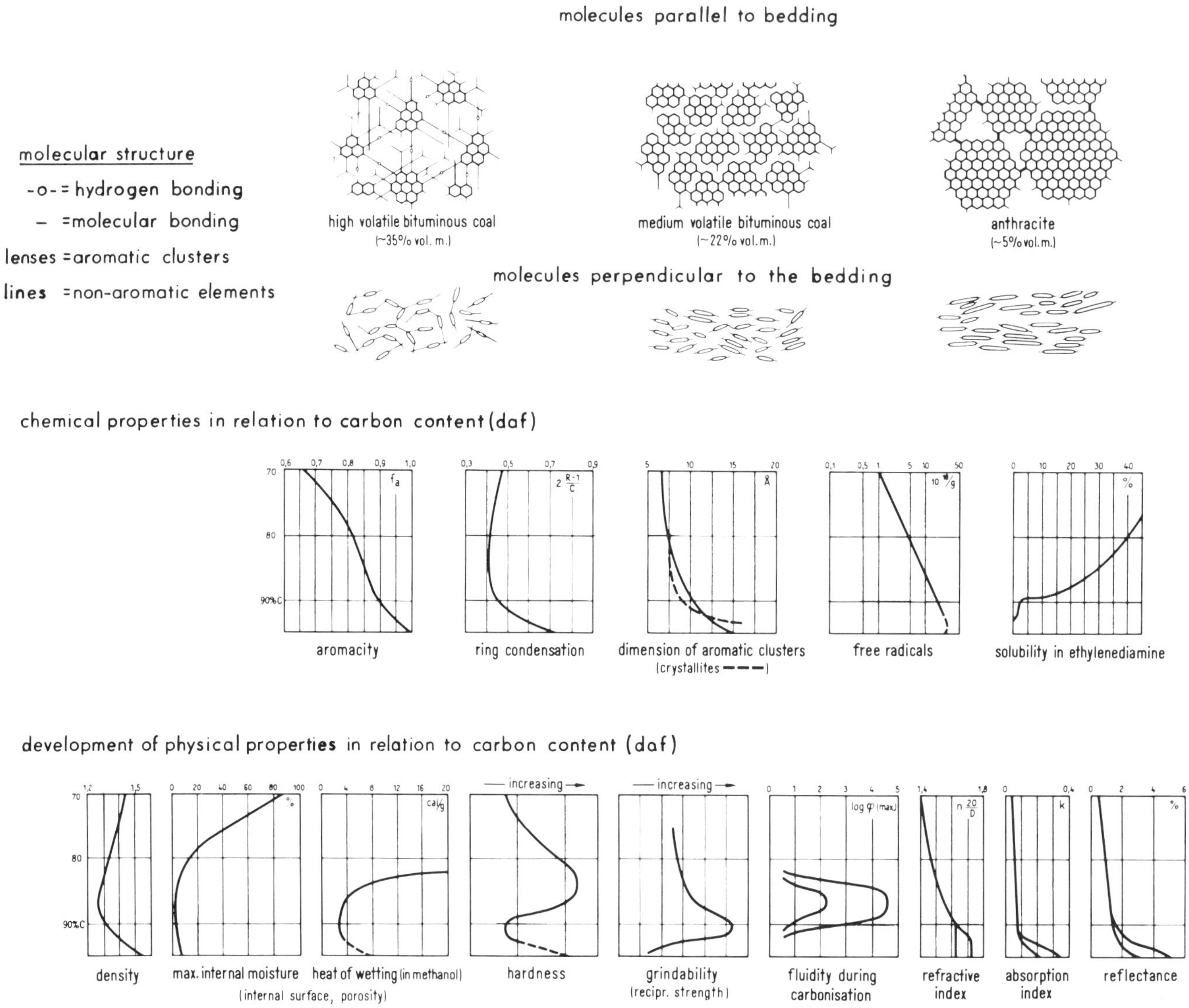

Figure 11. Changes in the chemical and physical properties of coal with increasing rank. Modified from Teichmüller and Teichmüller (1982).

The changes in the chemical and physical properties of the whole coal are the summation of changes to the coal constituents. During progressive coalification the three maceral groups (organic coal components), liptinite, inertinite, and huminite become enriched in carbon, but each component follows a predetermined and distinct coalification 'track.' On Figure 12 the coalification tracks of the liptinites, inertinites, and their derivatives are shown in terms of their hydrogen/carbon and oxygen/carbon ratios.

The chemical and petrographic properties of huminite and its hard coal derivative, vitrinite, alter most uniformly during coalification. With increasing coalification the oxygen and volatile contents decrease and carbon content increases. Hydrogen increases a minor amount and then decreases past the coking coal stage (Teichmüller, 1982b). The aromaticity progressively increases whilst the nonaromatic fraction de-

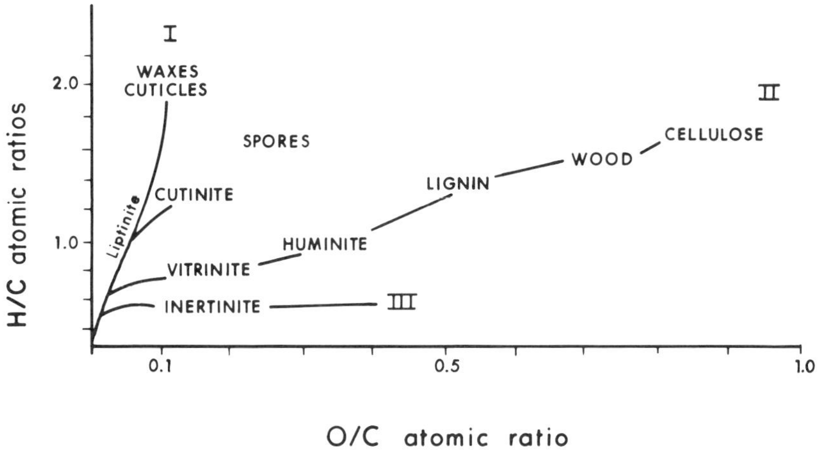

Figure 12. Van Krevelen diagram showing the coalification tracks of liptinite, inertinite, and huminite-vitrinite. Modified from Tissot and Welte (1978).

creases. Changes in the chemical and physical properties of the huminite-vitrinite are, however, not completely uniform and four coalification jumps have been recognized (Teichmüller, 1982b). The first coalification jump occurs during the high-volatile bituminous coal stage (80 per cent carbon), and corresponds with level of organic maturation at which oil generation begins from kerogen. The second jump occurs in the medium-volatile bituminous coal stage (87 per cent carbon) and corresponds to a marked reduction in oxygen and the beginning of methane generation and the death line for oil. The third jump occurs at the boundary between the semi-anthracite and anthracite stages (91 per cent carbon) and corresponds to the release of large amounts of hydrogen as methane and is accompanied by strong aromatization and ring condensation. The fourth jump occurs at the anthracite-meta-anthracite boundary and corresponds to a further release of hydrogen as methane and increased aromatization.

The petrographic properties of huminite-vitrinite change relatively uniformly with increasing rank (Plate 1). In reflected light the reflectance progressively increases as does the bireflectance. In transmitted light the organics become darker coloured and increasingly opaque. Cell structures and plant tissues, readily discernible in low rank coals, become increasingly difficult to recognize. At the low-volatile bituminous coal stage (88 per cent carbon) little cell structure can be recognized and in the semi-anthracite to anthracite stages cell structure can generally only be recognized by etching with acids and using special immersion oils.

Liptinite shows little or no evidence of coalification up to the subbituminous coal stage. At the boundary between subbituminous and bituminous coal liptinite experiences a coalification jump, resulting in a marked reduction in volatiles and an increase in carbon content, corresponding to the level of maturation of oil generation. At the medium-volatile bituminous coal stage liptinite undergoes a second jump resulting in a marked reduction of hydrogen and volatiles. In reflected light the liptinite is darker than huminite-vitrinite in low rank coals. At progressively higher coal ranks the liptinite becomes blue-grey and eventually blends into the vitrinite. At the low volatile coal stage of maturation liptinite assumes the optical (in reflected light) and chemical properties of vitrinite and can only be differentiated from vitrinite by etching (see Section 4.1.1). In reflected light the liptinite is darker than huminite-vitrinite in low rank coals. At progressively higher coal ranks the liptinite becomes blue-grey and eventually blends into the vitrinite. Under ultra-violet light excitation the liptinites show a progressive increase in maximum spectral fluorescence wavelength with increasing maturation and the total fluorescence intensity progressively declines (also see Section 7.3). In low-volatile bituminous coals the fluorescence intensity is so weak it cannot be routinely measured. In transmitted light the liptinite, initially pale yellow, becomes progressively darker and eventually black and opaque.

Plate 1. Changes in the petrographic appearance of coal with increasing rank. All photos are taken in reflected light and oil immersion except A which is taken in transmitted light.

A) Peat: Fraser River Delta; allochthonous and partially fusinized (dark black epidermis) *Pinus* needles (Pn) and grass stems (Gs) in a matrix of cell debris.

B) Lignite: Eureka Sound Formation, Ellsmere Island; humotelinite (h) composed of well-defined and little compacted woody tissues. L = liptinite.

C) Subbituminous coal: Eureka Sound Formation, Ellsmere Island; cell structure is readily visible, however, the cells have been compacted to a greater degree than in the lignite. Humotelinite (h), liptinite (l), inertinite (i).

D) High-volatile bituminous coal: Belly River Formation, southern Alberta; cell structure of vitrotelinite (v) is still visible but much less distinct than the lower rank coals. Carbonate (c), the bright specks are inertinite.

E) Medium-volatile bituminous coal: Mist Mountain Formation, Elk Valley, British Columbia; cell structure is no longer visible in the vitrinite (v) but liptinite (l) is still distinguishable.

F) Semi-anthracite: Mist Mountain Formation, Canmore, Alberta; the vitrinite (v) is homogeneous, liptinite is no longer distinguishable and the reflectance of vitrinite is close to that of inertinite (i).

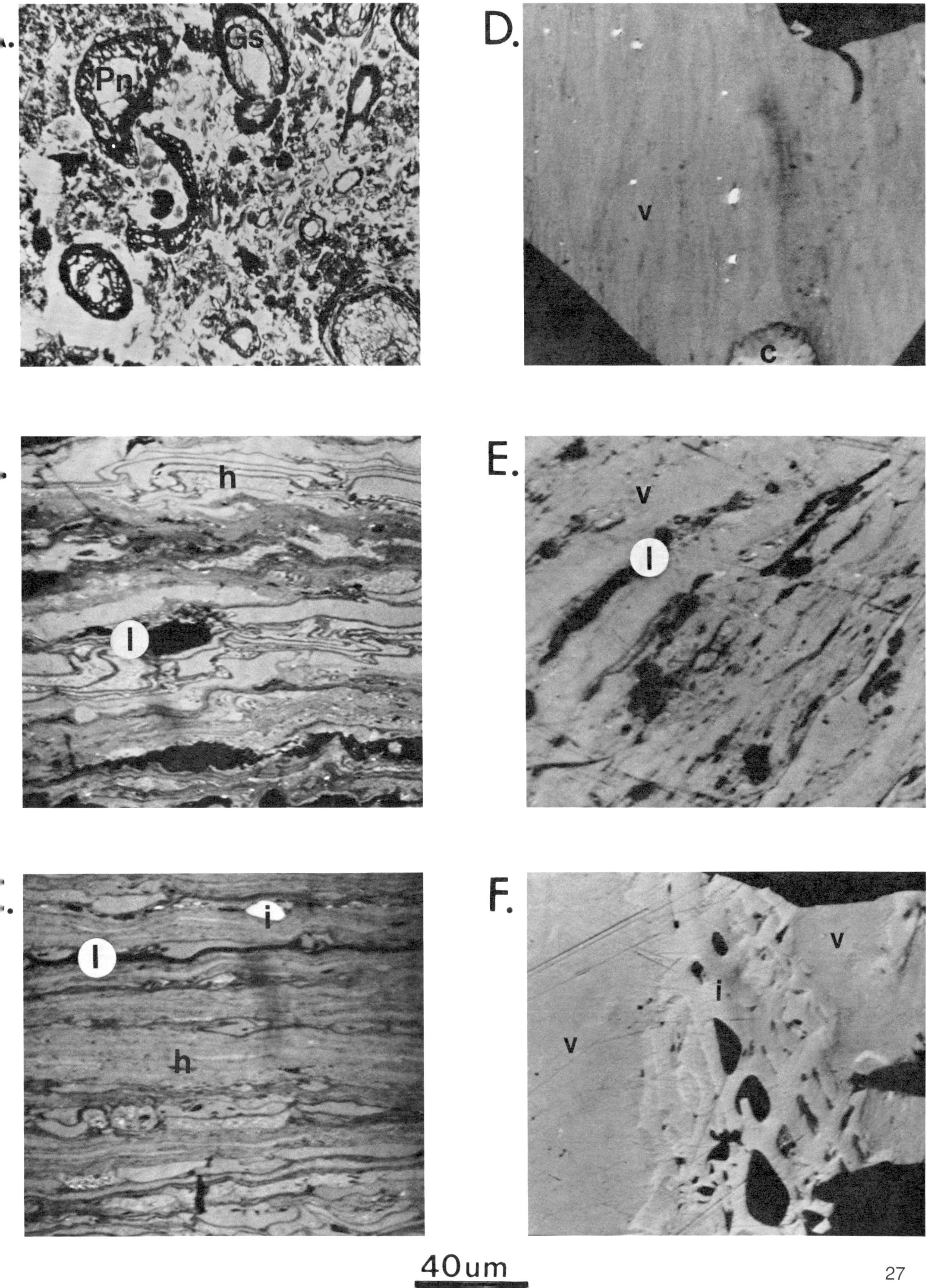

Gs
Pn
D.
v
c
h
l
E.
v
l
i
l
h
F.
v
i
v
40um

Plate 2. A and B. Cannel coal - A) reflected light. The coal is composed mainly of vitrinite (v) and liptinite (l) with minor inertinite (i). The liptinite component is mainly spore and pollen exines. Most cannel coals, such as this one, contain abundant vitrinite even though they are massive in hand specimen. B) Same sample as A but under blue-light excitation. The bright white material is the liptinite which fluoresces under blue light or ultra-violet light. The short stubby liptinite macerals are pollen exines, whereas the long thin liptinite are likely leaf cuticles. C and D. Boghead coal - C) reflected light. The coal is comprised mainly of liptinite which originated from algal bodies (*Botryococcus*) (1). The groundmass material is mineral matter. D) Same sample as C but under blue-light excitation. The bright white material is the algal bodies which fluoresce under blue light or ultra-violet light excitation.

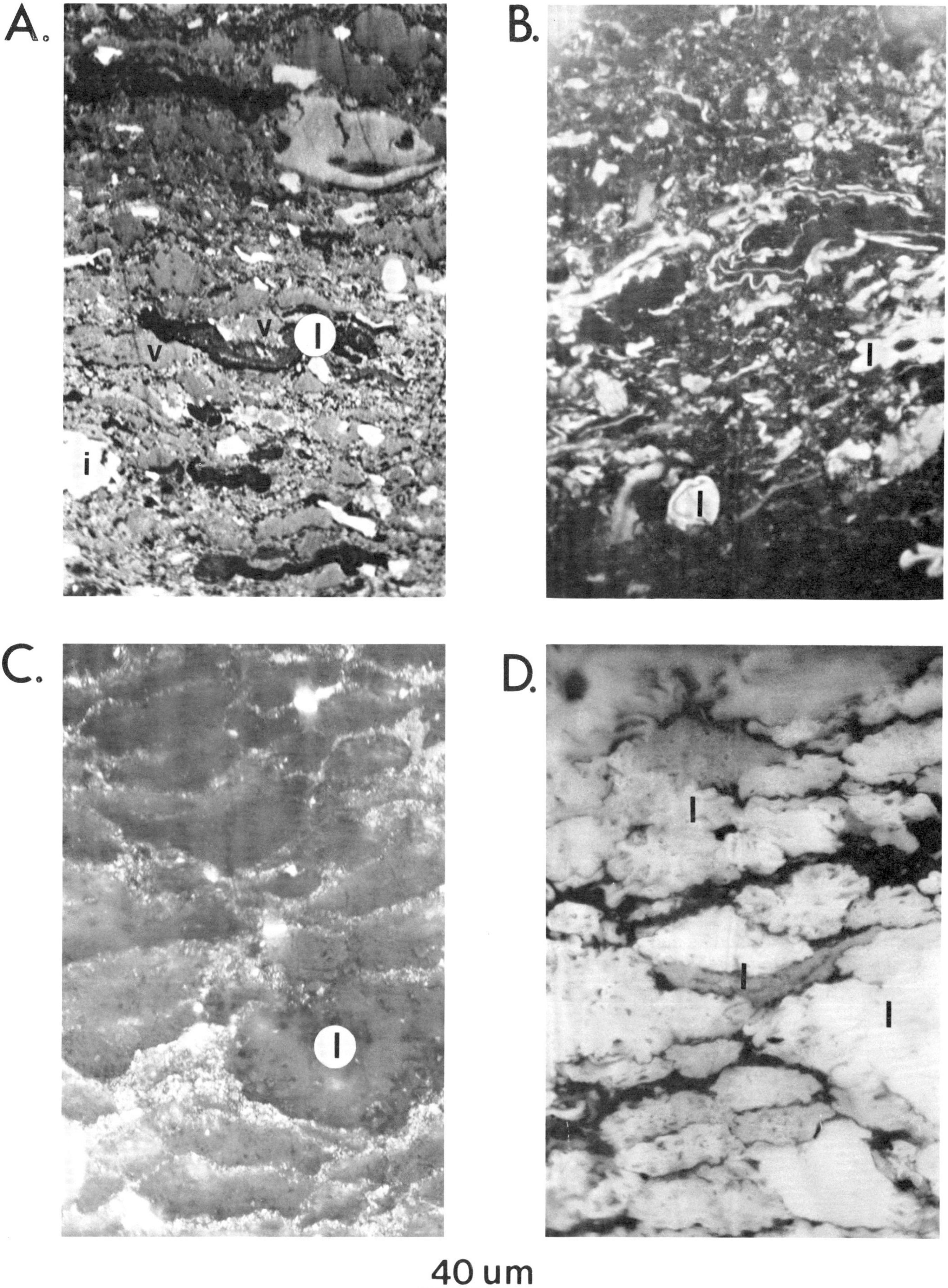
A.
v
v
l
i
B.
l
l
C.
l
D.
l
l
l
40 um

Inertinite is rich in carbon and strongly aromatized at deposition and thus undergoes little additional alteration. With increased levels of coalification any remaining oxygen and hydrogen are eliminated and the carbon content increases. The degree of later alteration depends on the initial degree of charring. The reflectivity of inertinite in incident light is initially high and only at advanced stages of coalification (semi-anthracite unward) does any increase in reflectivity occur. In transmitted light the inertinite is opaque throughout coalification.

During coalification sapropelic coals undergo alteration similar to that of the liptinite component of humic coals, but, in addition, components that would normally be destroyed under aerobic conditions are preserved (Plate 2; Teichmüller, 1982). Bacterial remains may also comprise a significant proportion of the organic material, contributing abundant protein, wax, and lipids. At the peat stage sapropelic coals are enriched in hydrogen relative to humic coals but at advanced stages of coalification (90 per cent carbon) the chemical composition of bogheads, cannels, and humic coals are similar. During coalification significant amounts of bitumen may be generated from sapropelic coals. Bogheads from Torbane Hill, Scotland, for example, yielded an average of 600 litres of oil per tonne during pyrolysis (Francis, 1961, p. 612).

2.5.3 *Causes of Coalification*

The level of coalification (maturation) of organic matter is a product of temperature and length of time of heating. Pressure, although important in the early stage of biochemical coalification, resulting in compaction and expulsion of water, is not considered to have a significant effect during the geochemical stage of coalification. Laboratory studies suggest confining pressure may actually retard maturation (Huck and Patteisky, 1964) but the effect has not been quantified. In most stratigraphic sequences increased temperature (and time) accompanies burial, and thus more deeply buried coals are of generally higher rank. This relationship was first documented by Hilt (1873), who observed a progressive decrease in volatile matter in coals with depth (Hilt's law). Exceptions exist where there is an extraneous source of heat such as igneous intrusion, circulation of hydrothermal solutions, frictional heating, or tectonic complications (see Section 9.2). In normal stratigraphic successions the rank increases with depth. The higher the geothermal gradient the greater the rate of change in coal rank with depth (coalification gradient).

The effect of temperature on coalification cannot be considered separately from the duration of heating. For longer heating durations at the same temperature, the resulting coal rank will be higher. The effect of heating duration on coalification was illustrated by Teichmüller and Teichmüller (1966) from a study of Miocene rocks from Louisiana (17 Ma) and Carboniferous rocks of northwest Germany (270 Ma). Coals from both areas had been exposed to a temperature of about 140 degrees Celsius but the Miocene coals were high-volatile bituminous, whereas the Carboniferous coals are low-volatile bituminous. The influence of time is greater at higher temperatures (Karweil, 1956). The rate of organic maturation is generally considered to double for every 10 degree Celsius rise in temperature as deduced from reaction kinetics (Waples, 1980). Very short durations at high temperatures may not, however, result in rank equilibrium. Also in rapidly subsiding basins the equilibrium temperature for a given depth and geothermal gradient may not be reached (Teichmüller and Teichmüller, 1982) resulting in coal of lower rank than predicted. Methods of resolving the time-temperature history from the study of the levels of coalification are described in Section 9.4.1.

3 SYSTEMS OF COAL RANK CLASSIFICATION

The classification of coal has never been a simple task. Because of the complexity of its chemical and physical properties and its varied uses, a multitude of classification schemes for coal have been proposed over the years using a variety of parameters as terms of reference. Elliott and Yohe (1981) list 14 of the more important approaches. Bases used for classification include: geological age, type (essentially the petrographic composition), grade (mainly content of impurities and heating value), behaviour toward solvents, oxidizability, friability and grindability, and rank.

Of these approaches to classification, rank is one of the more important. The standard rank system used in North America is the ASTM system (American Society for Testing and Materials, 1981). It is based on volatile matter and calorific value, with supplementary definition according to whether a coal is, or is not agglomerating. It does not include petrological parameters. Most people who are concerned with coal in North America are familiar with rank terms such as lignite, bituminous, and anthracite, but often do not know the exact criteria of designation. When terms such as anthracite, lignite, etc. are used in the ASTM classification sense, they mean something very specific as regards volatile matter content or calorific value. They may not, however, mean the same thing when used in the context of another country's classification. As a starting point, then, it is useful to review briefly the ASTM system.

The currently used version of this system is shown in Table 4. The coal rank spectrum is divided into classes by boundary lines which, in some instances, appear rather arbitrary. Although the coal series begins with peat at the low end of the scale, and progresses with increasing metamorphism through meta-anthracite to graphite at the high end, neither one is shown in the ASTM system. This spectrum of maturity or metamorphism is divided into four groups (lignitic, subbituminous, bituminous, anthracitic) and 13 classes. The grouping is based on dry, mineral-matter free volatile matter and moist, mineral-matter free calorific value. The reasons for using two parameters (volatile matter and calorific value) instead of one, and the reasons for using calorific value at the low end of the scale and volatile matter at the high end, are made clear on Figure 13, which shows variation in these two properties with rank. Below the rank of medium-volatile bituminous, volatile matter content changes only slightly and the changes are erratic, while above the medium-volatile - high-volatile bituminous boundary, volatile matter content decreases much more sharply

TABLE 4. ASTM CLASSIFICATION OF COALS BY RANK (FROM ASTM ANNUAL BOOK OF STANDARDS, 1981)

Class	Group	Fixed Carbon Limits, percent (Dry, Mineral-Matter-Free Basis)		Volatile Matter Limits, percent (Dry, Mineral-Matter-Free Basis)		Calorific Value Limits, Btu per pound (Moist,[B] Mineral-Matter-Free Basis)		Agglomerating Character
		Equal or Greater Than	Less Than	Greater Than	Equal or Less Than	Equal or Greater Than	Less Than	
I. Anthracitic	1. Meta-anthracite	98	...	...	2	...	...	nonagglomerating
	2. Anthracite	92	98	2	8	...	...	
	3. Semianthracite[C]	86	92	8	14	...	...	
II. Bituminous	1. Low volatile bituminous coal	78	86	14	22	...	...	commonly agglomerating[E]
	2. Medium volatile bituminous coal	69	78	22	31	...	...	
	3. High volatile *A* bituminous coal	...	69	31	...	14 000[D]	...	
	4. High volatile *B* bituminous coal	...	...	...	...	13 000[D]	14 000	
	5. High volatile *C* bituminous coal	...	...	...	...	11 500	13 000	
						10 500	11 500	agglomerating
III. Subbituminous	1. Subbituminous *A* coal	...	...	...	...	10 500	11 500	nonagglomerating
	2. Subbituminous *B* coal	...	...	...	...	9 500	10 500	
	3. Subbituminous *C* coal	...	...	...	...	8 300	9 500	
IV. Lignitic	1. Lignite *A*	...	...	...	...	6 300	8 300	
	2. Lignite *B*	...	...	...	...	...	6 300	

[A] This classification does not include a few coals, principally nonbanded varieties, which have unusual physical and chemical properties and which come within the limits of fixed carbon or calorific value of the high-volatile bituminous and subbituminous ranks. All of these coals either contain less than 48 % dry, mineral-matter-free fixed carbon or have more than 15 500 moist, mineral-matter-free British thermal units per pound.

[B] Moist refers to coal containing its natural inherent moisture but not including visible water on the surface of the coal.

[C] If agglomerating, classify in low-volatile group of the bituminous class.

[D] Coals having 69 % or more fixed carbon on the dry, mineral-matter-free basis shall be classified according to fixed carbon, regardless of calorific value.

[E] It is recognized that there may be nonagglomerating varieties in these groups of the bituminous class, and that there are notable exceptions in high volatile *C* bituminous group.

with increased metamorphism. The situation is just the reverse with calorific value, which increases more or less regularly at the lower end of the rank scale, reaches a plateau at the medium-volatile rank stage, then decreases slightly.

The ASTM system was originally designed, after much discussion, to provide a standard whereby coals could be compared on a common basis. The prescribed tests for the critical rank parameters are rigidly defined, so that results from different laboratories adhering to the standard can be accepted within relatively narrow tolerance limits. It has weaknesses, including the fact that volatile matter may be affected by the presence of carbonate minerals or abnormal amounts of either of the maceral groups liptinite or inertinite. Nevertheless over the years it has served the coal industry and the coal scientific community reasonably well, the former more so than the latter.

With its emphasis on volatile matter, ash, and moisture, the ASTM system relies heavily on proximate analysis. Other systems have been based on ultimate analysis, that is, the elemental composition of the coal, particularly its carbon content. One of the better known rank classifications based on ultimate analysis is that of Seyler which was first developed for British coals, but was extended later to other coals as well (Seyler, 1924 *in* Elliott and Yohe, 1981). Seyler's scheme was based on carbon and hydrogen contents. An advantage which it possessed in comparison with the present ASTM system was that coals with abnormal maceral distributions could be accommodated more easily in terms of their chemical composition. A disadvantage was the rather complex terminology which Seyler developed for his varieties of coal. According to Berkowitz (1979) the use of the Seyler classification for low rank coals (< per cent carbon) is of questionable validity.

Other parameters have been used or suggested to determine the level of coalification or rank. In peat, lignite and subbituminous coals, moisture content and behaviour with potassium hydroxide solutions and presence of free cellulose have been used. An International Standard for the classification of brown coals and lignites recommends the use of total moisture and low temperature tar yield (Montgomery, 1978). In bituminous and higher rank coals vitrinite reflectance has come to be recognized as one of the most reliable

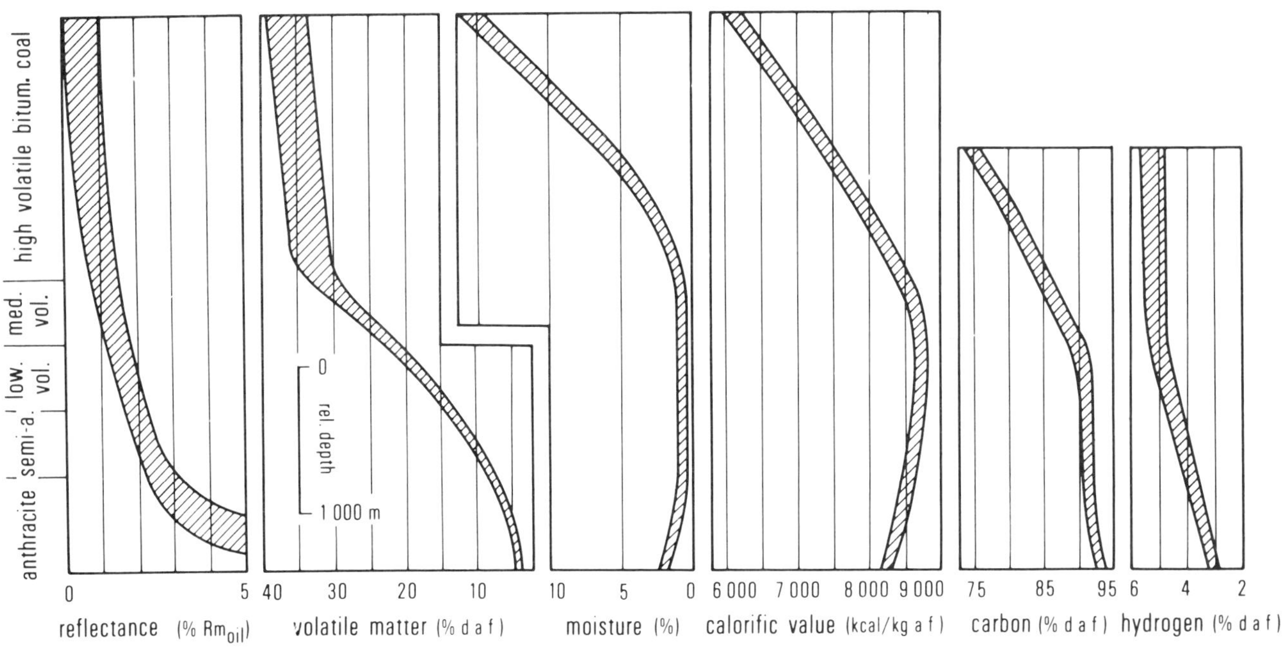

Figure 13. Variation of various chemical and physical parameters with rank. Note especially the changes in volatile matter and calorific value and compare with the way they are used in establishing the ASTM system (from Teichmüller, 1982).

rank criteria (Section 8.1). It is performed microscopically on one component and is thus independent of the relative abundance of the different coal constituents. There is some difference of opinion as to its validity for lower rank coals (McCartney and Teichmüller, 1972), though more recent literature suggests that it may have application for them as well (Castano and Sparks, 1975; Marchioni, in press). Fluorescence microscopy has some application in determining rank of coals lower than low-volatile bituminous by determination of the red/green quotient and maximum wave-length from the fluorescence spectra (Ottenjann *et al*., 1975), but as yet it is not a widely accepted method. Figure 10 (Section 2.5) compares the German (DIN) and North American (ASTM) systems of rank classification and summarizes the range of applicability of various rank-determining parameters.

A classification which involves a utilization parameter, in this case coking type, along with volatile matter was adopted by the National Coal Board in Britain (Table 5). In this scheme coals are divided into groups and classes which are identified by a three-digit code. An International Classification for hard coals (sub-bituminous A coals of ASTM system and higher) has been proposed, which has many similarities with the British (NCB) system. It identifies coal varieties with a three-digit code, characterizes rank by volatile matter or calorific value, and utilization behaviour by coking and caking indices (Berkowitz, 1979).

Recently Alpern (1981) proposed another system for coal classification and suggested it had more potential for universal application than existing classifications. The system embodies three main parameters: *Rank* (as measured by vitrinite reflectance), *Type* (as expressed by proportions of the three main maceral groups) and *Facies* (a term expressing ash content and cleaning aptitute).

TABLE 5. NCB (BRITAIN) COAL CLASSIFICATION SYSTEM (modified from Berkowitz, 1979)

Group	Class	Volatile matter (% dmmf)	Gray-King coke type	Description
100	101	6.1*	A	Anthracite
	102	6.1-9.0*	A	Anthracite
200	201	9.1-13.5	A-G	
	201a	9.1-11.5	A-B	Dry steam coals
	201b	11.6-13.5	B-C	
	202	13.6-15.0	B-G	
	203	15.1-17.0	B-G4	Coking steam coals
	204	17.1-19.5	G1-G8	
	206	9.1-19.5	A-B**	Heat-altered low-volatile bituminous coals
			A-D***	
300	301	19.6-32.0		
	301a	19.6-27.5	G4	Prime coking coals
	301b	27.6-32.0		
	305	19.6-32.0	G-G3	Heat-altered medium-volatile bituminous coals
	306	19.6-32.0	A-B	
400	401	32.1-36.0	G9	Very strongly caking coals
	402	>36.0		
500	501	32.1-36.0	G5-G8	Strongly caking coals
	502	>36.0		
600	601	32.1-36.0	G1-G4	Medium caking coals
	602	>36.0		
700	701	32.1-36.0	B-G	Weakly caking coals
	702	>36.0		
800	801	32.1-36.0	C-D	Very weakly caking coals
	802	>36.0		
900	901	32.1-36.0	A-B	Noncaking coals
	902	>36.0		

*To distinguish between classes 101 and 102, it is sometimes more convenient to use a hydrogen content of 3.35% instead of 6.1% volatile matter.

**For volatile matter contents between 9.1 and 15.0%.

***For volatile matter contents between 15.1 and 19.5%.

4 PETROGRAPHIC COMPOSITION OF COAL

4.1 Nature of Macerals

Coal is not a homogeneous substance but consists of various basic components analogous to the minerals of inorganic rocks. In coal these components are called macerals. In low rank coals in particular they can be further divided into maceral types and maceral varieties. The macerals are classified into three groups: vitrinite, liptinite, and inertinite either because of similar origin (liptinite group) or because of differences in preservation (macerals of vitrinite and inertinite groups). In low rank coals the term huminite is used instead of vitrinite, along with a greater variety of distinguishable macerals, submacerals, and maceral varieties (Table 6). Macerals of the liptinite and inertinite groups are used under the same names in low and high rank coals (Table 7).

Chemical and physical properties of the macerals such as elemental composition, moisture content, hardness, density, etc. and petrographic features differ widely and are also subject to changes in the course of diagenesis and coalification. These differences in chemical and physical properties are reflected in the technological behaviour of the macerals.

TABLE 6. CORRELATION OF THE HUMINITE MACERALS OF BROWN COALS AND LIGNITES WITH THE VITRINITE MACERALS OF BITUMINOUS COALS (AFTER INT. HANDBOOK OF COAL PETROGRAPHY, 1971, 1975, FROM STACH *ET AL*, 1982)

	Brown Coal — Lignite				Bituminous Coal		
Maceral Group	Maceral Subgroup	Maceral	Maceral Type	Maceral Variety	Maceral Type	Maceral	Maceral Group
huminite	humotelinite	textinite		A (dark) B (light)			vitrinite
		ulminite	texto-ulminite	A (dark) B (light)	telinite 1	telinite	
			eu-ulminite	A (dark) B (light)	telinite 2		
	humodetrinite	attrinite densinite				vitrodetrinite	
	humocollinite	gelinite	levi-gelinite: detrogelinite, telogelinite, eugelinite		desmocollinite telocollinite	collinite	
			porigelinite		gelocollinite		
		corpo-huminite	phlobaphinite pseudo-phlobaphinite		corpocollinite		

TABLE 7. MACERALS OF LIPTINITE AND INERTINITE GROUPS

Maceral Group	Maceral	Maceral Type	Maceral Variety
Liptinite	Sporinite		Tenuisporinite Crassisporinite Microsporinite Macrosporinite
	Cutinite Resinite		
	Alginite		Pila-Alginite Reinschia-Alginite
	Liptodetrinite Fluorinite Bituminite Exsudatinite Suberinite		
Inertinite	Micrinite Macrinite Semifusinite		
	Fusinite	Pyrofusinite Degradofusinite	
	Sclerotinite	Fungosclerotinite	Plectenchyminite Corposclerotinite Pseudocorposclerotinite
	Inertodetrinite		

In Table 8 are summarized, in a simplified manner, the classification of macerals, their origin, their microscopic characteristics, and their technological behaviour in coals up to the rank of low volatile bituminous.

TABLE 8. COAL MACERALS: CLASSIFICATION, ORIGIN, AND PROPERTIES (AFTER DAVIS *ET AL.*, 1976)

COAL MACERALS: CLASSIFICATION, ORIGIN, AND PROPERTIES

		ORIGIN		APPEARANCE				
MACERAL GROUP	MACERAL	SOURCE	PROCESS	COLOR IN THIN SECTION	REFLECTED LIGHT	CHEMISTRY	CARBONIZATION	LIQUEFACTION
VITRINITE	TELINITE (CELL WALLS) COLLINITE (CELL FILLINGS, STRUCTURELESS VITRINITE)	WOODY TISSUES, BARK, LEAVES, ETC.	MUMMIFICATION	RED-BROWN	INTERMEDIATE GRAY	INTERMEDIATE HYDROGEN CONTENT AND VOLATILES	PRINCIPAL REACTIVE CONSTITUENT IN COKING COAL	SUSCEPTIBLE TO LIQUEFACTION
LIPTINITE	RESINITE SPORINITE CUTINITE ALGINITE	RESINS SPORE EXINES CUTICLES ALGAE	RESISTANT	YELLOW	DARK GRAY	HIGHER HYDROGEN CONTENT AND VOLATILES; MORE ALIPHATIC	REACTIVE DURING CARBONIZATION	SUSCEPTIBLE TO LIQUEFACTION
INERTINITE	MICRINITE (MASSIVE AND GRANULAR)	PLANT MATERIALS	DEGRADATION PRODUCTS	OPAQUE	WHITE, YELLOWISH, LIGHT GRAY	LOWER HYDROGEN CONTENT AND VOLATILES; MORE AROMATIC	INERT DURING CARBONIZATION	RESISTANT TO LIQUEFACTION
	FUSINITE	WOODY TISSUES, ETC.	FIRE; OR BIOCHEMICAL OXIDATION					
	SEMIFUSINTE	WOODY TISSUES; ETC.	INTERMEDIATE BETWEEN VITRINITE AND FUSINITE					
	SCLEROTINITE	FUNGAL SCLEROTIA, HYPHAE	RESISTANT REMAINS					

AFTER DAVIS ET AL. (1976)

4.1.1 *Petrographic Properties*

The petrographic properties of coals are described either by means of transmitted light or incident light microscopy. Transmitted light can be used readily up to the rank of low volatile bituminous because at higher rank levels the organic material becomes opaque. Incident light microscopy has the advantage that coals over the complete range of rank from peat to anthracite and graphite can be analysed both qualitatively and quantitatively. The main parameters used to distinguish the macerals in incident light are:

(1) Reflectance, anisotropy;
(2) Morphology, relief, size;
(3) Etching; and
(4) Fluorescence.

(1) Reflectance and Anisotropy
In low to intermediate rank coals (lignites to low-volatile bituminous), the three maceral groups can be distinguished by different grey levels or reflectivity as exemplified on Figure 14 where:

Liptinites = dark grey, low reflecting;
Vitrinites = medium grey, medium reflecting; and
Inertinites = light grey to white, high reflecting.

With increase in rank the differences in reflectance between macerals diminish due to an overall convergence of chemical and physical properties (Figure 15).

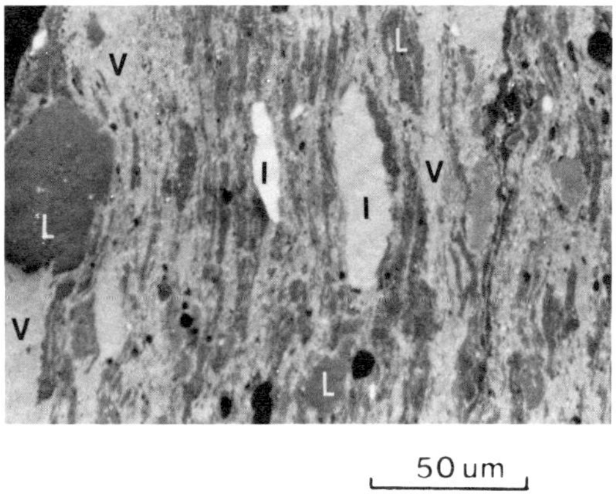

Figure 14. Grey levels of liptinite, vitrinite, and inertinite macerals in a bituminous coal; l = liptinite, v = vitrinite, i = inertinite.

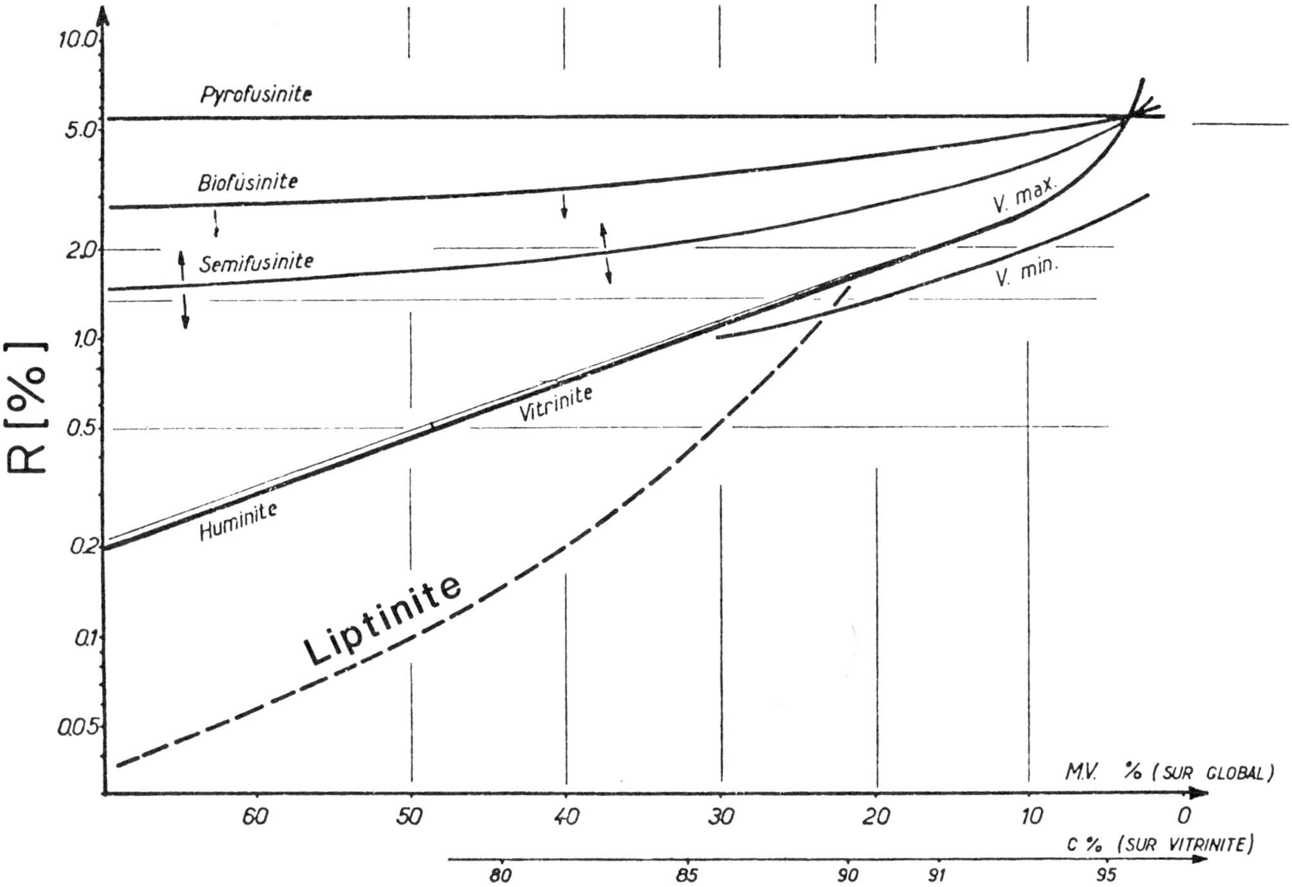

Figure 15. Changes in reflectances of liptinite, vitrinite, and inertinite macerals with increasing rank (modified from Alpern and Lemos de Sousa, 1970).

With increase in rank, coal becomes distinctly anisotropic due to increasing aromatization and orientation of the molecular structure. This effect is measured in terms of maximum reflectance and minimum reflectance and is expressed as:

Anisotropy = Maximum reflectance - Minimum reflectance.

As can be seen from Figure 15 anisotropic effects (Vmax, Vmin) in vitrinite can be observed in high volatile bituminous and higher rank coal. Macerals of the liptinite and inertinite groups are in general isotropic.

(2) Morphology, Relief, Size

Shape (form), size, and internal structures are used to distinguish macerals where reflectance is similar but morphology is different. (Examples are given on Plates 3 to 5). Besides differences in morphology and size, the macerals are characterized by differences in relief when observed in incident light. This effect is caused by variations in hardness. Liptinite macerals, such as alginite, sporinite, and cutinite, tend to show positive relief as compared to the vitrinite matrix, whereas others such as resinite or exsudatinite do not show any significant change in relief as compared to vitrinite. Macerals of the inertinite group are in general characterized by positive relief.

(3) Etching

Etching the polished surface of coal can give better insight into the nature of the macerals, and can be used to enhance differences among macerals, in particular by comparison of etched and unetched portions of the same maceral (see Figure 16). The etching procedure is essentially an oxidizing process. The surface of the coal is exposed to a strong oxidizing agent, for example, a mixture of $KMNOz_4$, H_2SO_4, and H_2O. The duration of exposure to this agent depends on the rank of the coal (details of the procedure are given in Stach, 1982). The etched macerals are prefixed Crypto-, for example, a corpocollinite after treatment would be called cryptocorpocollinite.

(4) Fluorescence

Fluorescence properties of the macerals are described in Section 7.3.

Brief descriptions of the more common macerals follow. For further details see International Handbook of Coal Petrology (1971, 1975) and Stach's Textbook of Coal Petrology (1982).

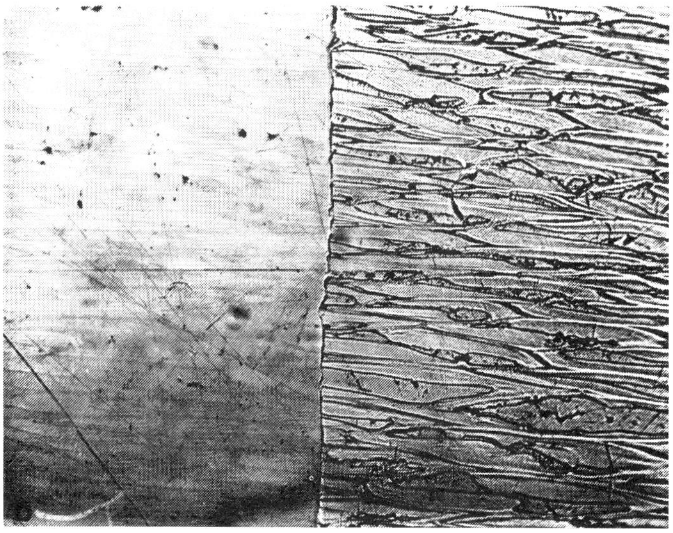

Figure 16. Comparison of etched (right) and unetched (left) portion of a vitrinite (from Stach, 1982).

4.1.1.1 **Huminite/Vitrinite Group**

The huminite/vitrinite group is commonly the most abundant group in coals. The macerals have been derived from woody tissues and bark of trees. The cell structure can often still be found, in particular in low rank coals. The term huminite applies to this group when used in the description of lignites and sub-bituminous coals (see Table 6).

The main characteristics of the huminite macerals are listed below:

Maceral	Origin	Petrological Features
textinite	woody tissue	primary cell wall structure still distinguishable, cell lumina mostly open
ulminite	woody tissue	higher degree of humification; *texto-ulminite* = cell wall structure still visible, *eu-ulminite* = no visible cell wall structure, cell lumina mostly closed
attrinite	finely comminuted	particle size <10 micron degradation product of the other huminite macerals
densinite	same as attrinite	tighter packed than attrinite
gelinite	derived from collodal humic solutions which migrate into existing cavities and precipitate as gels	secondary cell filling
corpohuminite	condensation products of tannins, characteristic for bark tissues	in cross-section globular to tabular shaped

Increase of rank leads to homogenization of the macerals of the huminite/vitrinite group. In bituminous coals the term telinite is used to describe vitrinite characterized by faint cell structure. The term collinite is used to describe homogeneous structureless vitrinite (see Table 6). The term pseudovitrinite is applied to a special form of vitrinite, found and described first in Appalachian coking coals by Benedict *et al.* (1968). The petrological characteristics of pseudovitrinite are: reflectance at least 0.025 per cent higher than normal vitrinite; slit-like openings; presence of cell structure; serrated boundaries of grains.

4.1.1.2 **Liptinite Group**

This group of macerals can be divided into two subgroups. Macerals of the first subgroup can be distinguished by means of normal reflected light supplemented by observations in the fluorescent mode. It comprises coalified exines of pollen and spores (sporinite), resins (resinite), cuticles (cutinite), algae (alginite), and suberinite, a liptinitic cell-wall material typically found in bark tissues. In the lower part of the rank scale these liptinite macerals are characterized by a much lower reflectance than the corresponding vitrinite of the same rank. With increase in rank, the reflectance increases slowly up to the stage of medium-volatile bituminous coals. At this point (ca. 1.1 per cent vitrinite reflectance) the reflectance of the liptinites increases rapidly and reaches the reflectance of the vitrinite (second coalification jump) when the rank of low-volatile bituminous is attained (ca. 1.35 to 1.5 per cent). Macerals of the second group can only be detected and distinguished from mineral matter and mounting medium by fluorescent light. It comprises the macerals fluorinite, bituminite, and exsudatinite. The petrological properties of liptinite macerals are summarized as follows:

Plate 3. Macerals of the huminite/vitrinite group (reflected light, oil immersion).

Figure 1. *Textinite* characterized by wide open cell-lumina and cell-wall layering associated with *corpohuminite* (C) in a lignite from Greece, Tertiary in age.

Figure 2. *Texto-ulminite*, most of the cell-lumina are compressed. Lignite, Areece, Tertiary in age.

Figure 3. *Eu-ulminite* associated with *corpohuminite* (C) in a lignite from Greece, Tertiary in age.

Figure 4. *Eu-ulminite* and *corpohuminite* (C) in a subbituminous coal. Drumheller area, Alberta, Late Cretaceous in age.

Figure 5. *Telinite* in a high-volatile A bituminous coal. Peace River Coalfield, British Columbia, Early Cretaceous in age.

Figure 6. *Attrinite* in Rhenish brown coal, Cologne area, West Germany, Tertiary in age (from Hagemann, RWTH, Aachen).

Figure 7. *Densinite* (D) and *gelinite* (G) in a subbituminous coal. Drumheller area, Alberta, Late Cretaceous in age.

Figure 8. *Desmocollinite* (Dc) associated with macerals of the liptinite and inertinite groups. Sclerotinite (Sc), inertodetrinite (I), and cutinite (Cu). Subbituminous coal, Drumheller area, Alberta. Late Cretaceous in age.

Figure 9. Homogeneous band of *telocollinite* (Tc) and *desmocollinite* (Dc) associated with macerals of liptinite and inertinite groups. Sporinite (S), semifusinite (Sf), and inertodetrinite (I). High-volatile A bituminous coal, Peace River Coalfield, British Columbia, Early Cretaceous in age.

Figure 10. *Telocollinite* (Tc) associated with macerals of the inertinite group, semifusinite (Sf) and inertodetrinite (I). Low-volatile bituminous coal, Peace River Coalfield, British Columbia, Early Cretaceous in age.

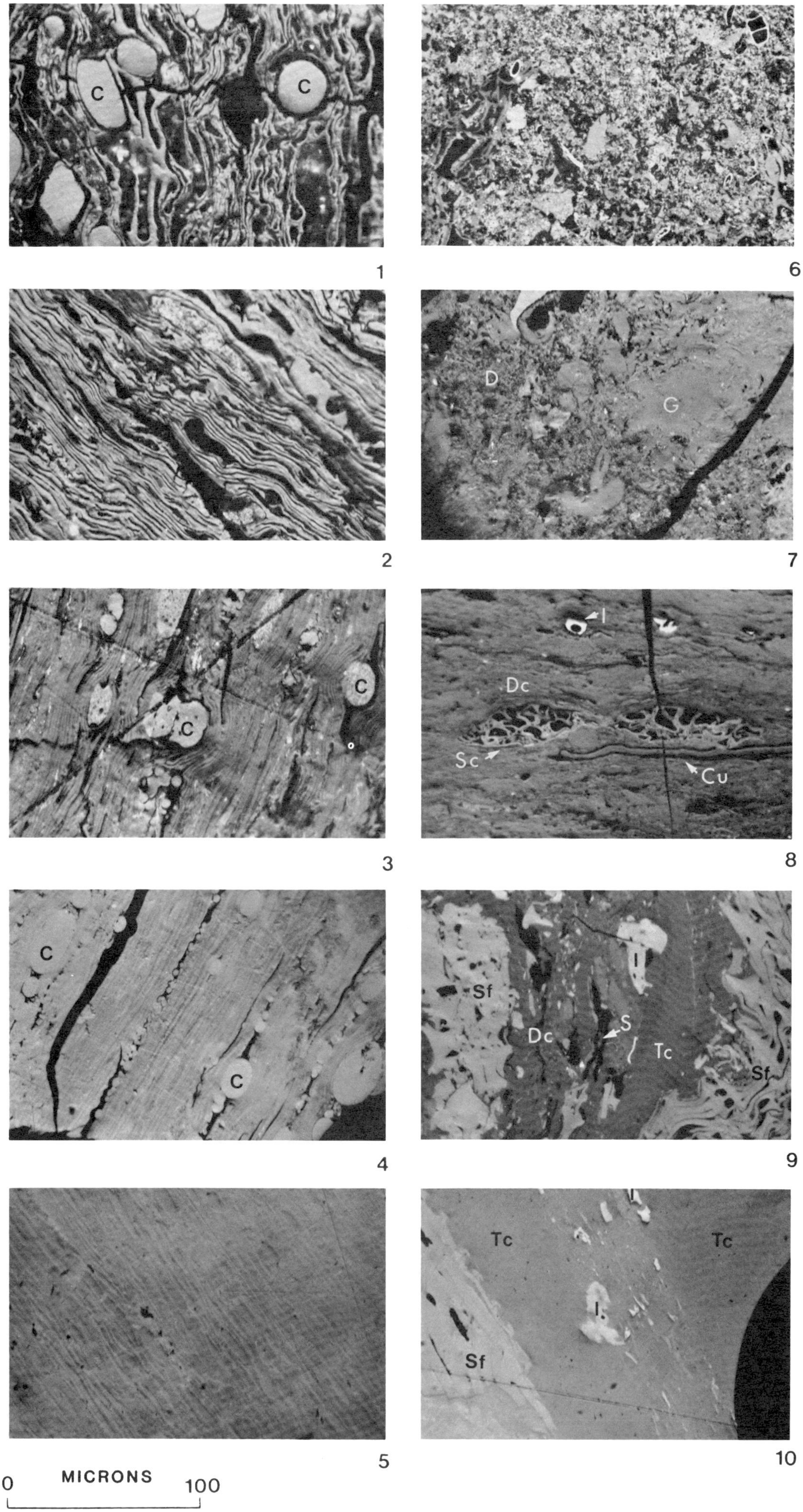
1
C
C
2
3
C
C
C
4
C
C
5
6
7
D
G
8
I
Dc
Sc
Cu
9
Sf
I
S
Dc
Tc
Sf
10
Tc
Tc
I
Sf
0 MICRONS 100

Figure 11. Dark *eu-ulminite A* (A) and light *eu-ulminite B* (B) in a Greek lignite, Tertiary in age.

Figure 12. *Eu-ulminite A*? (dark) and *eu-ulminite B*? (light) in a lignite from Saskatchewan, Paleocene in age.

Figure 13. *Desmocollinite* (Dc) associated with thin walled cutinite (Cu). Medium-volatile bituminous coal, Peace River Coalfield, British Columbia, Early Cretaceous in age.

Figure 14. *Desmocollinite* (Dc) associated with macerals of the liptinite group; thick walled cutinite (Cu) and small fragments of liptinite macerals (liptodetrinite). High-volatile A bituminous coal, Peace River Coalfield, British Columbia, Early Cretaceous in age.

Figure 15. Highly detrital coal. *Vitrodetrinite* associated with inertodetrinite and liptodetrinite. Medium-volatile bituminous coal, Peace River Coalfield, British Columbia, Early Cretaceous in age.

Figure 16. *Phlobaphinite* (P) and eu-ulminite (E) in a longitudinal section of a bark tissue. The black cell-wall material between the phlobaphinite bodies consists of the liptinite maceral suberinite. Lignite from Greece, Tertiary in age.

Figure 17. *Pseudovitrinite,* characterized by serrated boundary of the grain, slit-like openings and the presence of faint cell-structure. High-volatile A bituminous coal, Peace River Coalfield, British Columbia, Early Cretaceous in age.

Figure 18. *Telocollinite* (Tc) associated with the liptinite maceral resinite (R). High-volatile A bituminous coal, Peace River Coalfield, British Columbia, Early Cretaceous in age.

Figure 19. *Desmocollinite* associated with macerals of liptinite and inertinite groups. Sporinite (S), fragments of cutinite (Cu), macrinite (M), and inertodetrinite (I). High-volatile subbituminous coal, Peace River Coalfield, British Columbia.

Figure 20. *Resino-vitrinite*? (R) associated with desmocollinite (Dc) and macerals of liptinite and inertinite groups. Sporinite (S), megasporinite (M), and semifusinite (Sf). High-volatile A bituminous coal, Sydney Coalfield, Nova Scotia. Carboniferous in age.

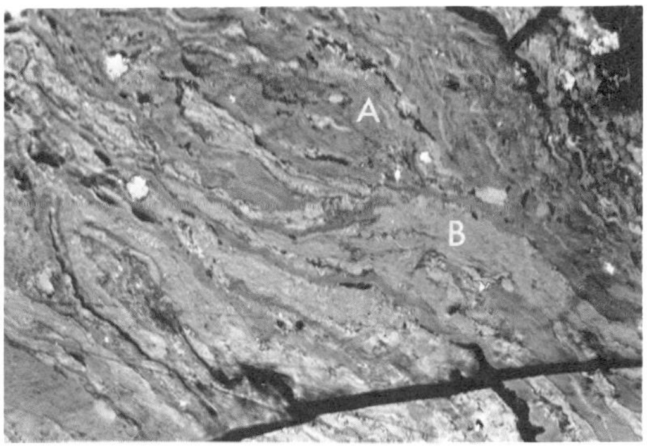

11

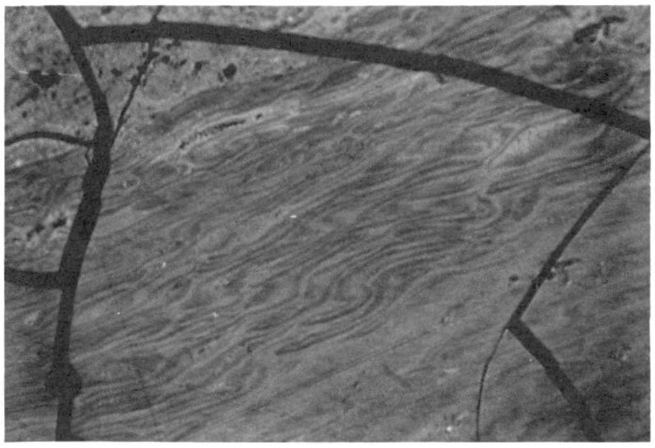

12

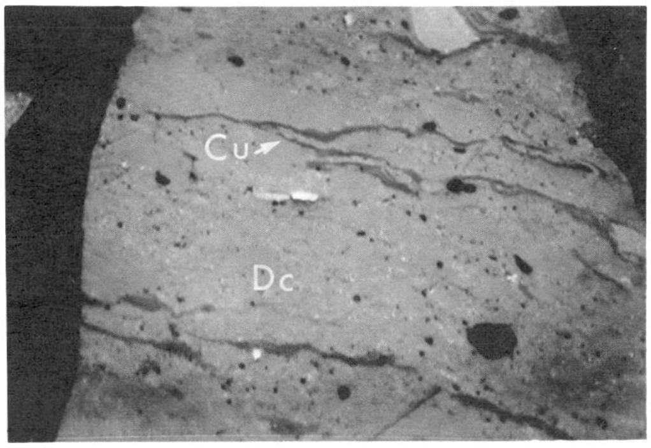

13

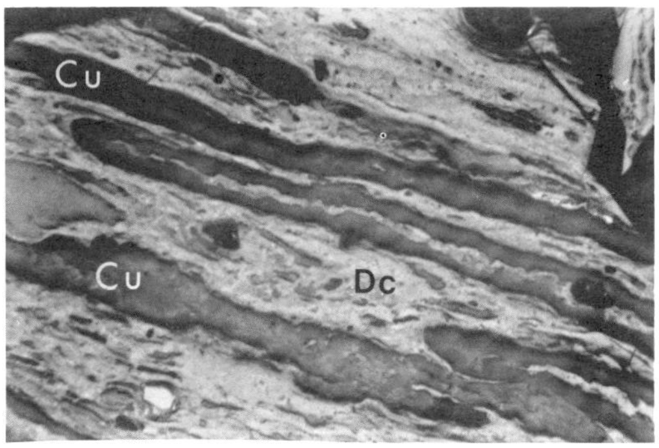

14

15

0 MICRONS 100

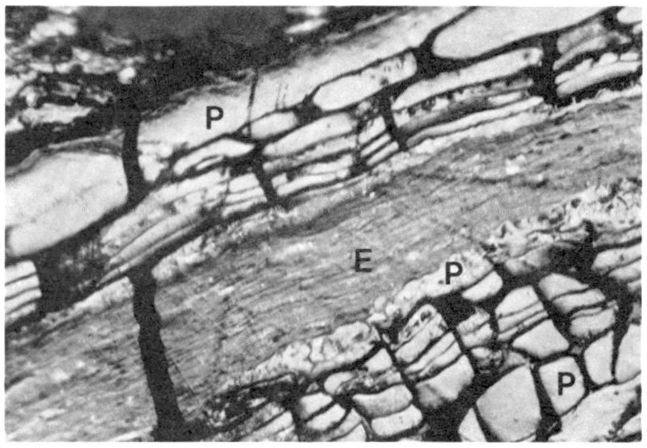

16

17

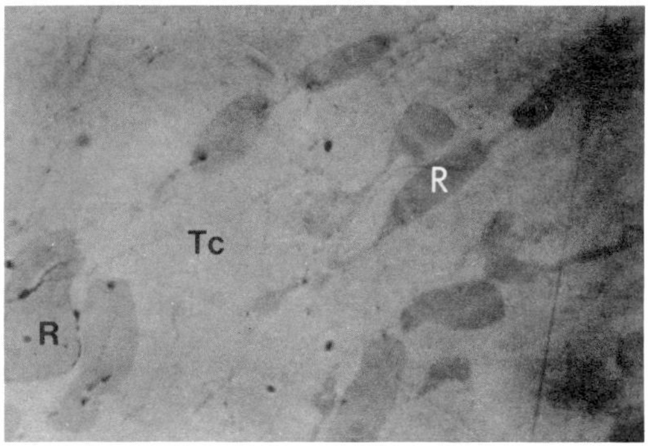

18

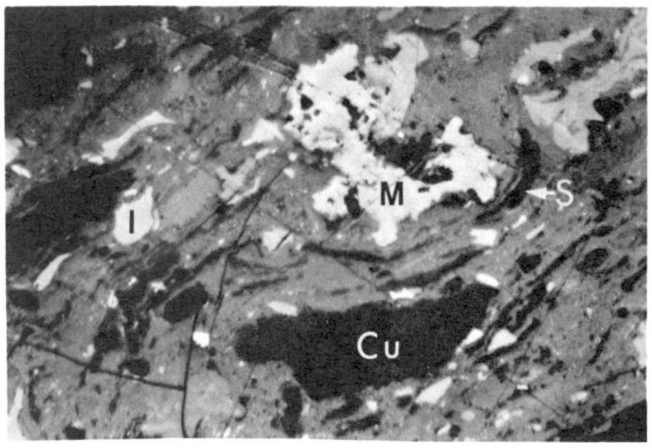

19

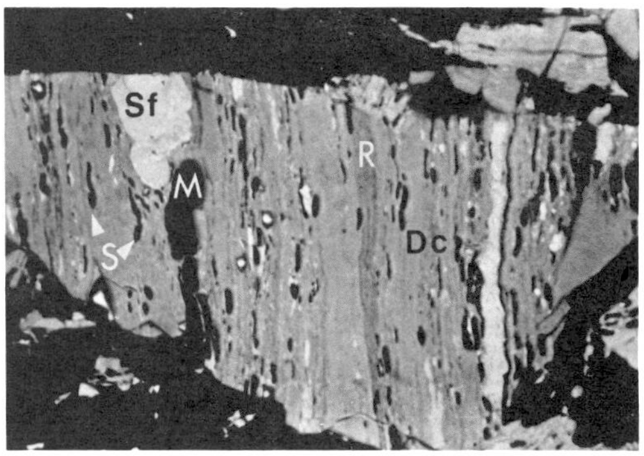

20

Plate 4. Macerals of the liptinite group.

Figure 1a. Normal reflected light; phlobaphinite (P) in attrinitic peat matrix. Peat from Villeroy-Lotbiniere district, Quebec, Pleistocene in age.

Figure 1b. Same section after blue-light excitation; the large amount of liptinite macerals contributing to the peat matrix becomes now visible. *Suberinite* (Su) displays greenish-yellowish fluorescence colours. In the centre a greenish-yellowish fluorescing *pollen grain* (P).

Figure 2a. Normal reflected light; *resinite* (R) characterized by reddish to brown colours caused by internal reflections of the incident light. Greek lignite, Tertiary in age.

Figure 2b. Same section after blue-light excitation; *resinite* (R) displays greenish-yellowish fluorescence colours.

Figure 3a. Normal reflected light; *sporinite* (S) associated with desmocollinite (Dc) and inertodetrinite (I). High-volatile B bituminous coal, Sydney Coalfield, Nova Scotia, Carboniferous in age.

Figure 3b. Same section after blue-light excitation; *sporinite* displays yellow fluorescence colour. Desmocollinite and inertodetrinite do not fluoresce.

Figure 4a. Normal reflected light; *cutinite* (Cu) characterized by serrated edges associated with desmocollinite (Dc). High-volatile A bituminous coal, Peace River Coalfield, British Columbia, Early Cretaceous in age.

Figure 4b. Same section after blue-light excitation; *cutinite* displays orange-brownish fluorescence colour. Desmocollinite does not fluoresce.

Figure 5a. Botryococcus-type *alginite* (A) associated with desmocollinite (Dc) and inertodetrinite (I). Alginite-rich coal from an oil shale zone in the Pictou Coalfield, Nova Scotia; high-volatile A bituminous in rank, Carboniferous in age.

Figure 5b. Same section after blue-light excitation; in contrast to normal reflected light, shape and internal structure of the *alginite* (A) show up much better under blue-light excitation. In fact it is the only safe way to determine whether the dark area in Figure 5a consists of liptinitic material, mineral matter or epoxy resin. At this rank alginite displays a yellow-brownish fluorescence colour. Desmocollinite and inertodetrinite do not fluoresce.

1a

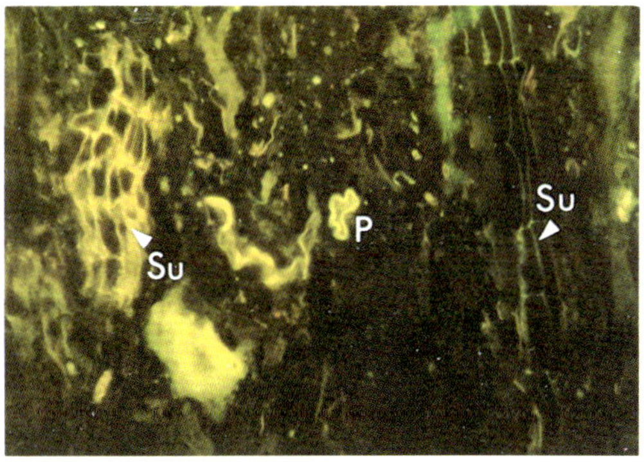

1b

2a

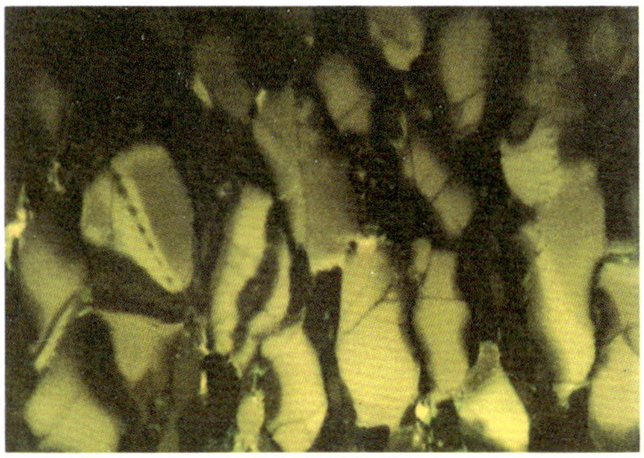

2b

3a

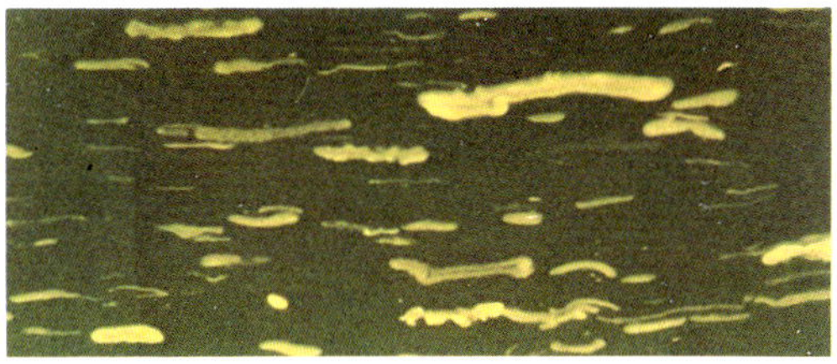

3b

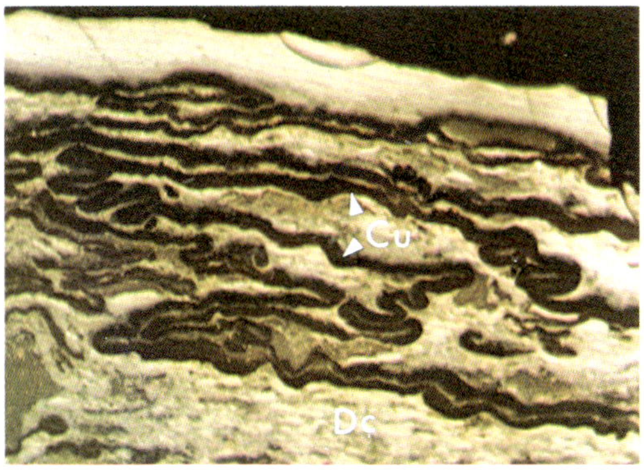

4a

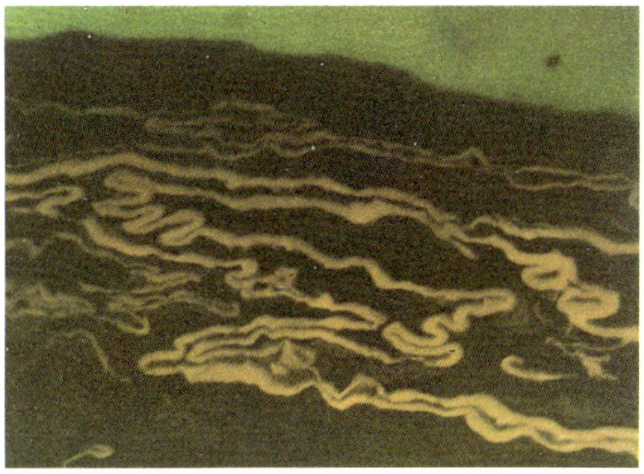

4b

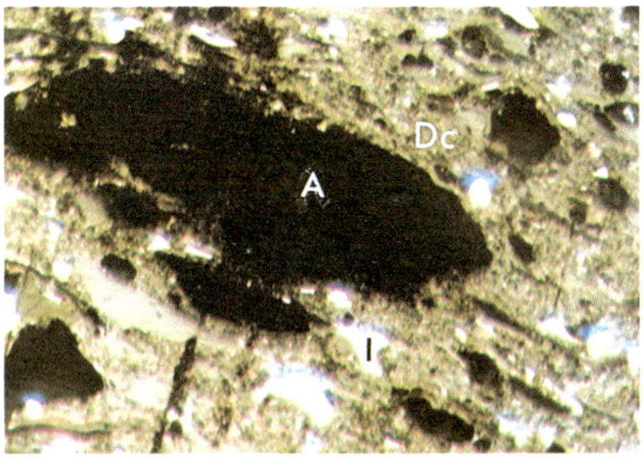

5a

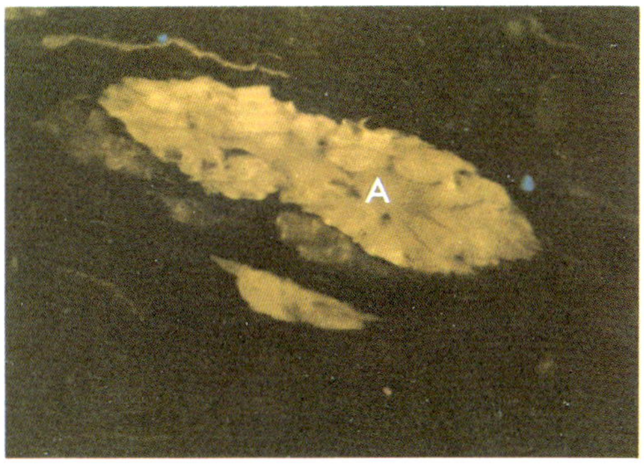

5b

MICRONS

0 100

Plate 5. Macerals of the inertinite group (reflected light, oil immersion).

Figure 1. *Fusinite* in a medium-volatile bituminous coal, Peace River Coalfield, British Columbia, Early Cretaceous in age.

Figure 2. *Fusinite* showing characteristic "Bogen-structure." High-volatile A bituminous coal, Peace River Coalfield, British Columbia; Early Cretaceous in age.

Figure 3. *Semifusinite* (Sf) associated with desmocollinite (Dc) and *inertodetrinite* (I). Medium-volatile bituminous coal, Peace River Coalfield, British Columbia; Early Cretaceous in age.

Figure 4. *Semifusinite* (Sf) associated with *inertodetrinite* (I), desmocollinite (Dc), and liptodetrinite (Ld). High-volatile A bituminous coal, Peace River Coalfield, British Columbia; Early Cretaceous in age.

Figure 5. High reflecting *macrinite* (M) associated with desmocollinite (Dc), *inertodetrinite* (I), and sporinite (S). High-volatile A bituminous coal, Peace River Coalfield, British Columbia; Early Cretaceous in age.

Figure 6. Low reflecting *macrinite* (M) associated with desmocollinite (Dc) and *semifusinite* (Sf). High-volatile A bituminous coal, Peace River Coalfield, British Columbia; Early Cretaceous in age.

Figure 7. Lenses of *micrinite granules* (Mi) associated with desmocollinite and telocollinite (Tc) and mineral matter (MM). Low-volatile bituminous coal, Peace River Coalfield, British Columbia; Early Cretaceous in age.

Figure 8. Occurrence of massive concentration of *micrinite* granules (Mi) associated with desmocollinite. Low-volatile bituminous coal, Peace River Coalfield, British Columbia, Early Cretaceous in age.

Figure 9. Fungal *sclerotinite* (Sc) associated with telocollinite (Tc) and desmocollinite (Dc). Subbituminous coal, Hat Creek deposit, British Columbia; Tertiary in age.

Figure 10. Plectenchymatous *sclerotinite* (Sc) in a subbituminous coal. Hat Creek deposit, British Columbia; Tertiary in age. Photo F. Goodarzi.

Figure 11. Fungal *sclerotinite* (Sc) in a subbituminous coal. Drumheller area, Alberta. Late Cretaceous in age.

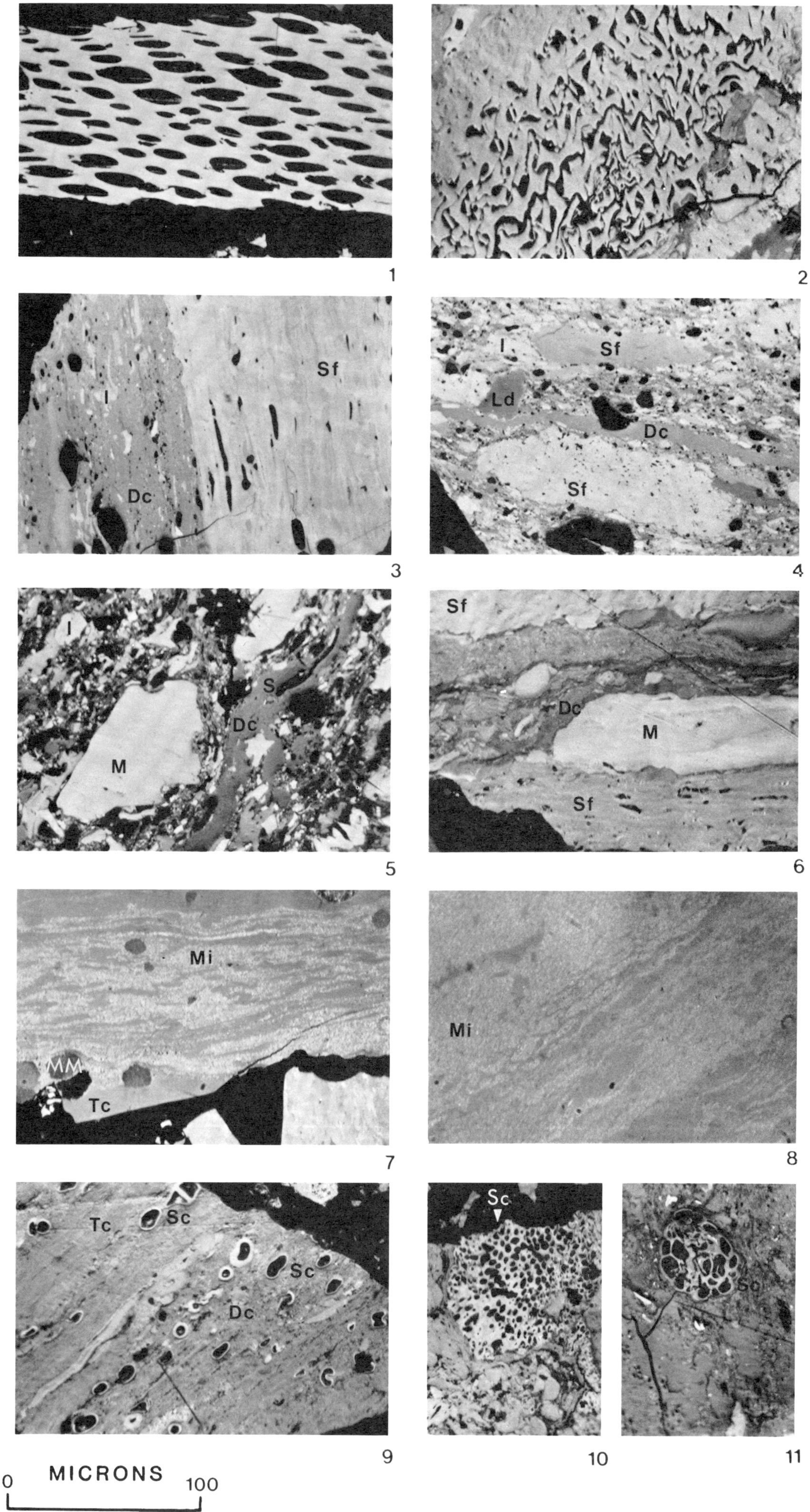
1
2
I
Sf
Dc
3
I
Sf
Ld
Dc
Sf
4
I
S
Dc
M
5
Sf
Dc
M
Sf
6
Mi
MM
Tc
7
Mi
8
Tc
Sc
Sc
Dc
9
Sc
10
Sc
11
0 MICRONS 100

Maceral	Origin	Petrological Features
sporinite	pollen, spores	individual bodies with distinct cell walls, high relief
cutinite	cuticles	serrated edges, high relief
alginite	algae	individual bodies or colonies, high relief
resinite	resin, waxes	secondary cell-filling, often zonation and internal reflections
suberinite	bark tissues	cell-wall material associated with phlobaphinite
liptodetrinite	degradation product of liptinitic macerals	
exsudatinite	secondary maceral, formed from lipid constituents of liptinites and huminites	filling veins and cell lumina, strong fluorescence intensities of varying colours, black in normal reflectance light
fluorinite	uncertain, probably from plant oil	ng fluorescence, black in normal reflected light, no definite shape
bituminite	uncertain, probably decomposition product of algae, animal plankton, and bacterial lipids	no definite shape, ground-mass for other macerals, weak fluorescence, strong alteration after UV-excitation

4.1.1.3 **Inertinite Group**

The inertinite group comprises macerals which in general are characterized by higher reflectances than the corresponding vitrinite of the same rank. The processes which produce some of the inertinite macerals, for example, fusinite and semifusinite, are different from those which produce vitrinite from the same parent material (see Table 8). During the course of coalification the petrographic properties of the inertinites vary very little due to the fact that aromatization has taken place before or at a very early stage after deposition (charring, oxidation, mouldering, fungal attack). Exceptions are the maceral semifusinite, which represents an intermediate stage between vitrinite and fusinite and whose properties are believed to change considerably during the course of coalification, and the maceral micrinite, which is considered to be formed as a secondary maceral at the transition between subbituminous to bituminous coals.

Maceral	Origin	Petrological Features
fusinite	woody tissue	high reflectance, colour white to yellowish, open cell lumina, thin cell walls, 'bogen' structure
semifusinite	woody tissue	reflectance between vitrinite and fusinite, smaller cell lumina often closed, often cloudy appearance, anisotropic
macrinite	uncertain, probably formed through oxidation of gelified plant materials	amorphous bodies of irregular shape, high reflectance
micrinite	secondary maceral, genesis in part related to occurrence of liptinite macerals	small size, ca. 1 micrometre granular appearance, high reflectance
sclerotinite	fungal mycelia, fungal spores	cellular structure, high reflectance
inertodetrinite		fragments of other inertinitic macerals (normally <30 micrometres)

4.2 **Lithotypes and Microlithotypes**

Macerals are the basic building blocks of coal seams, but they do not occur in isolation. They are associated with one another in different proportions and with variable amounts of mineral matter the visible result of which is the characteristic banded or striped aspect of coal (Plate 6A). We distinguish these associations as lithotypes and microlithotypes. The difference between them is a matter of scale; lithotypes are distinguished macroscopically in material of hand specimen size, while microlithotypes, as the name implies, are identified microscopically.

4.2.1 *Lithotypes*

Variations vertically and laterally in the banded characteristics of a coal seam, that is, in its lithotype composition, reflect changes in the plant communities which gave rise to different layers in the parent peat, as well as reflecting changes in the chemical and physical conditions which affected the preservation of plant material in the swamp and the input of mineral matter. This has been amply documented by studies on the vertical changes in composition of layers in modern peat beds (Spackman *et al.*, 1974; see Section 9.3).

4.2.1.1 **Definition**

The ICCP Handbook (ICCP, 1963) defines lithotypes as 'the macroscopically recognizable bands of humic coals', and describes four of them, namely, vitrain, clarain, durain, and fusain. Stach (1982) gave essentially the same definition except that he called lithotypes 'macroscopically recognizable bands of coal seams,' and included the two lithotypes of sapropelic coal (cannel and boghead) in his definition. According to Stach the most conspicuous macroscopic features of sapropelic coals are their dull lustre, homogeneous texture, lack of stratification, and high strength. Fracture is often conchoidal, especially in cannels. Cannels and bogheads are difficult to differentiate macroscopically though boghead may be somewhat more brown than cannels and have a brownish streak. The proper distinction can be made only microscopically and the term boghead is reserved for coals rich in the maceral alginite. Both cannel and boghead are relatively rare. Plate 6B is a macroscopic photograph of a cannel.

4.2.1.2 **Stopes-ICCP System**

The four lithotypes described in the ICCP Handbook are essentially the 'four visible ingredients in banded bituminous coals' of Stopes (1919). She did not originate this four-fold breakdown; that had been done earlier by Fayol (1887), though he used different terms to designate his 'lithotypes.' The characteristic properties of the Stopes-ICCP lithotypes are summarized on the following pages.

The application of the lithotype approach to the description of coal has always been attended with problems. One of the most important of these involves the relation between lithotype description and maceral or microlithotype composition as determined microscopically. Stopes in her 1919 paper provided good macroscopic descriptions of her 'ingredients', and she supplemented these with microscopic observations. From this it was implied that the ingredients or lithotypes consisted of preferred associations of microscopically identifiable constituents, that is, macerals, though that term was unknown at the time. She especially stressed the importance of spore exines (liptinite) as a component of durains. Such suggested relationships had some confusing consequences later on, because some petrographers began to define lithotypes in terms of microscopic composition. A related difficulty concerns the coals which Stopes examined to make her initial descriptions. They were bituminous coals of Carboniferous age. Later studies of coal of other ranks and ages showed the occurrence of clarain and durain bands whose microscopic compositions did not correspond closely to those described by Stopes. A good discussion of some of these difficulties is given by Davis (1978b). Nevertheless the lustre and surface texture of coal are dependent on maceral composition and for this reason many people over the years have tried to use macroscopic description as an indicator of maceral composition. As a first approximation this is valid but it should not be regarded on the same level of precision as microscopic analysis.

A second problem with the use of lithotypes concerns band or layer thickness. There is no precisely defined minimum thickness for lithotypes. In Germany the practice has been to use 10 millimetres as a minimum, but in other countries minimum thicknesses from 3 to 10 millimetres are used. This creates confusion in comparing data from different investigators or from different laboratories.

Plate 6. Macroscopic appearance of coal lithotypes.

A) Naturally broken cleat face of Pennsylvanian age bituminous coal (Harbour seam, Sydney Coalfield, Nova Scotia) showing examples of three lithotypes; vitrain (V), clarain (C), and durain (D). Note the extremely rapid lateral thickness change in some of the vitrain layers.

B) Naturally broken surface of cannel coal of Pennsylvanian age from West Virginia. Note the uniformly dull appearance of this lithotype; upper right-hand corner of block shows typical conchoidal fracture and greasy luster.

A
C
D
C
V
C
V
C
D
C
B

VITRAIN*

I - Origin of Term

The term vitrain was introduced by M. C. Stopes in 1919 to designate the macroscopically recognizable very bright bands of coals.
Synonyms: Glanzkohle, charbon brillant.

II - Description

Very bright bands or lenses, usually a few mm (3-5) in width: thick bands are rare. Clean to the touch. In many coals the vitrain is permeated with numerous fine cracks at right angles to stratification, and consequently breaks with a conchoidal fracture. In other coals the vitrain is crossed by only occasional perpendicular cracks.
In the macroscopic description of seams only the bands of vitrain having a thickness of several mm are usually noted.
Examination with the microscope shows vitrain to consist of microlithotypes very rich in vitrinite.

III - Occurrence

After clarain, vitrain is the most widely distributed and common macroscopic constituent of humic coals.

CLARAIN*

I - Origin of Term

The term clarain was introduced by M. C. Stopes in 1919 to designate the macroscopically recognizable bright lustrous constituent of coal, which, in contrast to vitrain, is intrinsically striated by dull intercalations. Nowadays the term is used to describe all finely striated bands of coal which have an appearance intermediate between vitrain and durain.
Synonyms: Halbglanzkohle, charbon semibrillant.

II - Description

Bands of variable thickness having a lustre between that of vitrain and durain.
Conventionally the thicknesses of the fine, bright and dull striations must be less than 3 mm; - if greater than this they should be considered as thicknesses of separate lithotypes:
In macroscopic descriptions of seams, only bands having a thickness of several mm are usually recorded as clarain. Examination with the microscope shows clarain may have variable maceral composition from the three groups, vitrinite, exinite and inertinite.

III - Occurrence

Clarain is, as a general rule, the most widely distributed and common macroscopic constituent of all humic coals.

*Modified from ICCP Handbook, 1963.

FUSAIN*

I - Origin of Term

The term fusain was introduced by Grand'Eury in 1882 to designate the black silky, lustrous bands, recognized macroscopically in coal.
Synonyms: Faserkohle - charbon fibreux.
Analogous terms: mineral charcoal - mother of coal - Russkohle.

II - Description

Fusain is recognized macroscopically by its black or grey-black colour, its silky lustre, its fibrous structure and its extreme friability. It is the only constituent of coal which marks and blackens objects with which it comes in contact. Fusain may include a high proportion of mineral material, which strengthens it and reduces its friability; it retains, however, its silky lustre. In macroscopic descriptions of seams, only those bands having a thickness of several mm are recorded. Microscopic examination shows that fusain consists in the main of fusite.

III - Occurrence

As fine bands and lenses in almost all humic coals. Widely distributed, but not abundant.

DURAIN*

I - Origin of Term

The term durain was introduced by M. C. Stopes in 1919 to designate the macroscopically recognizable dull bands in coals.
Synonyms: Mattkohle, charbon mat.
Analogous terms: splint coal, dull coal.

II - Description

Bands of durain are characterized by their grey to brownish black colour and rough surface with dull or faintly greasy lustre: reflection is diffuse: they are markedly less fissured than bands of vitrain, and generally show granular fracture.
Wandless and Macrae (1934) distinguished two types of durain, - grey durain and black durain.
In macroscopic descriptions of seams, only bands having a thickness of several mm are usually recorded. Examined under the microscope, durain is shown to consist of microlithotypes rich in exinite or inertinite. By macroscopic examinations alone durain may sometimes be confused with carbonaceous shale and dirty (mineral-rich) coal.

III - Occurrence

In humic coals, durain occurs in bands up to many cms in thickness. Widely distributed, but with exceptions not abundant.

*Modified from ICCP Handbook, 1963.

Another difficulty involves the definition of clarain. From the lithotype descriptions in the Stopes-ICCP system, it is clear that clarain is composed of alternating layers of vitrain, durain, and fusain which are not thick enough to be considered lithotypes on their own (depending on the minimum thickness which is adopted). Because of the descriptive constraints on vitrain, durain, and fusain, it should not be surprising to find that the bulk of most coals is composed of clarain (60 per cent or more). To get around this problem investigators have proposed subdivisions of clarain. One such breakdown defined *bright clarain* - more bright (vitrainic) layers than dull (durainic or fusainic); *clarain* - equal proportions bright and dull; and *dull clarain* - less bright than dull (Mackowsky, 1982). However, this practice is not presently used, at least in Germany. Another subdivision suggested by Cady (1942), proposed the term clarodurain for the duller varieties of clarain, roughly corresponding to the 'dull clarain' described above. In 1971 Cameron proposed a textural discrimination of clarain types based on the dominant thickness of the contained bright (vitrainic) layers, and in this way described the clarain fraction of the Harbour seam in Nova Scotia in terms of fine, medium, and coarsely banded clarain. However, none of these schemes for subdividing clarain are entirely satisfactory.

Finally there are problems concerning the subjectivity of description when comparing results of two or more persons, problems relating to the environment in which descriptions are made (mine or laboratory), and problems related to the degree of shearing or weathering to which the coal has been subjected. Shearing in extreme cases, may completely obliterate the banded appearance of coal. Changes in rank also affect the lustre of coal, and can produce problems in lithotype identification. Lithotypes probably show up best in high- volatile bituminous coal. As rank increases contrasts in lustre between bands in the coal tend to diminish; in anthracite they are very much more difficult to identify than in bituminous coal. At the other extreme of rank, layers in lignite which would undoubtedly give rise to vitrain in higher rank coals have a much duller, less glassy appearance, although they can still be identified.

Figure 17 illustrates a macroscopic profile developed according to the Stopes approach for a high-volatile bituminous coal in the Illinois Basin. In this case a detailed description was possible because a polished column sample was examined in the laboratory. An attempt was made to differentiate between bright and dull clarain. On the basis of the initial description the seam was divided into petrographic intervals (indicated by Roman numerals). These intervals are characterized by the predominance of one or more lithotypes. On a rough coal surface, as in a mine, it is the larger layers represented by the intervals, which would most likely be recognized.

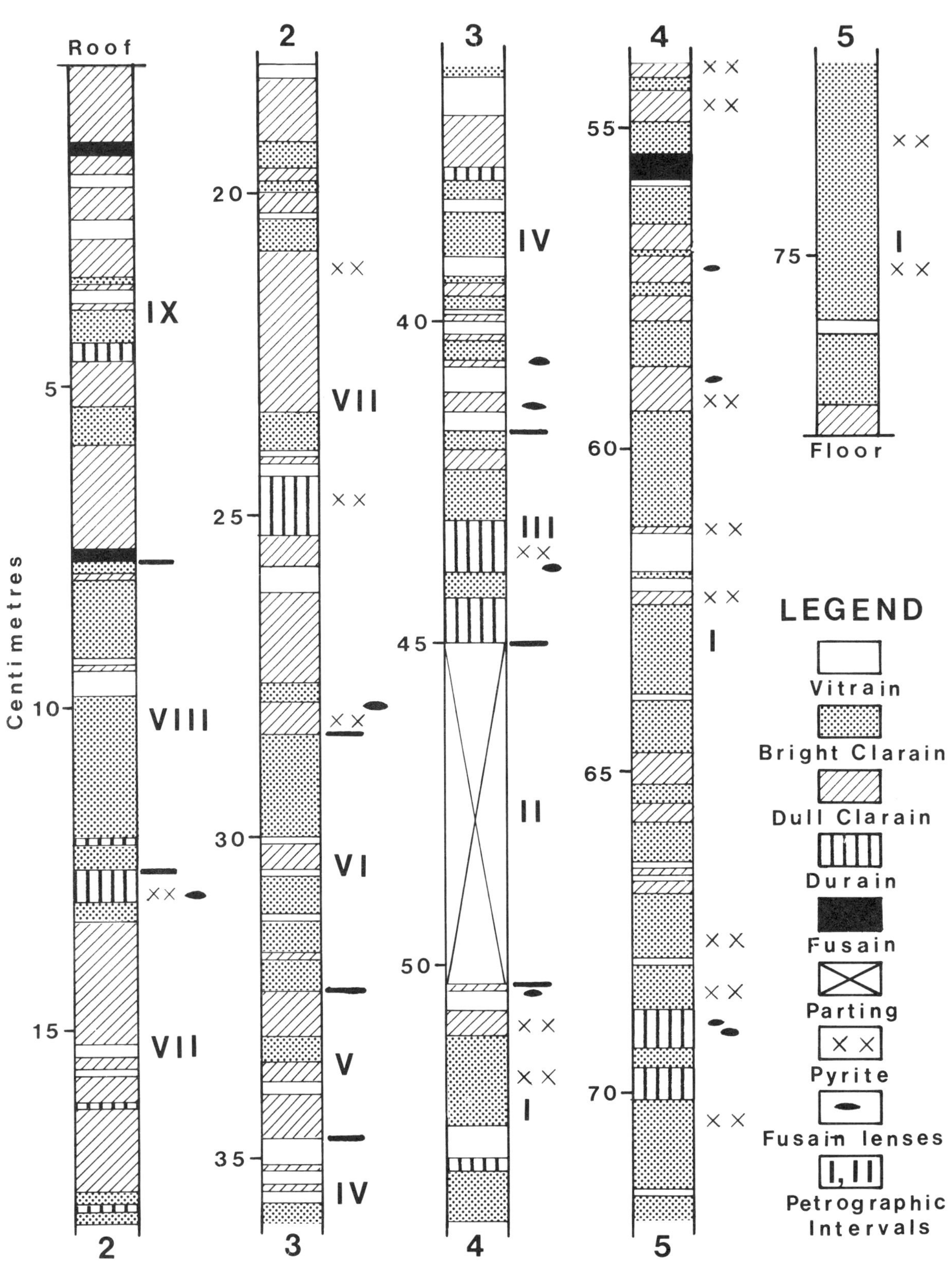

Figure 17. Macroscopic profile of Dekoven seam (Pennsylvanian age) from southern part of Illinois Basin (from Cameron, 1978).

4.2.1.3 **Other Lithotype Approaches**

Other systems for the macroscopic or lithotype evaluation of coal have been described, two of which will be reviewed briefly. Both represent an attempt to make the descriptive procedure more quantitative than is the case with the Stopes-ICCP system.

One of these descriptive methods was proposed by J. M. Schopf of the United States Geological Survey (Schopf, 1960). It was intended mainly, though not solely, for field descriptions of coal seams. He proposed no minimum thickness for his layers, and his main criteria for delineation were that layers should be lithologically distinct and have some lateral continuity. The delineation of such layers on a coal face is the first step in the Schopf procedure. In addition to identifying layers of dirty coal and rock partings, he proposed that the layers of clean coal be described in terms of proportions of three constituents, namely, vitrain, fusain, and attrital coal, with the banded characteristics or texture of a layer being determined on the thickness and frequency of occurrence of vitrain bands and, to a lesser extent, fusain. He recognized vitrain and fusain according to the Stopes-ICCP descriptions. In addition, Schopf proposed a five-fold breakdown of attrital coal according to its lustre, namely bright, moderately bright, medium, moderately dull, and dull. Numerical limits for his thickness and frequency classes of vitain are given in Table 9. He suggested that the fusain component in a given layer be similarly classified. Thus in a given seam a clean coal zone might be described as 'moderately thin banded - moderately dull - sparse thin banded,' meaning a layer with 15 to 30 per cent vitrain, dominated by the occurrence of the vitrain in bands 0.5 to 2.0 millimetres thick, with a moderately dull lustre to the attrital matrix, and with sparse (less than 15 per cent) fusain in the thin-banded mode (0.5 to 2.0 millimetres). If, in this example, the same amount of vitrain is equally distributed in two different thickness categories, then one might use a composite terminology such as 'moderately thin-medium banded,' meaning equal amounts in the thin and medium categories. This procedure has merit in that it establishes numerical thresholds, but in actual practice it might be somewhat cumbersome. Another problem is the subjective evaluation of lustre in the attrital coal, though Schopf stated that between two observers there is usually only a one-step discrepancy.

Some of Schopf's suggested criteria have been included in a method being used by geologists at U.S. Steel (Goscinski *et al.*, 1978). Figure 18 (from Goscinski *et al.*) illustrates macrosopic description of a coal core made according to their procedures. Note that the thicknesses of vitrain bands and lustre are included following a somewhat modified version of Schopf's recommendations. Note also that specific gravities are determined on the various lithotypes.

An alternative and somewhat simpler quantitative approach for macroscopic description has been developed in Australia, where coal is regarded as being a mixture of 'bright' and 'dull' components, and lithotypes are defined according to the proportions of the basic ingredients in a given layer. 'Bright' and 'dull' have been used deliberately instead of the terms in the Stopes system because, in contrast to the latter, they carry no compositional implication (Davis, 1978b). Several versions of such schemes have been proposed. The nomenclature for one of them described by Diessel (1965) is presented in Table 10. Diessel's terminology for his clean coal lithotypes are: bright, banded bright, banded, banded dull, dull, and fibrous. Note that for the first five, threshold proportions have been established; bright and dull are treated as end

TABLE 9. BOUNDARIES FOR THICKNESS OF VITRAIN BANDS AND FREQUENCY OF OCCURRENCE IN SCHOPF CLASSIFICATION (AFTER SCHOPF, 1960)

(1) THICKNESS OF VITRAIN BANDS	
Designation	Range of Thickness
Thin bands*	0.5 to 2.0 mm (about 1/50 to 1/12 inch)
Medium bands	2.0 to 5.0 mm (about 1/12 to 1/5 inch)
Thick bands	5.0 to 50.0 mm (about 1/5 to 2.0 inches)
Very thick bands**	more than 50.0 mm (over 2 inches thick)
(2) FREQUENCY OF OCCURRENCE	
Designation	Range of Frequency
Sparse	comprising less than 15% of the layer
Moderate	comprising from 15 to 30% of the layer
Abundant	comprising from 30 to 60% of the layer
Dominant	comprising more than 60% of the layer

*Vitrain bands less than 1/2 mm (0.5 mm) thick are regarded as a part of attrital coal.
**Very thick vitrain bands are not common; lenses of very thick pre-vitrain occur more commonly in lignite.

Thickness, mm		Rock Type	Luster	Vitrain Thickness, Amount	Organic Accessories	Inorganic Accessories	Sp.Gr.	Wgt.
Cum.	Layer							
		1100'0" - Top of Coal						
1701	87	Coal	Int-Br	Thin to Med-Moderate	Fusain Lenses		1.40	
1614	75	Coal	Int-Br	Thin-Moderate	Fusain Lenses		1.45	
1539	03	Fusain					1.55	
1536	94	Coal	Br	Thin to Thick-Abundant			1.35	
1442	65	Coal	Br	Thin to Thick-Abundant			1.35	
1377	130	Gray Shale					Sk.1.70	520.0 g
1247	75	Bone					Sk.1.70	
1172	49	Coal	Int-Br	Thin-Moderate	Fusain Lenses	Calcite in Cleat	1.45	
1123	65	Coal	Int-Br	Thin-Moderate	Fusain Lenses		1.45	
1058	40	Coal	Int-Dull	Thin - Sparse	Fusain Lenses	Pyrite Lenses	1.60	
1018	06	Pyrite					Sk.1.70	
1012	50	Coal	Int	Thin - Sparse	Fusain and Bone Lenses	Pyrite in Cleat	1.50	
962	12	Bone					Sk.1.70	
950	52	Coal	Int-Br	Thin-Moderate	Fusain Lenses	Calcite in Cleat	1.45	
898	70	Coal	Br	Thin to Thick-Abundant		Calcite in Cleat	1.35	
828	58	Coal	Int.Br	Thin to Med-Moderate		Calcite in Cleat	1.40	
770	125	Black Shale					Sk.1.70	500.7
645	25	Gray Clay					Sk.1.70	101.7 g
620	50	Coal	Int.Dull	Thin-Sparse	Fusain and Bone Lenses		1.55	
570	08	Bone					Sk.1.70	
562	70	Coal	Int-Br	Thin-Moderate	Fusain Lenses		1.40	
492	80	Coal	Int-Br	Thin-Moderate	Fusain Lenses		1.40	
412	101	Gray Mud Shale					Sk.1.70	410.5g
311	27	Coal	Int	Thin-Sparse	Fusain and Bone Lenses		1.50	
284	04	Fusain					1.50	
280	60	Coal	Int-Br	Thin-Moderate	Bone Lenses	Calcite in Cleat	1.45	
220	105	Coal	Br	Thin to Med-Abundant		Calcite in Cleat	1.35	
115	35	Coal	Br	Thin-Abundant	Fusain Lenses	Calcite in Cleat	1.40	
80	05	Pyrite					Sk.1.70	
75	25	Coal	Br	Thin-Abundant		Calcite in Cleat	1.35	
50	50	Coal	Int-Br	Thin-Moderate	Fusain Lenses	Calcite in Cleat	1.40	
		1105'7" - Bottom of Coal						

Figure 18. Macroscopic log of coal seam prepared according to method of Goscinski *et al.* (from Goscinski *et al.*, 1978).

members of a series, while the other three are distinguished according to varying proportions of the end members. Note also on this table the correlation between the Stopes lithotypes and those of the Australian scheme. Diessel suggested a minimum band width of 5 millimetres for the layer to be considered a separate lithotype. In addition to the above, carbonaceous shale or dirty coal and true rock partings are also identified. A somewhat similar system has been used by the Commonwealth Scientific and Industrial Research Organization (CSIRO) in Australia (Table 11). Unlike the system proposed by Diessel, no numerical boundaries for lithotypes are given and apparently have not been established.

TABLE 10. THE CORRELATION BETWEEN LITHOTYPES AFTER STOPES AND THOSE DISTINGUISHED IN N.S.W. (AUSTRALIAN) COALS. (MODIFIED FROM DIESSEL, 1965)

Divisions after Stopes	Divisions as used for Australian coals	Description
Vitrain	Bright coal	Subvitreous to vitreous lustre, or conchoidal fracture <10 per cent dull
	Banded bright coal	Bright coal with some thin dull bands 10-40 per cent dull
Clarain	Banded coal	Bright and dull coal bands in equal proportion 40-60 per cent dull
	Banded dull coal	Dull coal with some thin bright bands 10-40 per cent bright
Durain	Dull coal	Matt lustre, uneven fracture <10 per cent bright
Fusain	Fibrous coal	Satin lustre, friable

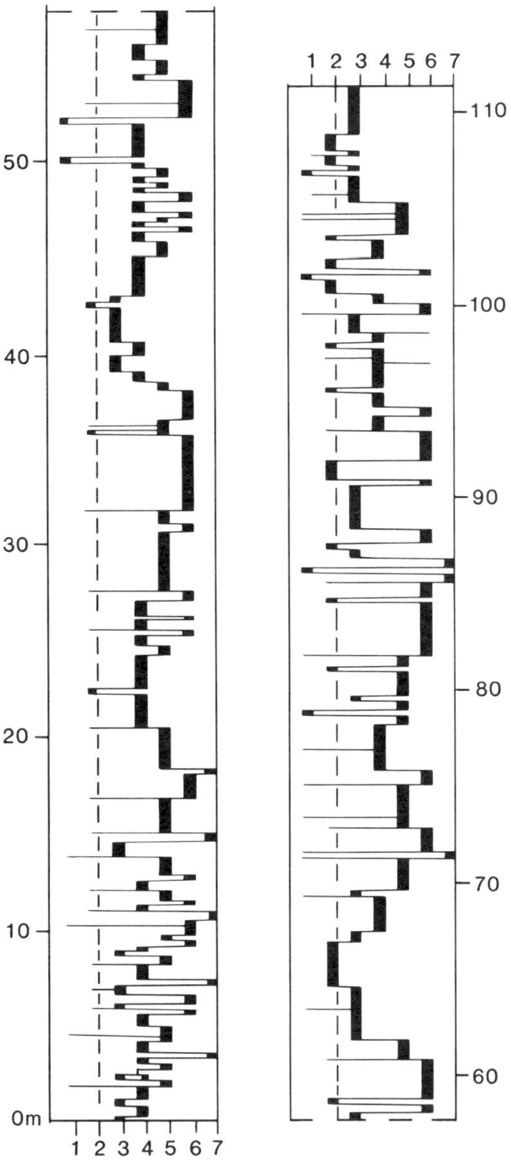

Figure 19. Macroscopic profile of "D" coal zone, Hat Creek Coalfield, British Columbia, described according to Australian system. Numbers on the abscissa correspond to lithotypes as follows: 1 = shale, 2 = carbonaceous shale, 3 = dull, 4 = banded dull, 5 = banded, 6 = banded bright, 7 = bright (from Marchioni, in press).

TABLE 11. CSIRO (AUSTRALIA) TERMS FOR THE MACROSCOPIC DESCRIPTION OF COAL (AFTER DAVIS, 1978b)

Bright coal
Interbanded dull and bright coal
Mainly dull, finely interbanded with bright coal
Dull coal

Figure 19 is an example of a lithotype profile prepared according to Diessel's system. The seam in question is the D zone from the Hat Creek deposit, British Columbia. In this example, because of seam thickness (over 110 metres), minimum lithotype thickness has been increased to 5 centimetres.

Macroscopic descriptions can be used to give an *approximation* of maceral composition. Data from Diessel (1965) show that mean values of maceral composition of a number of lithotypes can be projected to samples for which only lithotype data are available and used to calculate maceral data. The relevant data for lithotype distribution and maceral content of lithotypes are given in Tables 12 and 13 respectively. Table 14 shows a comparison of calculated maceral data and that actually determined by analysis on the coals listed in Table 12. Table 14 shows surprisingly good correlation between calculated and analysed maceral contents.

Figure 20 further illustrates the relationship between macroscopic petrography and maceral composition. This figure shows a plot of the vitrain content of intervals of the No. 10 seam in the Fernie Basin, British Columbia, plotted against the vitrinite content of the same intervals. There is a fairly good positive relationship of one to the other.

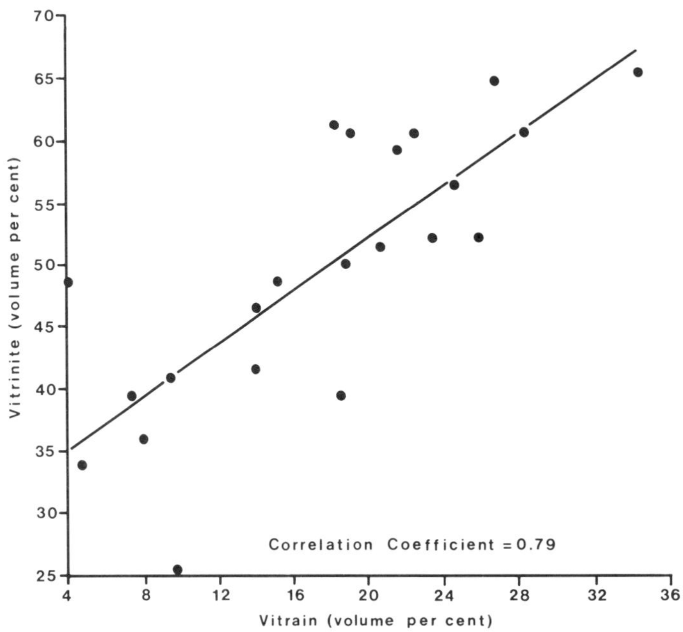

Figure 20. Relationship between vitrain and vitrinite contents of petrographic intervals in No. 10 seam, Sparwood, British Columbia (from Cameron, 1978).

TABLE 12. DISTRIBUTION OF LITHOTYPES IN SOME COKING COALS OF THE ILLAWARRA COAL MEASURES, AUSTRALIA (MODIFIED FROM DIESSEL, 1965)*

Seam	Bright coal	Banded bright	Banded coal	Banded dull	Dull coal	Carbar-gilite	Fibrous coal	Carb. sh. & sh.
Wongawilli	9.0	35.3	22.2	14.5	8.9	3.7	0.2	6.2
	8.8	34.7	25.0	15.2	10.0	2.1	0.3	3.9
	7.8	37.0	19.7	13.9	10.0	2.5	0.3	8.8
Mean	8.5	35.7	22.3	14.5	9.6	2.8	0.3	6.3
Bulli	1.0	9.2	7.9	45.6	33.4	2.6	0.3	0.0
	6.7	5.8	13.1	47.5	21.0	0.7	0.2	0.0
	5.2	5.3	11.5	45.5	31.6	0.7	0.0	0.0
	4.4	7.3	25.1	33.2	29.0	1.0	0.0	0.0
	5.9	9.5	13.7	35.0	34.7	0.6	0.0	0.6
	5.5	4.6	11.2	34.2	43.3	0.2	0.0	1.0
Mean	4.8	7.0	14.6	40.1	32.1	1.0	0.1	0.3

*Data in volume per cent; study encompassed three columnar sections of Wongawi seam and six from Bulli seam.

TABLE 13. NOMINAL MACERAL CONTENTS OF VARIOUS LITHOTYPES DISTINGUISHED IN SOME COALS OF ILLAWARRA COAL MEASURES, AUSTRALIA (MODIFIED FROM DIESSEL, 1965)

Lithotype	Vitrinite		Liptinite		Inertinite		Minerals	
	W.S.*	B.S.**	W.S.	B.S.	W.S.	B.S.	W.S.	B.S.
Bright coal	93***	91	1	1	3	7	3	3
Banded bright coal	80	65	2	1	12	30	6	4
Banded coal	72	53	2	2	20	41	6	4
Banded dull coal	57	39	2	2	31	55	10	4
Dull coal	45	30	2	2	38	61	15	7
Carbargilite	30	20	1	1	26	36	43	43
Fibrous coal	0	0	0	0	81	91	19	9
Carb. sh. & sh.	3	3	0	0	4	4	93	93

*Wongawillie seam
**Bulli seam
***Data in volume per cent

TABLE 14. CALCULATED AND ANALYSED MACERAL CONTENTS IN SAMPLES OF WONGAWILLI AND BULLI SEAMS, ILLAWARRA COAL MEASURES, AUSTRALIA (MODIFIED FROM DIESSEL, 1965)

Seam	Vitrinite		Liptinite		Semi-inertinite plus inertinite		Minerals	
	Calc.	Anal.	Calc.	Anal.	Calc.	Anal.	Calc.	Anal.
	66.1	66.4	1.7	0.8	17.8	19.3	14.4	13.5
Wongawilli	68.0	65.3	1.7	0.7	18.6	18.9	11.7	15.1
	64.5	61.5	1.6	0.4	17.4	19.3	16.5	18.8
Mean	66.2	64.4	1.7	0.6	17.9	19.2	14.2	15.8
	45.5	44.0	1.8	0.2	47.0	51.3	5.7	4.5
	44.3	53.3	2.1	0.4	47.9	42.4	5.7	3.9
Bulli	44.1	49.8	1.8	1.1	48.7	44.3	5.4	4.8
	43.0	41.6	1.8	1.1	49.2	51.4	6.0	5.9
	40.5	41.0	1.9	2.4	51.2	51.8	6.4	4.8
	39.2	34.6	1.8	2.9	52.7	55.1	6.3	7.4
Mean	42.7	44.1	1.9	1.3	49.5	49.4	5.9	5.2

4.2.1.4 **Lithotypes in Low Rank Coals (Lignites-Subbituminous)**

Most of the above discussion on lithotypes applies to coals of bituminous rank or higher though Schopf indicates that his classification could be extended to lignites. However, an internationally acceptable system of lithotypes for low rank coals does not exist as yet. Some methods have been proposed: those of Sontag and Suess (1976) in East Germany, Malan (1971) in Czechoslovakia, Allardice *et al.* (1977) in Australia, and Hagemann (1978) in West Germany. The last one will be discussed here because it was applied by Hagemann to Saskatchewan lignites and lignite-subbituminous coal in the Hat Creek, British Columbia deposit (Hagemann, 1980).

An important criterion in Hagemann's classification is the relative proportion of groundmass and woody (xylitic) remains plus the relative abundance of mineral impurities. Groundmass is used somewhat in the sense of the attrital coal in Schopf's definition in that it comprises more finely comminuted particles of varied origin, too small to be identified separately. Another important criterion relates to the texture or banded characteristics. Finally, intensity and hue of colour, degree of gelification, and the presence or absence of inclusions, such as resin rodlets and fusain lenses, are all incorporated into the descriptive system. Based on these five main criteria a hierarchy of lithotype classes and varieties is proposed (Table 15).

TABLE 15. MACROPETROGRAPHIC CLASSIFICATION OF LIGNITE (FROM HAGEMANN, 1980)

MACROPETROGRAPHIC CLASSIFICATION OF LIGNITES (A AND B)						
categories	lithotype-classes	lithotypes	lithotype-varieties			
reading column matrix	10	20	30	40	50	
quality parameter	structure; presence of xylite and/or contamination by mineral matter	texture: ratio between groundmass and plant remains	intensity and hue of colour	evidence of gelification	inclusions	additional features
macropetrographic description	1 pure coal, non-xylitic	1 unbanded coal	1 pale-yellow	1 gelified groundmass (+,++,+++)	1 resin bodies (+,++,+++)	a) surface texture extensive cracking moderate cracking no cracking
	2 pure coal, xyl. fibrous xylite brittle xylite - - - - - - - - tree stumps tree trunks etc.	2 moderately banded coal	2 medium light-yellow	2 gelified tissues (+,++,+++)	2 cuticles (+,++,+++)	b) fracture even imbricated
	3 impure coal, non-xylitic clayey coal sandy coal calcareous coal iron sulphides etc.	3 banded coal	3 pale-brown	3 microgranular humic gel particles (+,++,+++)	3 charcoal (+,++,+++)	c) size degradation coarse fragmentation fine fragmentation crumbling
	4 impure coal, xylitic	4 highly banded coal	4 medium light-brown	4 -	3 -	
			5 dark-brown			
			6 dark-black			

Following the format of Table 15, lithotype identification proceeds by steps from left to right, that is, the lithotype class (column 10) is first identified, then the lithotype (column 20). These are the only two stages of the operation which can be carried out in the field, because the observations for columns 30, 40, and 50 must be carried out on dried material. The suggested technique for drying is freeze-drying. This accentuates the original colour of the specimen and preserves its structure. The laboratory examinations are still carried out macroscopically or with the use of a low power stereo microscope.

Hagemann stated that this classification was really designed for coals of lignite B rank (ASTM classification). Many of the Saskatchewan coals belong to the lignite A rank group, while the Hat Creek coals are even higher in rank. Therefore he suggests that the classification should be modified, particularly in application of the 'colour/hue' and 'gelification' criteria (Hagemann, 1980). As will be explained below, only some of the possible variations in these two properties were observed in the Saskatchewan coals. The resulting data can be recorded as a numerical code or in graph form (Figure 21).

Figure 21 shows the lithotype profile of the Souris coal seam from the Estevan area, Saskatchewan. The figure is plotted in the form of several columns arranged according to Table 15. Note that the various layers identified in the seam are fairly thick, most being greater than 10 centimetres. The procedure in the field is to first visually establish and mark off such layers. An important qualification is that an identified layer should have a certain degree of lateral continuity. As an example of the numerical indexing of a lithotype, the layer from 137 to 146 centimetres woud be described as 12-23-36-41(+)-42(++)-53(+); that is, a pure coal (xylitic), banded, dark/black in hue, sparse amount of gelified groundmass, moderate content of gelified plant remains, and sparse charcoal (fusain).

Hagemann (1980) emphasized that there is not a strong correlation between the macroscopic petrography and microscopic data, though certain microlithotypes and macerals have a preferred association with particular lithotypes. Hagemann and Hollerbach (1979) investigated the relations between lithotypes and chemical composition.

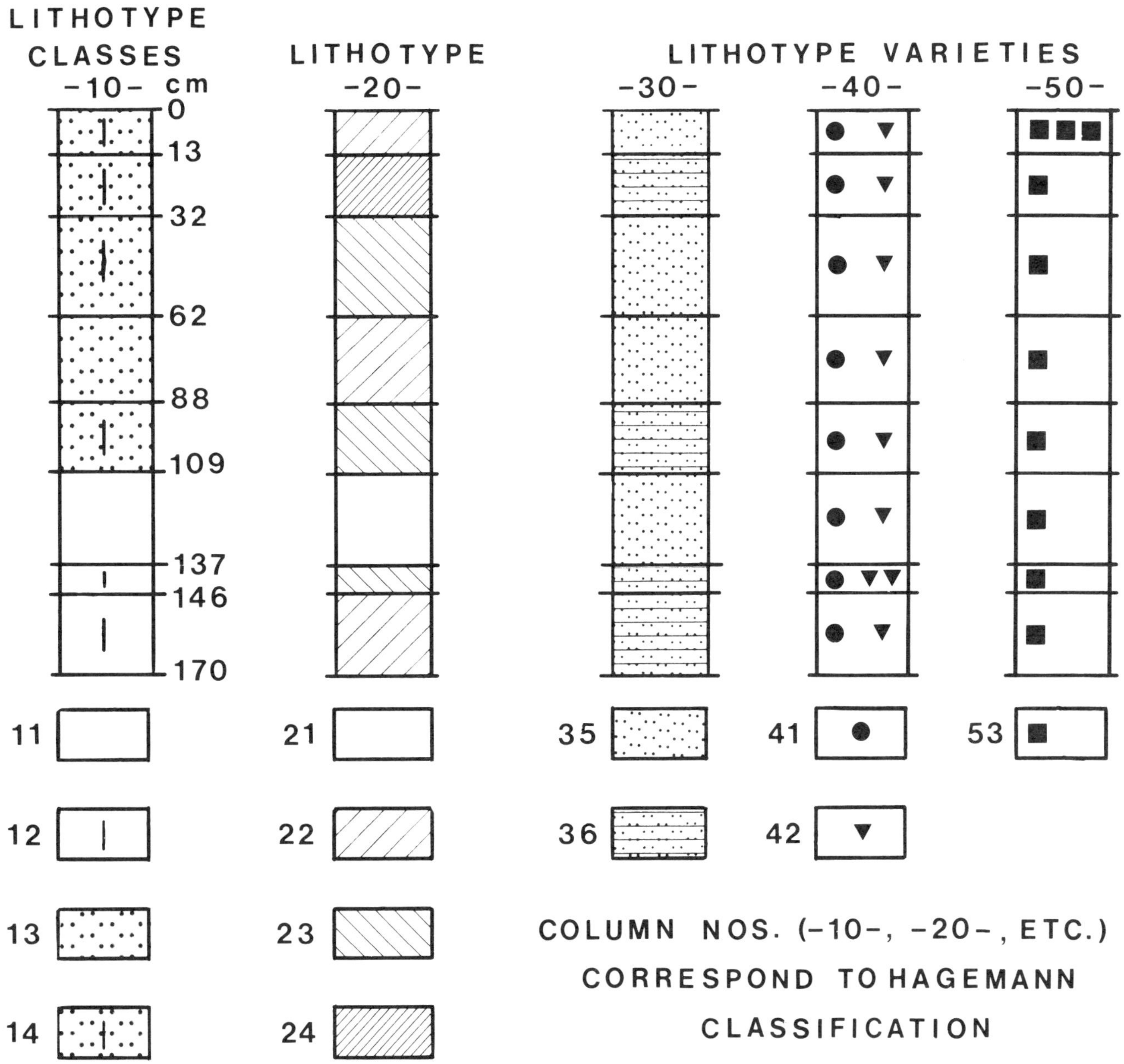

Figure 21. Macroscopic profile of lignite bed (Souris seam), Estevan area, Saskatchewan. Numbers and symbols represent subdivision in various columns of Table 15, for example, 11 and 12 represent, respectively, "pure coal, non-xylitic" and "pure coal, xylitic" in column 10; symbols 35 and 36 represent colour classes "dark brown" and "dark black" in column 30 (from Hagemann, 1980).

4.2.2 *Microlithotypes*

Microlithotypes are defined in the ICCP Handbook (ICCP, 1963) as 'typical associations of macerals with a minimum band width of 50 micrometres. As the name implies, microlithotypes can only be identified microscopically. The three main groups of microlithotypes are designated monomaceralic, bimaceralic, and trimaceralic according to whether they contain representative macerals of one, two, or three of the maceral groups. In addition, a fourth microlithotype group called carbominerites can be identified. This group is made up of mixtures of various minerals and macerals. In contrast to lithotypes, the maceral contents and, where applicable, mineral contents of microlithotypes are rigidly defined following what appear to be arbitrary boundaries.

In addition to the 50 micrometres minimum band width, the ICCP has adopted another rule regarding microlithotypes. That is, the so-called 5 per cent rule. This states that the monomaceralic microlithotypes may contain up to 5 per cent of accessory macerals, for example, the monomaceralic microlithotype 'vitrite' may contain up to 5 per cent liptinite and/or inertinite in addition to at least 95 per cent vitrinite. A summary of microlithotype classification is given in Table 16 and is presented diagrammatically on Figure 22.

TABLE 16. TABLE OF MICROLITHOTYPES (FROM ICCP HANDBOOK, 1963 and STACH, 1982)

Microlithotype Constituent Macerals	Principal Groups of Composition	Maceral	Group
Vitrite	Vitrinite (V)	>95% V	Monomaceralic
Liptite	Liptinite (L)	>95% L	Monomaceralic
Inertite	Inertinite (I)	>95% I	Monomaceralic
Clarite	Vitrinite + Liptinite	>95% V + L	Bimaceralic
Durite	Inertinite + Liptinite	>95% I + L	Bimaceralic
Vitrinertite	Vitrinite + Inertinite	>95% V + I	Bimaceralic
Duroclarite	Vitrinite + Liptinite + Inertinite	V,L,I each >5%, V I,L	Trimaceralic
Vitrinertoliptite	Vitrinite + Liptinite + Inertinite	V,L,I, each >5%, LI,V	Trimaceralic
Clarodurite	Vitrinite + Liptinite + Inertinite	V,L,I, each >5%, IV,L	Trimaceralic

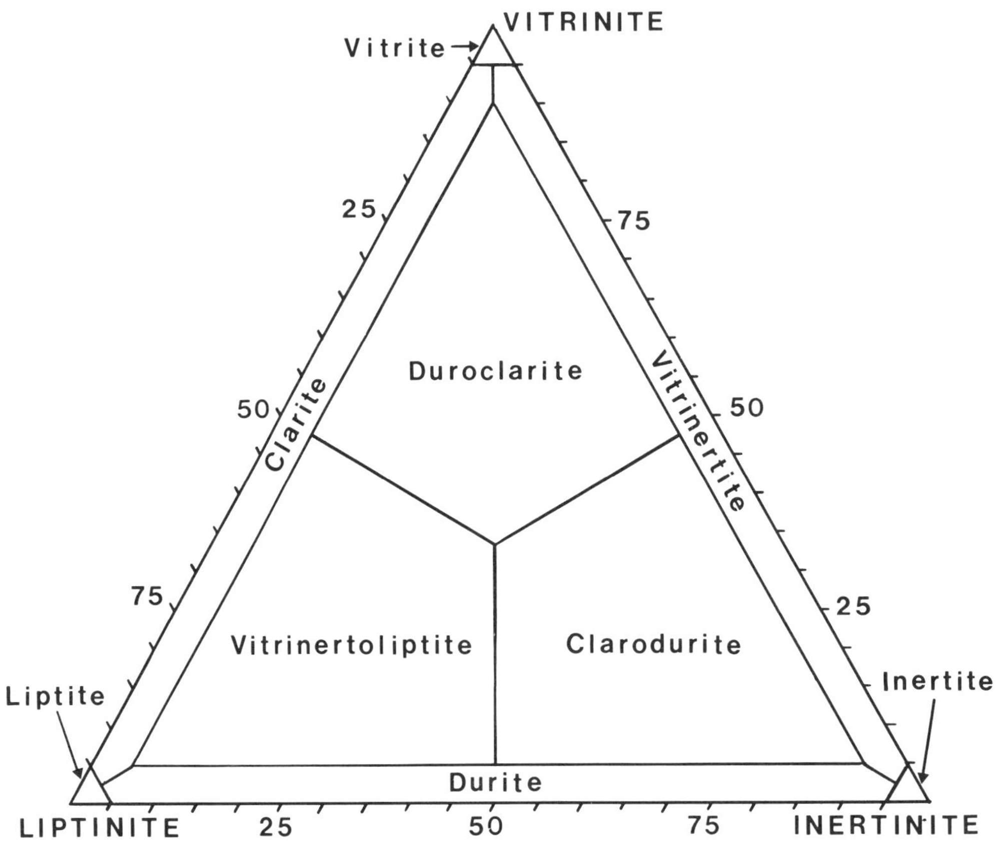

Figure 22. Diagrammatic representation of microlithotype classification shown in Table 16.

Note that the microlithotype names given in the table refer to groups. Just as there are maceral groups which are subdivided into individual macerals, so is it possible to subdivide the microlithotypes according to their contained macerals. Thus the microlithotype group 'inertite' can include the microlithotypes 'fusite' and 'inertodetrite' signifying microlithotypes with at least 95 per cent inertinite, being, in one case predominantly fusinite, and in the other, inertodetrinite. Plate 7 shows representative photomicrographs of microlithotypes.

The carbominerite group can similarly be subdivided according to the minerals which they contain. Thus 'carbargilite' is a mixture of coal and clay minerals, 'carbopyrite' is coal and pyrite, etc. Associations of coal and minerals with a mineral volume percentage of 20 to 60 per cent fall into the carbominerite category. If higher percentages of mineral matter occur, the association is simply identified as shale or rock. These percentages relate to separation densities often used in coal washing. It has been found that coal and mineral intergrowths with mineral matter contents of 20 to 60 per cent (that is, carbominerites) most often report in the 1.5 to 2.0 density fraction, that is, the middlings in most wash plants, while associations with more than 60 per cent by volume of mineral matter report in the reject. Thus microlithotype analyses can give considerable information on the behaviour of a coal during washing. In Plate 7 are shown examples of carbominerites.

A microlithotype analysis can give an indication of the texture of a coal. If two coals have equal overall contents of vitrinite and one has a higher vitrite content than the other, it is likely that the vitrinite occurs in thicker bands in the former relative to the latter. This in turn may have some bearing on the preparation characteristics of a coal. Microlithotype analysis gives an indication of the size distribution of masses of inertinite and this may be important in coking behaviour. It is for this reason that Brown *et al.* (1964) chose to use microlithotype data to characterize their coking coals.

4.3 **Mineral Matter in Coal**

Mineral matter in coal occurs both as finely disseminations and as discrete partings. Macroscopic identification of mineral matter in coal is sometimes possible, as tonsteins (clay bands), coal balls and siderite concretions, for example, are distinguishable in outcrop or mine wall. More concise identification of the inorganic constituents of coal, however, requires detailed analysis.

There are three basic approaches to identification of mineral matter in coal: chemical analysis of high-temperature (combustion) ash; petrography; and X-ray and SEM analysis of low-temperature ash. The first is the most readily carried out, but its major drawback is that original minerals are destroyed and so mineral matter must be expressed in terms of oxides. Petrographic analysis is difficult to carry out, mainly due to the generally fine-grained nature of coal mineral matter. The X-ray and SEM method requires expensive analytical equipment, but definitely provides the most accurate picture of the true mineral matter content.

Knowledge of the minerals and of their distribution in a coal has many useful applications (Mackowsky, 1968), including:

(1) Certain authigenic minerals form readily only under certain physical and chemical conditions, and therefore their presence provides an indication of the conditions of formation of the coal (Sections 2.3.2.3 and 9.3).

(2) A coal's primary mineral matter content, including tonsteins, may be characteristic, allowing for seam identification and correlation (Section 9.1).

(3) Washability of coal is affected by the degree of intergrowth of mineral matter and coal (Section 10.1).

(4) Mineral matter affects heating value of a coal, its ash fusion point, and the tendency for deposits and corrosion to occur on surfaces of heating chambers (Section 10.2).

(5) Mineral matter composition affects the makeup of mine dust.

Plate 7. Representative photomicrographs of microlithotypes. All photos taken with reflected light and oil immersion; the 20-point graticule is superimposed on all. At the proper magnification the area covered by the graticule on the specimen should be 50 x 50 micrometres; in this plate the size of the graticule has been exaggerated.

V = vitrinite, L = liptinite, I = inertinite, P = pyrite, C = clay minerals

Figure 1. Vitrite - all points on vitrinite

Figure 2. Liptite - all points on liptinite

Figure 3. Inertite (graticule on left side) - all points on inertinite
Vitrinertinite (graticule on right side) - one point on inertinite, remainder on vitrinite

Figure 4. Clarite - one point on liptinite, remainder on vitrinite

Figure 5. Durite - 10 points each on inertinite and liptinite

Figure 6. Duroclarite - one point on liptinite, three on inertinite, remainder on vitrinite

Figure 7. Clarodurite - two points on vitrinite, three on liptinite, remainder on inertinite

Figure 8. Vitrinertoliptite - 10 points on liptinite, five points each on inertinite and vitrinite

Figure 9. Carbopyrite - two points on pyrite, remainder on vitrinite and liptinite

Figure 10. Carbargilite - nine points on clay minerals, remainder on vitrinite

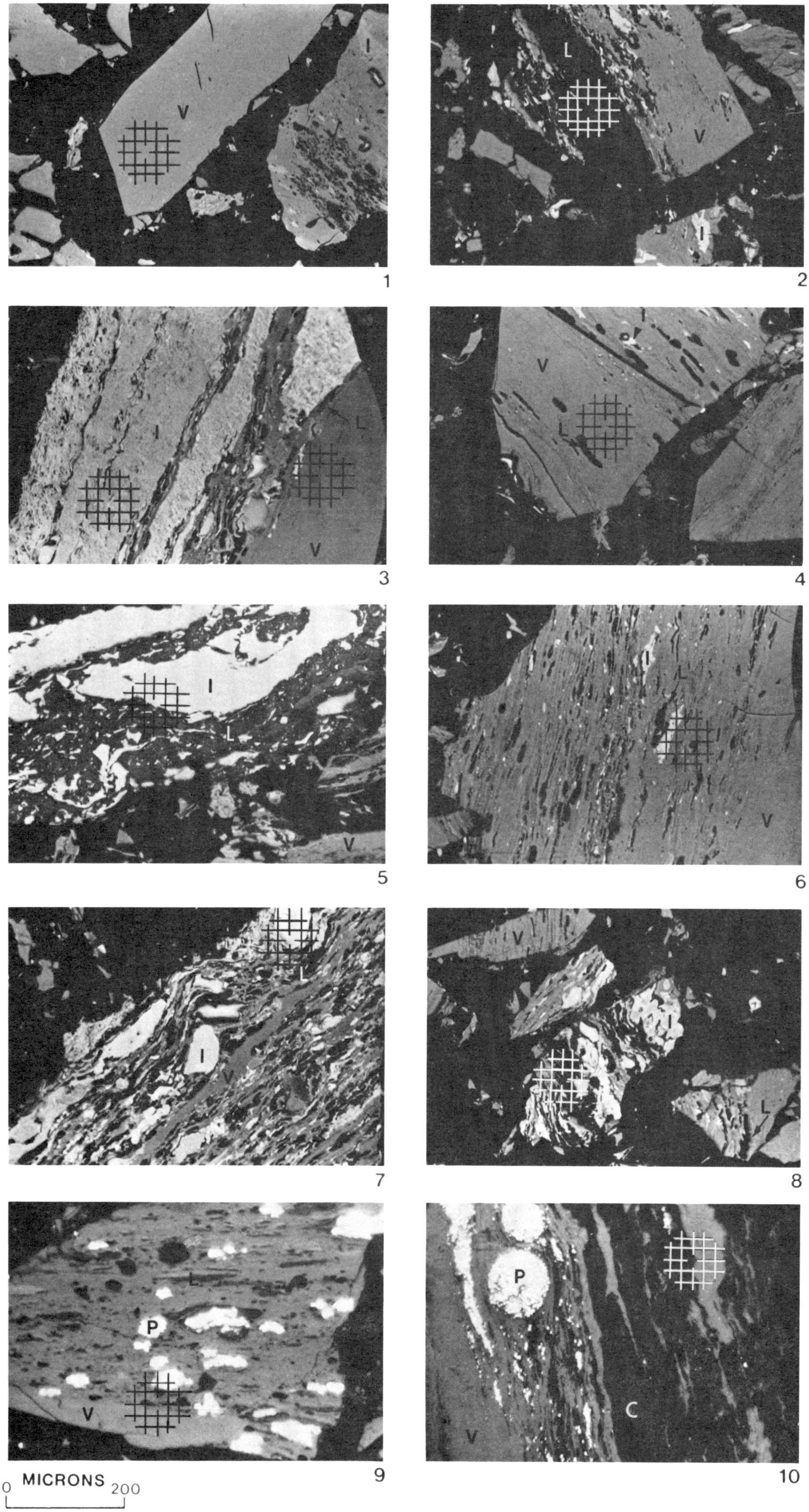

1
2
3
4
5
6
7
8
9
10
0 MICRONS 200

This chapter will cover mineral matter in coal, excluding shale bands and country rock. Mineral matter in coals may be divided into three groups (Mackowsky, 1982; ICCP, 1963; Rao and Gluskoter, 1973):

(1) mineral matter contained in original plants (inherent ash);

(2) primary mineral matter, emplaced during or shortly after deposition, including both detrital and authigenic minerals; and

(3) secondary mineral matter, formed at a later stage.

The terms syngenetic and epigenetic are also used for (2) and (3), respectively, and both may be referred to as extraneous ash.

The first group is generally not recognized petrographically, except perhaps by SEM, and will not be discussed at any length. The second and third groups are usually recognizable. Primary mineral matter is generally intimately intergrown with the coal, whereas secondary mineral matter tends to be coarsely intergrown and associated with cleat, fractures, and cavities. The latter may also, however, represent replacement of primary mineral matter and may therefore appear similar to primary minerals. Primary minerals are generally the more common.

The major groups of minerals include the clays, carbonates, iron sulphides, and silicas. Other groups, which occur more rarely, include oxides, hydroxides, other sulphides, phosphates, silicates, and sulphates.

Plates 8 and 9 illustrate examples of mineral matter in coal. Table 17 summarizes the mineral groups in coal and categorizes them with respect to their type of formation. Write-ups by Mackowsky (1968, 1982) provide good summations of minerals in coal and were referred to for much of the following information.

TABLE 17. MINERALS IN COAL (ADAPTED FROM MACKOWSKY, 1982 AND ICCP, 1963)

	PRIMARY FORMATION		SECONDARY FORMATION	
	Detrital	Authigenic	Deposited in cleat, fractures and cavities	Transformation of primary minerals
Clays	kaolinite, illite, mixed-layer clays,	sericite, smectites		illite, chlorite (from other clays)
Carbonates		siderite, dolomite, ankerite, calcite,	ankerite, dolomite, calcite	
Sulphides		pyrite, marcasite, melnikovite	pyrite, marcasite, sphalerite, galena, chalcopyrite	pyrite (from siderite)
Silicas	quartz	quartz, chalcedony	quartz, chalcedony	
Oxides and hydroxides	rutile	hematite limonite	goethite	
Phosphates	apatite	(phosphorite) apatite		
Silicates	zircon, feldspars, tourmaline, micas			
Sulphates			hydrated iron sulphates, gypsum (oxidation products)	

4.3.1 *Clays*

Clays are the most common mineral group in coal, generally comprising 60 to 80 per cent of total mineral matter. They are generally primary, representing material introduced to the basin by water or wind, and authigenic material. Their distribution within coals is partly related to original chemical conditions within the swamp. Staub and Cohen (1979), for example, found that in the modern Snuggedy Swamp of South Carolina acid water conditions within peat swamps have led to rapid flocculation of detrital clays, and resultant concentrations of clay deposition near the edges of the swamp.

The most common clay minerals in coal are kaolinite and illite, with sericite, mixed-layer clays, and smectites being less common (Table 17). Chemical conditions at the site of deposition also influence the type of clay minerals associated with coal. In particular, freshwater swamps, with their low pH, tend to favour in situ alteration of smectites, illite, and mixed-layer clays to kaolinite (Berner, 1971). Generally illite is dominant in coals with marine roofs, whereas kaolinite is dominant in nonmarine influenced coals.

Secondary clays are generally produced by tranformation of primary clays. With increasing depth of burial, the proportion of kaolinite decreases while that of illite rises. Chlorite can be expected to occur with greater pressure and temperature. Clay minerals occur in coal in two fundamental associations, either as finely dispersed inclusions in macerals or as tonsteins (clay bands). Origin of tonsteins, which are composed predominantly of kaolinite, is not completely understood. Detrital and authigenic origins have been postulated in different examples (Mackowsky, 1968; Moore, 1968). Tonsteins of apparent volcanic origin (ash falls or tuffs) are also known. These usually contain kaolinite, smectites, and mixed-layer clays (ICCP, 1963), along with accessory sanidine and zircon. Kaolinite in tonsteins occurs in two distinctive forms: (1) as graupen, anisotropic spherules of fine kaolinite (Plate 9), and (2) as vermicules, clearly authigenic growths of crystalline kaolinite (Plate 8).

Petrographically, coal containing up to 20 per cent clay can be termed contaminated, while that containing 20 to 60 per cent clay can be referred to by its microlithotype name, carbargilite. Actual identification of clay minerals is not practical under normal petrographic conditions. One exception is recognition of kaolinite as vermicules and graupen in tonsteins. However, X-ray diffraction should be considered a necessary step in positive identification of clay minerals.

4.3.2 *Carbonates*

The carbonate minerals are emplaced as both primary and secondary precipitations (Table 17). Primary carbonates occur as concretions or infillings of cell lumens. Siderite, which forms under reducing conditions as concretions and nodules (Plate 8), is generally considered to be of primary origin (Gould and Smith, 1979), but the other common carbonates in coal (calcite, ankerite, and dolomite) are not so restricted. Primary dolomite is thought to reflect marine roof conditions, and may occur in idiomorphic form or as a constituent of coal balls. Secondary carbonates generally take the form of sparry replacements or fracture and cavity fillings, the last including fillings of cell lumens.

Carbankerite is the microlithotype term for coal containing 20 to 60 per cent carbonate.

4.3.3 *Iron Sulphides*

Iron sulphides, dominantly pyrite, but including marcasite and melnikovite, are of primary and secondary origin in coal (Table 17). Primary iron sulphides occur as both fine crystals and fine nodules or globules (Plate 8), and as infillings of cell openings, and are especially abundant in seams with marine roofs. Secondary iron sulphides are deposited in fractures and cleat (Plate 9), and may also impregnate cell openings in fusinite and semifusinite, or replace primary minerals such as siderite (Smyth, 1966).

Petrographically, associations of coal with 5 to 20 per cent iron sulphides are called carbopyrites.

Plate 8. Mineral matter in coal (reflected light oil immersion, except 5b).

Figure 1. Kaolinite graupen (G) in A-seam (medium-volatile) from the Kootenay Group, Crowsnest Coalfield, Sparwood, British Columbia.

Figure 2. Kaolinite vermicules in tonstein band from subbituminous coal of the Horseshoe Canyon Formation, Drumheller, Alberta.

Figure 3. Siderite concretion in subbituminous coal of the Horseshoe Canyon Formation, Drumheller, Alberta.

Figure 4. Rare occurrences of elemental sulphur (S) filling cell lumina in fusinite from high-volatile A coal from the Gething Formation, north of Williston Lake, Peace River Coalfield, British Columbia.

Figure 5a. Quartz in 10-seam (medium to low-volatile) from the Kootenay Group, Crowsnest Coalfield, Sparwood, British Columbia.

Figure 5b. Same field of view as 5a. Dry lens.

Figure 6. Clays (M.M.) and pyrite in high-volatile coal from the Bickford Formation, Peace River Coalfield, British Columbia. Note occurrence of large pyrite framboid.

Figure 7. Pyrite in high-volatile coal from the Bickford Formation, Peace River Coalfield, British Columbia. Note stringers of pyrite aligned with liptinite macerals.

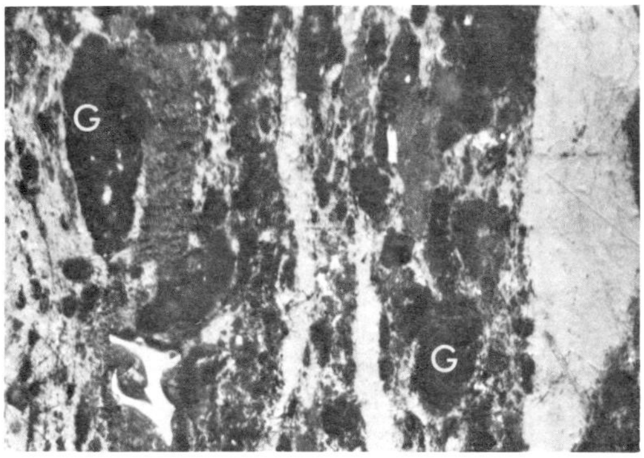

1

2

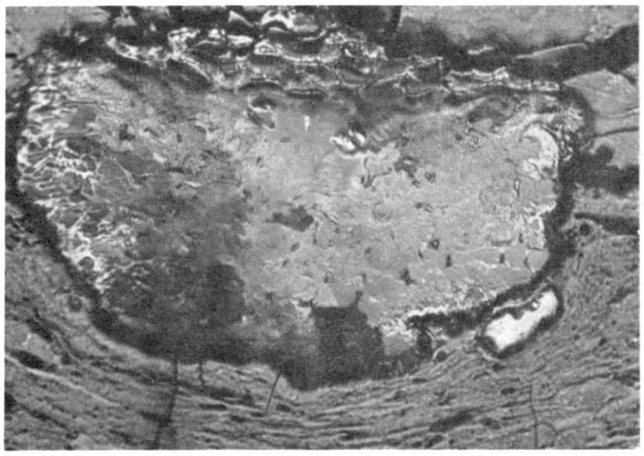

3

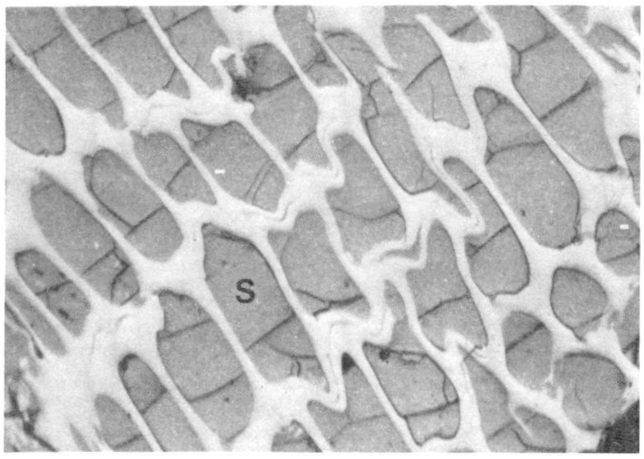

4

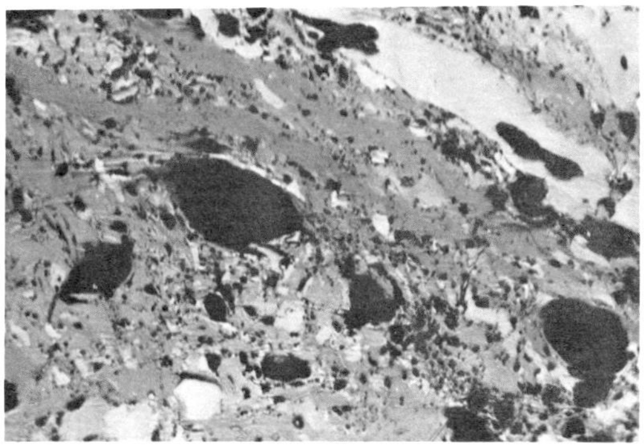

5a

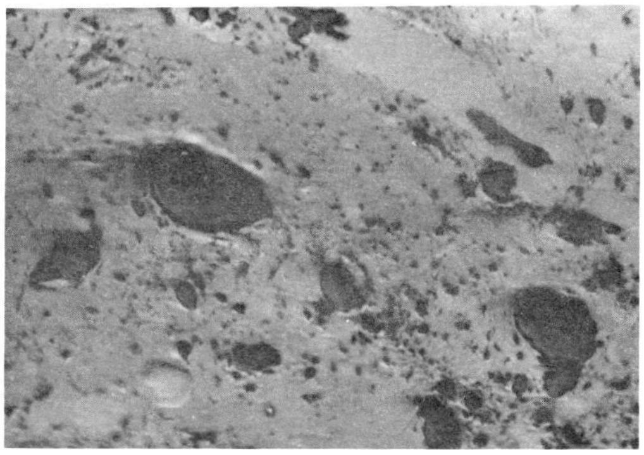

5b

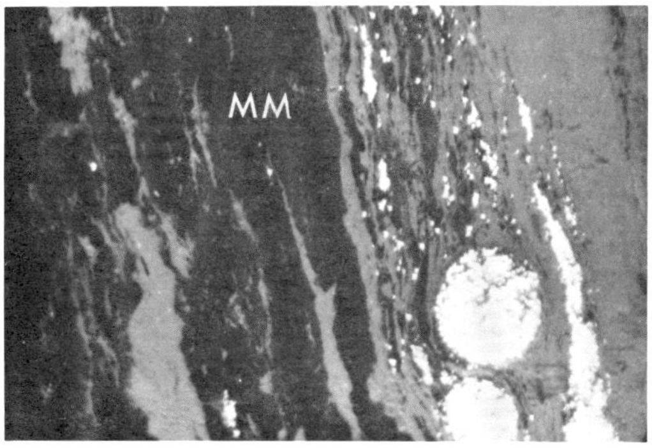

6

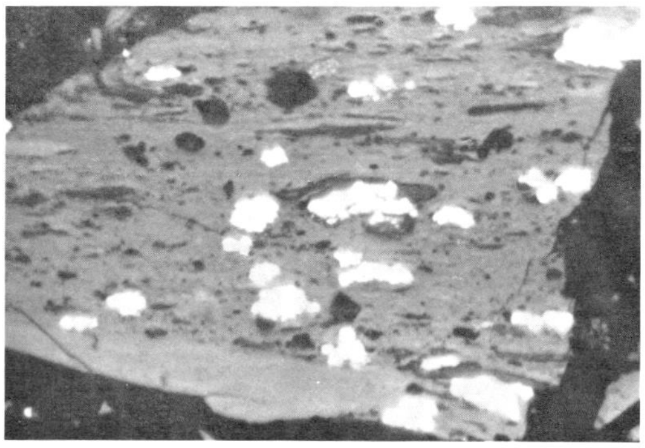

7

MICRONS

0 100

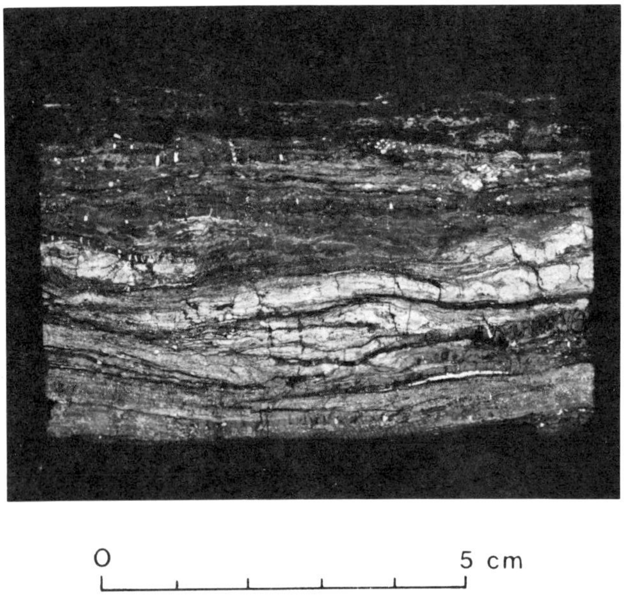

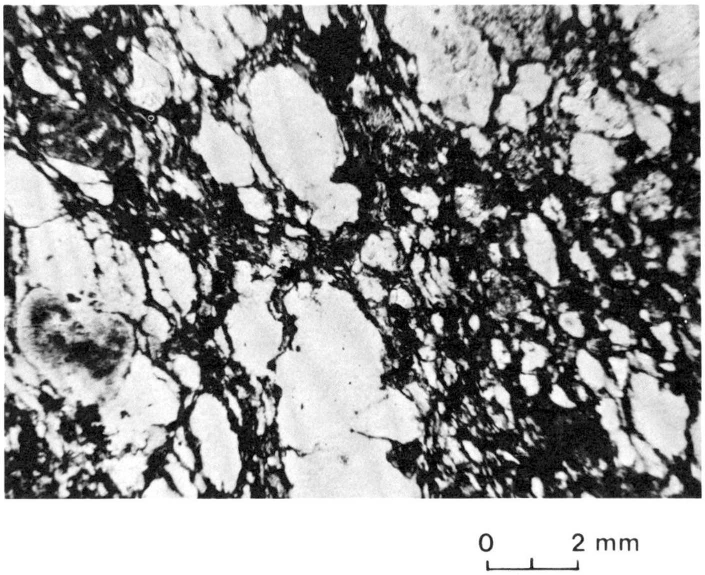

Plate 9. Mineral matter in coal.

Figure 1. Secondary pyrite stringers parallel to banding in a polished block of semi-anthracite coal from the Groundhog Coalfield, British Columbia.

Figure 2. Thin section of graupen tonstein parting from coal seam in the Kootenay Group, Elk Valley Coalfield, British Columbia.

4.3.4 *Silicas*

Detrital quartz grains and irregular quartz blebs precipitated as fine intergrown bodies in coal are examples of primary quartz in coal (Plate 8). Both are associated with clays in coal and carbargilite. Also of probable primary origin are quartz or chalcedony infillings of cell lumens in fusinite and semifusinite. Secondary quartz takes the form of quartz or chalcedony in veins and fracture fillings.

4.3.5 *Other Mineral Groups*

Oxides and hydroxides include hematite (primary), limonite (primary), and goethite (secondary). Other sulphides, which are generally of hydrothermal (secondary) origin in coal, include galena, sphalerite, and chalcopyrite. Phosphate minerals (generally apatite) are primary in coal, and may represent detrital or authigenic material. Silicates include mainly detrital accessory minerals such as zircon, tourmaline, feldspars, and mica, which are found in coal, partings, and tonsteins. Sulphates occur mainly in oxidized coals as late alteration products. The most common are a series of hydrated iron sulphates (including melanterite and coquimbite) which result from weathering of pyrite (Rao and Gluskoter, 1973). Gypsum also occurs in oxidized coals (Pearson and Kwong, 1979).

4.3.6 *Trace Elements in Coal*

Trace elements cannot be routinely determined petrographically. A recent study by Finkleman (1980) applied scanning electron microscopy with an energy dispersive detector (SEM-EDS) to elucidate trace element content and distribution in coals. Finkleman determined that many trace elements (including lead, zinc, copper, cadmium, zircon, hafnium, thorium, and uranium) occur in micron-sized accessory mineral grains scattered throughout the macerals. Others (including arsenic and mercury) are in solid solution in pyrite lattices. Another group of trace elements (including titanium, selenium, bromine, and uranium) may be bound in the organic structure of the coal.

5 OXIDATION

Exposure of coal to the atmosphere or oxygenated waters in the subsurface results in oxidation of the organic and inorganic constituents. Oxidation markedly alters the physical and chemical properties of coal and reduces the coal quality. In particular, the lycophobic character of coal particles is destroyed, resulting in loss of floatability during washing, the calorific value is lowered, and the tar yield during caking and total yield on coking are both reduced. Due to the detrimental effects of oxidation on coal quality, there has been considerable research into determining the chemical and physical characteristics of oxidized coals and determining the degree of oxidation (Elliott, 1981; Crelling *et al.*, 1979; Alberts *et al.*, 1974; Wachowska *et al.*, 1974; and others).

The chemistry of oxidation, although still poorly understood, is considered to result in a decrease in the nonaromatic fraction of the coal and a relative increase in the aromatic fraction (van Krevelan and Schuyer, 1957). Initially, oxidation is characterized by chemisorption of oxygen on aromatic and nonaromatic sites by formation of acidic functional groups such as -COOH, -CO, and phenolic -OH. Advanced stages of oxidation are characterized by formation of humic acids and progressively smaller molecules (Berkowitz, 1979). The most pronounced change evident in routine chemical analysis of coal are a progressive increase in oxygen and a decrease in fixed carbon, hydrogen, and sulphur with degree of oxidation (Crelling *et al.*, 1979).

Changes in the surface chemistry of coal particles occur at the onset of oxidation and so even mild oxidation affects the lycophobic character of the coal and the coking properties. A decrease in lycophobic character results in poor froth flotation such that separation of ash and fine oxidized coal is difficult and much of the fine coal may thus be unrecoverable. The loss of caking properties and fluidity during coking may be evaluated if the FSI, dilatometry, or fluidity of the coal is determined. With accompanying loss of caking power, bulk density, strength of the coke, and tar yields are reduced. Presently, attempts are being made to develop methods of restoring the lycophobic and caking properties of coal by pretreatment with various chemicals. On the other hand, in particular types of fluid beds and hot coal transport cyclones, it may be desirable to pre-oxidize the coal to prevent caking, which is detrimental to the operation of the machinery.

Weathering of coal, which generally accompanies oxidation, results in physical breakdown to fine particles and promotes hydrolysis and hydration. Comminution accompanying weathering further enhances oxidation by increasing surface area. If the coal is tectonically sheared, and thus composed of fine particles, the extent of oxidation will be greater.

The susceptibility of coal to oxidation is determined by its maceral and mineral matter contents. Vitrinite is the most readily oxidized maceral, and if pyrite or other sulphides are present, they will readily oxidize to sulphates. Coals of all ranks are susceptible to oxidation; however, no quantitative comparisons between coal of different ranks have been made. Other external factors such as climate, physiography, particle size of the coal, and hydrology also govern the rates and degree of oxidation.

A number of different methods for quantifying oxidation based on chemical and physical properties have been developed (see Gray *et al.*, 1976). These include infrared spectroscopy, pyrolysis gas analysis, phase inversion, coal slurry pH, alkali solubility, electrostatic properties, dilatometry, fluidity, and FSI.

5.1 **Petrography**

The petrographic characteristics of oxidized coal vary with the temperature of oxidation and local oxidizing conditions (Chandra, 1962, 1982; Leythaeuser, 1973). Although all the macerals are susceptible to oxidation, in a given sample the vitrinite invariably shows greater evidence of oxidation than the inertinite or liptinite macerals.

Natural oxidation results in the formation of micropores, microfissures, and oxidation rims on the macerals (Chandra, 1962; Crelling *et al.*, 1979; Bustin, 1982). Oxidation in coal piles, gob areas of mines or in the laboratory (at elevated temperatures) have similar surface features, but the oxidation rims have higher reflectances than those of the core of the particles. In most naturally oxidized coals the oxidation rims generally have a lower reflectance, although Marchioni (1983) has found higher reflecting oxidation rims in some subbituminous coals from Drumheller, Alberta. Pearson and Kwong (1979) and Bustin (1982) both found that natural oxidation of medium and low-volatile bituminous coals in the southeastern Canadian Cordillera coalfields reduces the vitrinite reflectance.

As oxidation progresses there is a corresponding change in the petrographic features of the coal (Plate 10). In very slightly oxidized coals evidence of oxidation is generally manifest as a slight darkening of the rim and along natural fractures in the vitrinite. At advanced stages of oxidation the vitrinite develops micropores and microfissures and takes on a granular appearance and a high relief, and the inertinite macerals develop oxidation rims.

Oxidation affects the surface of coal particles (acts topochemically), and thus the degree of oxidation of the particles is proportional to the surface area. In any coal sample there may be a complete range of oxidized particles; the fine size fraction may be highly oxidized, whereas the coarse fraction may be only mildly oxidized. For coals which have undergone tectonic shearing or have otherwise become comminuted, the degree of oxidation will be greater.

Methods of quantification of the degree of oxidation by petrographic procedures have been suggested by Gray *et al.* (1976) and Gray (*in* Marchioni, 1983). Polished pellets prepared for maceral analysis are first etched in KOH, then dipped in safranin 'O' in alcohol, and then washed in water. By this method highly oxidized coal particles are stained a dark olive green, whereas less oxidized particles are stained brighter green. Unoxidized particles are unchanged (Gray *et al.*, 1976). Point counting the pellet stained in this fashion allows quantification of the amount of oxidation and enables a semiquantitative evaluation of the degree of oxidation. In Plate 11, a series of stained samples representing various degrees of oxidation from the southeastern Canadian Cordillera are shown.

5.2 **Depth of Oxidation**

Because oxidation markedly reduces the quality and value of coal, it is important to predict the extent of oxidation. In any deposit the depth of oxidation and the degree of oxidation will be a function of the climate, groundwater conditions, extent of fracturing of the rocks and coal, and the particle size and composition of the coal. If the coal is above the groundwater table, making it accessible to atmospheric oxygen, the degree of oxidation will be greater. If the coal is below the groundwater table, oxidation will be less, but, where the coal seam is an aquifer, oxidation may be pronounced (Johnson, personal communication, 1979). Fracturing of the coal and adjacent rocks and particle size of the coal have a notable effect on degree of oxidation.

In the southeastern Canadian Cordillera finely sheared coal adjacent to faults are more oxidized than adjacent unsheared coal, probably as a result of both the finer particle size of the sheared coal and because the faults commonly act as conduits for groundwater flow. Although it is known that vitrinite and pyrite-rich coal are more susceptible to oxidation, the composition of the coal is almost everywhere a secondary factor affecting the depth of oxidation. An example of changes in the degree of oxidation with depth in an adit (No. 6 seam) from Tent Mountain in the southeastern Canadian Cordillera is shown on Figure 23. At this locality there is evidence of oxidation up to 10 metres below the surface. A depth of 10 metres from the present weathering horizon is approximately the extent of oxidation observed in most adits in the southeastern Canadian Cordillera. If the coal is sheared or acts as an aquifer, oxidation will be more extensive, and oxidized coal has been observed at depths of 100 metres.

5.3 **Predicting Fresh Coal Characteristics**

In many areas undergoing coal exploration only surface samples may be available for analysis. Such samples will invariably be oxidized to some degree, and routine coal analysis will yield erroneous results. Although there is no substitute for fresh coal samples, petrographic studies of the oxidized coal may yield valuable data. The most coarse size fraction available from the surface should be collected and pellets prepared using only this fraction. Routine petrographic analysis and vitrinite reflectance determination can then be made. During vitrinite reflectance determinations only the central portions of unoxidized grains

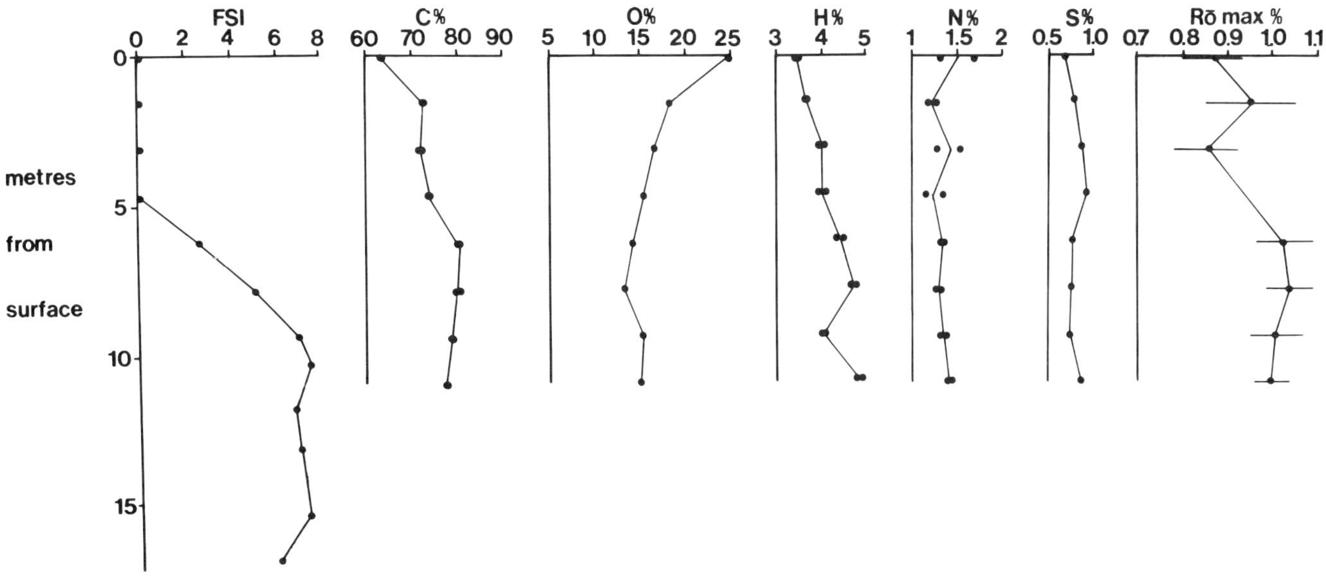

Figure 23. No. 6 seam adit. Variation in Free Swelling Index (FSI), elemental analysis (dmmf), and vitrinite reflectance with increasing distance from the modern weathering surface. Oxygen determined by difference. Replicate analyses are shown for carbon, hydrogen, nitrogen. The mean (•) and standard deviation (-) of the maximum reflectance values are shown (modified from Bustin, 1982).

should be measured. The petrographic results of such a study will give some information as to the composition of the seam, although it is unlikely that it will characterize the seam. The vitrinite reflectance should yield a minimum reflectance so as to allow determination of the minimum rank.

An alternative, purely chemical method for evaluating fresh coal chemistry has been outlined by Francis (1961) and Berkowitz (1979). Naturally and artificially oxidized coal and pure hydrocarbons were shown by Francis (1961) to progressively change composition along predictable trends when plotted on a ternary diagram with oxygen, carbon, and hydrogen as end members. Berkowitz (1979) has suggested that fresh coal composition can be estimated by plotting the composition of analysed oxidized samples on such a diagram and extrapolating a line parallel to established oxidation tracks, back to the coal band of Seyler (Figure 24). On Figure 24 the analysed samples from the No. 6 seam adit are plotted. With the exception of the sample at 9 metres from the weathering horizon, the analysed samples define a relatively straight oxidation track. The oxidation track defined by the No. 6 seam adit samples is oblique to that reported by Francis (1961) but clearly demonstrates progressive change in chemistry with oxidation. Because the extent of oxidation is at least in part a product of coal rank and maceral composition, it is to be expected that different coals will give different oxidation tracks. Nevertheless, such a method of predicting fresh coal composition from analysis of weathered samples could provide a good estimate of the fresh coal composition, particularly if the individual oxidation tracks of different seams are established in the area of study.

5.4 Spontaneous Combustion of Coal

Spontaneous combustion of coal is a result of oxidation in which temperatures capable of ignition are reached. Spontaneous combustion can only occur under conditions in which heat released by exothermic oxidation is greater than that carried away. These conditions exist where there is poor ventilation such as in gob areas or coal stockpiles. As the temperature of the coal rises the rate of oxidation increases further. The factors promoting spontaneous combustion of coal are poorly understood (Mackowsky, 1982, pp. 474-476). The presence of vitrinite (which oxidizes more readily than other macerals), fine particle size, pyrite, and moisture are all considered to enhance oxidation, which, together with poor ventilation, may result in spontaneous combustion. Studies by Munzner (1972, *in* Mackowsky, 1982, p. 475) have demonstrated that, if the coal is wetted, the reactivity of the coal is doubled. If finely dispersed pyrite is present, the reactivity is increased ten-fold. Spontaneous combustion thus appears to be more likely with fine, wet, vitrinite-rich coal in which finely dispersed pyrite is present.

Plate 10. Photomicrographs of oxidized coal. All photographs are taken in reflected light with oil immersion.

A) Slightly oxidized vitrinite (v) in low-volatile bituminous coal. Oxidized coals such as this one polish poorly.

B) Highly weathered and oxidized low-volatile bituminous coal. The vitrinite (v) has developed microfractures and micropores. Inertinite (i).

C) Highly oxidized low-volatile bituminous coal. A well-developed oxidation rind (darkish tint) has developed along the edge and along fractures of inertinite (i). Semifusinite (s) which is completely enclosed in oxidized vitrinite (v) shows little evidence of oxidation.

D) Highly weathered and oxidized anthracite. Oxidation rinds are well developed along the border of the vitrinite and pervade the smaller vitrinite particles. The centre of the large vitrinite particle, however, is unoxidized and vitrinite reflectance thus could be measured.

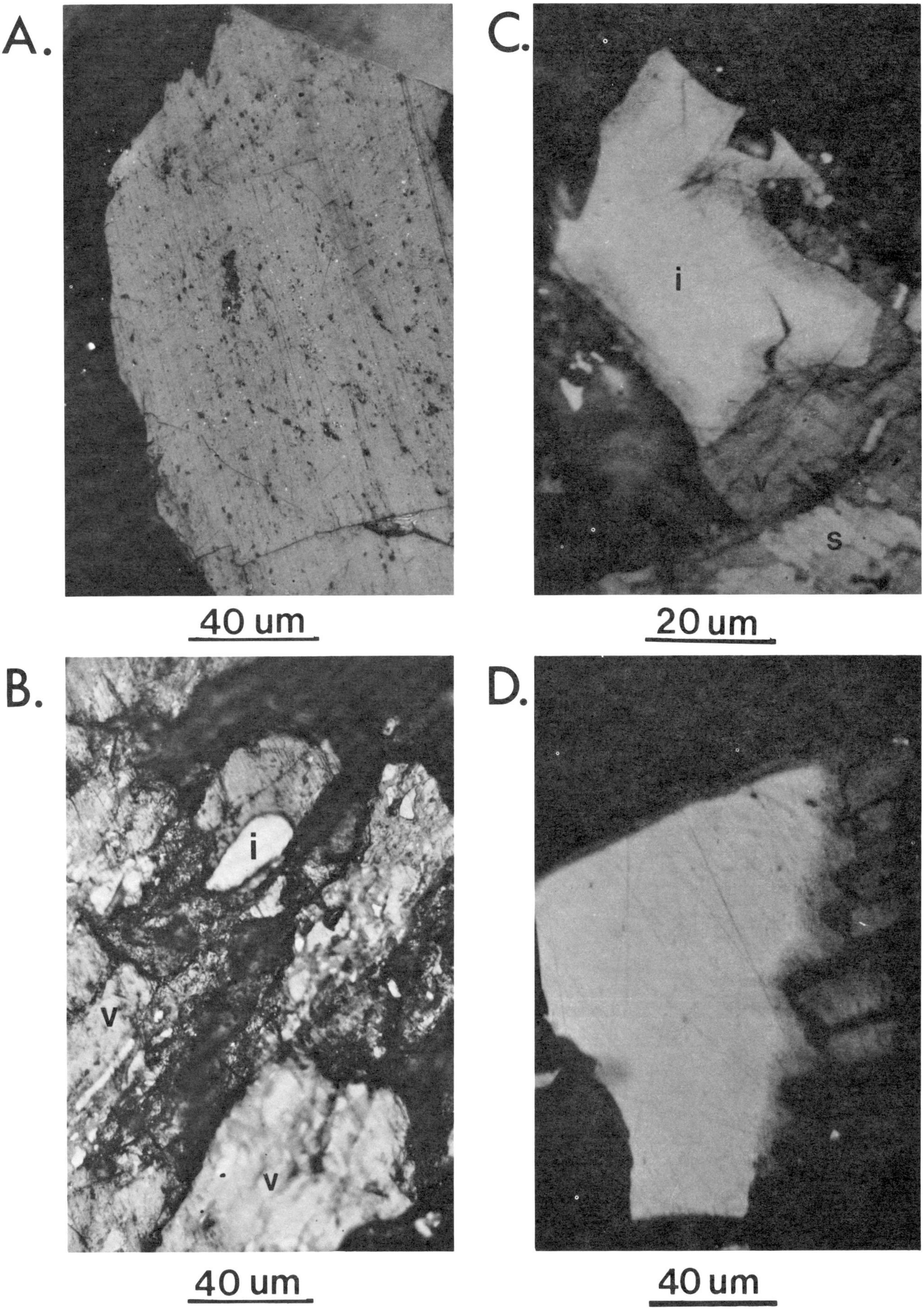
A.
40 um
C.
i
v
s
20 um
B.
i
v
v
40 um
D.
40 um

Plate 11. Coal stained with safranin "O." The four samples shown are from an adit in the southeastern Canadian Cordillera. The coal is a low-volatile bituminous, coking coal. All photographs with reflected light oil immersion.

A) Sample 21 metres from the present weathering surface. Evidence of oxidation (green stain) is restricted to small particles and the rims of some larger particles of vitrinite (v).

B) Sample from 38 metres from the present weathering horizon. This sample is more oxidized than the sample from 21 metres because it is adjacent to a fault which promoted oxidation by allowing circulation of oxygenated waters and/or atmospheric oxygen. Vitrinite (v), inertinite (i).

C) Sample from 38 metres from the present weathering horizon. Note how the smaller particles of vitrinite are much more oxidized (stained) than the larger particles of vitrinite (v), semifusinite (s), or inertinite (i).

D) Sample from 10 metres from the present weathering horizon. The vitrinite (v) is much more oxidized (stained) because of its proximity to the weathering horizon than the sample from 21 metres.

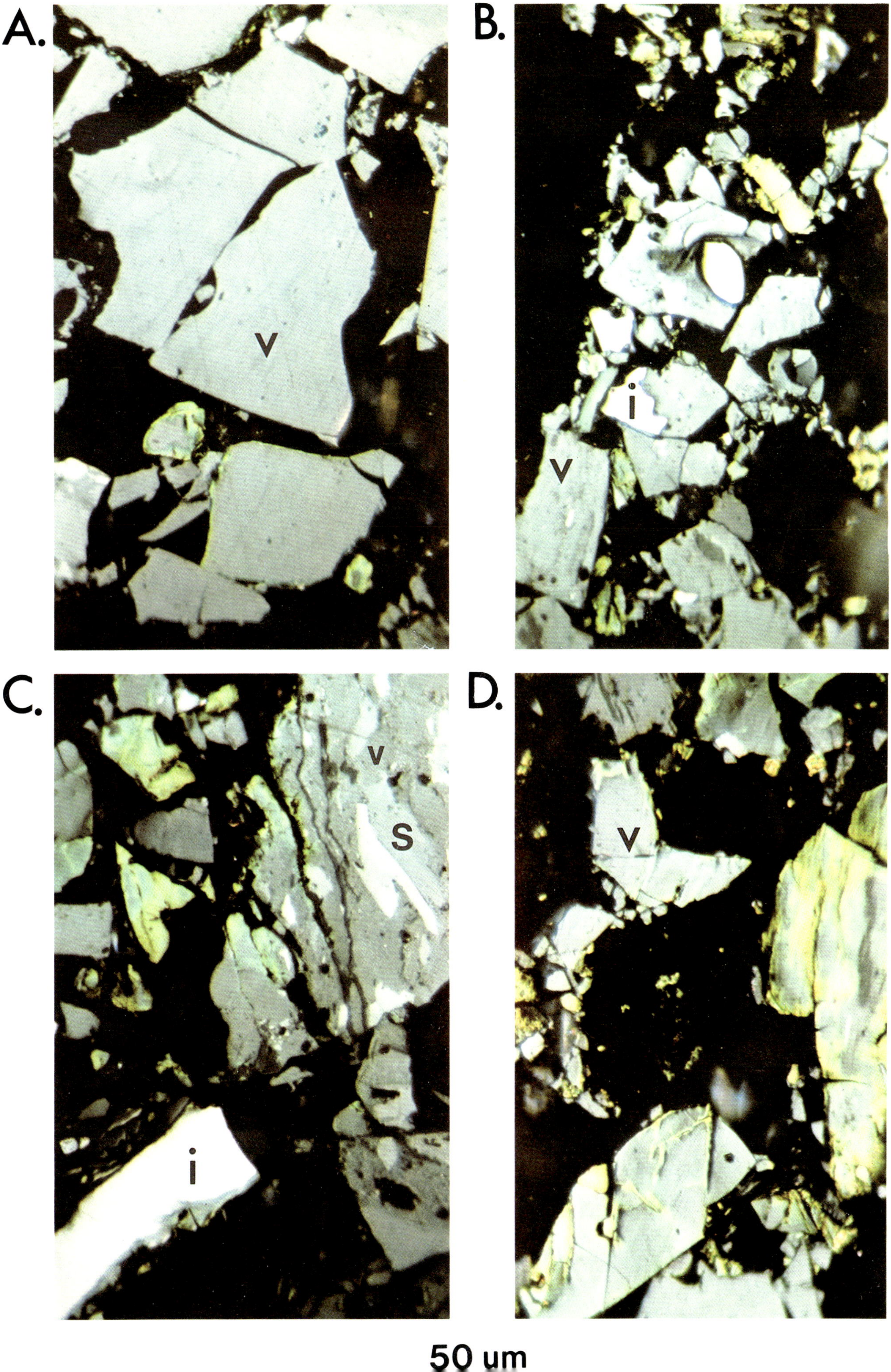
A.
v
B.
i
v
C.
v
S
i
D.
v
50 um

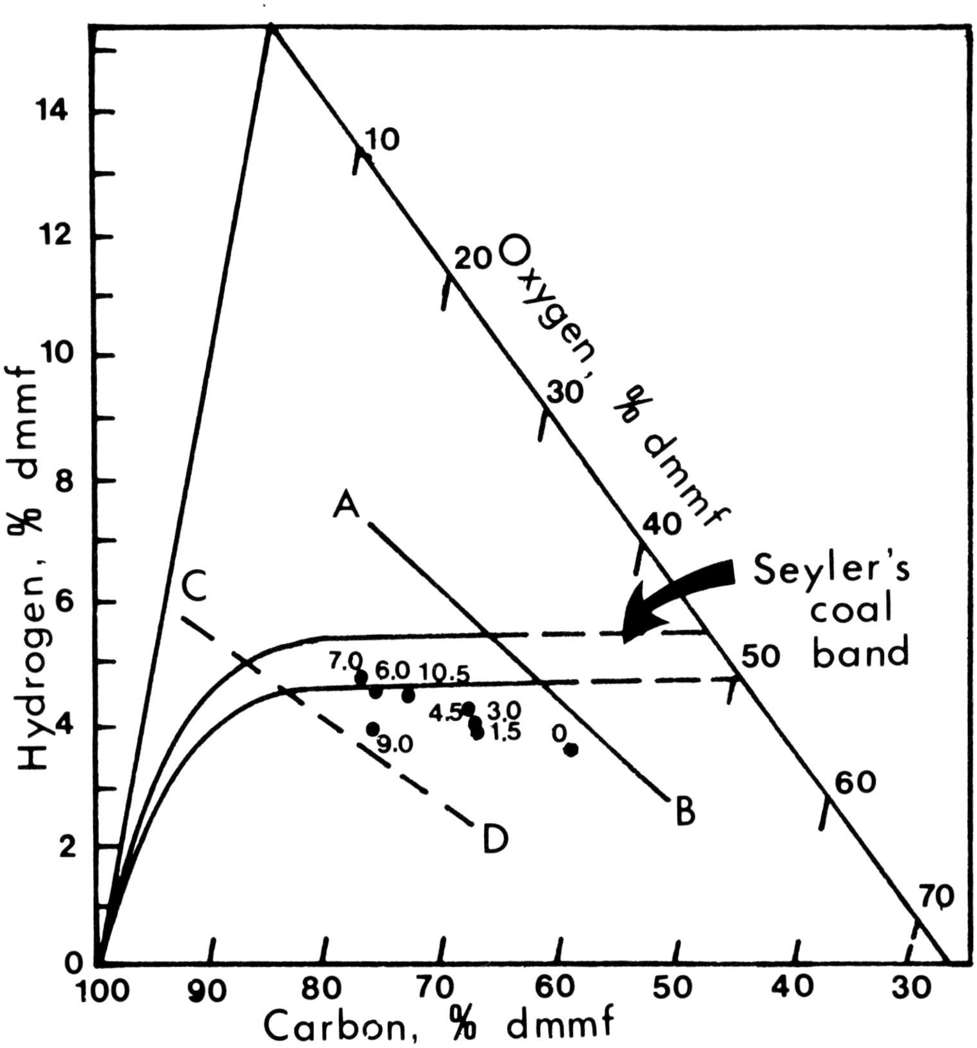

Figure 24. C-H-O diagram showing the shifts in elemental composition of coal with progressive oxidation. Lines A-B and C-D are typical oxidation tracks of pure hydrocarbons (modified from Francis, 1961). Normal coal samples are expected to plot within Seyler's band of "normal coal" and progressively shift, with oxidation, along tracks parallel to that of pure hydrocarbons and towards oxygen as shown by Francis (1961). Plotted are the No. 6 adit samples (•); the adjacent numbers indicate distance, in metres, from the modern weathering surface.

6 SAMPLING AND SAMPLE PREPARATION

6.1 Sampling Problems and Methods

The reliability of any analysis depends on sample quality. The type of sample collected should be determined by the nature of the analytical data desired. The degree of detailed information which a sample provides cannot exceed the weight, area, or dimension constraints put upon a sample no matter how carefully it is collected. For example, a carefully collected whole-seam channel sample cannot provide the kind of vertical variation information within a seam which a column sample yields. Therefore before a project begins it is always necessary to determine the kind of information required and to tailor sample collecting procedures to it.

Coal samples come from many sources; mines, exploration adits and trenches, borehole cores and cuttings, outcrops, old mine dumps, and a host of coal utilization and handling facilities including wash plants, coke-making installations, stockpiles, ships, and trains. Each of these situations presents its own unique set of conditions relevant to the basic premise that a sample should be a representative part of a population. Specifications for the collection of samples from industrial installations as well as channel samples in mines for coking or combustion tests, washability studies, and chemical properties are described in ASTM Standard D-2234-76 (ASTM, 1981). Of course splits from such samples are frequently used for petrological studies as well.

The collection of an initial representative sample is a most difficult step and, probably, the greatest single source of variation in the analysis of a coal population. The variability can be evaluated if it is feasible to collect and analyse multiple samples from the parent population of coal. Such techniques are readily applied in evaluating coal for technological purposes where coal is commonly pre-crushed, where sampling is done by automatic samplers, or where the parent population of the coal is limited, such as in a coal feed stockpile. If the purpose of the study is to characterize a coal seam, bench or layer samples of the coal must be collected through the entire thickness of the seam. It is necessary to collect duplicate bench samples from the same location, and from different locations, in order to evaluate not only the sampling but also the lateral variation in composition of the coal. If an attempt is made to determine whether variation in samples collected at one site is greater or less than the variation between sites, a large number of duplicate samples may be required depending on the levels of confidence demanded. Collecting of duplicate samples from coal seams is relatively easy if a drill core is available; however, if only surface exposures are available, it is a tedious task and great care is required to sample the entire seam equally.

Three types of coal samples are collected in the field: grab samples, column samples, and channel samples. The least representative type is the grab sample which may come from an undetermined level in a coal seam. This is the kind often collected by geologists working in remote areas with neither time nor equipment to excavate partially covered coal seams. It must be assumed that grab samples do not give an adequate measure of the maceral content of a coal. They may, however, provide rank data, which in many cases is invaluable information. Column samples represent the total thickness of a seam and show the natural association of layering and structure within the seam. They are usually comprised of an unbroken series of blocks representing the seam from roof to floor. Such samples are taken by cutting two parallel channels in the seam at least 20 centimetres apart and then loosening and removing the 'pillar' in the intervening space. Usually this comes out in the form of blocks which must be carefully labelled as to position in the seam and orientation with respect to roof and floor. These blocks may be wrapped with masking tape to prevent breakage during handling and transportation. This is especially important if the coal tends to be friable. If it is extremely friable, it is the practice in some places to impregnate the coal surface with resin prior to removal from the face.

Another technique which can be used with friable coals is to take a series of mini-channels which together represent the total thickness of the seam. It is best if these channels represent macroscopically defined layers or utilize natural breaks, such as partings, for layer boundaries. Otherwise some arbitrary layer thickness such as 25 or 50 centimetres might be employed for channel height. Many of the very friable western Canadian coking coals are collected in this way if detailed examination of the layering within a seam is an objective of the study.

Like column samples, whole-seam channels also represent the total thickness of a seam but unlike columns the internal arrangement of bedding is lost because they are collected as broken coal. In the collection of such channel samples it is recommended that partings of 1 centimetre thickness or greater be excluded. Procedures for collecting such samples are described by Schopf (1960). He suggests that 6 pounds be collected for each foot of coal in the seam, roughly 9 kilograms per metre.

Drill cores provide a ready made column sample if recovery is good. Because a number of washability or chemical tests may be planned for a core, a column sample for petrology may require lengthwise sawing of the core or some other form of splitting. If recovery is poor the sample may have to be treated as a channel although these will be increasingly unrepresentative as recovery deteriorates.

Drill cuttings can also be a source of coal samples, although here the possibility of contamination from material cut higher in the hole has to be considered. This problem may be largely overcome by the use of reverse circulation equipment (Irvine *et al.*, 1974). Coal samples may also be recovered from oil well cuttings although here too, the problem of contamination from caving may be quite significant. This is also a factor to be considered in the study of kerogen samples from oil wells. Methods for detection of contamination will be discussed later under microscopic techniques (see Section 10.5).

One problem noted at the Institute of Sedimentary and Petroleum Geology, with coal cuttings from an oil well, revealed itself microscopically by the presence of grains which had obviously been heat affected, some to the point of coking. The cause was believed to be excessive localized heating of coal particles during the drying of the cuttings. The reflectance measurements in some cases were seriously affected.

If possible, samples should be obtained from fresh coal, however, material from outcrop can also be studied provided weathering is not too severe. The removal of the highly weathered surface at outcrop to a depth where coherent pieces of coal begin to appear will usually provide material, which though it still shows weathering microscopically, may yet be quite adequate for petrological examination (see Chapter 5).

Coal samples collected for petrology are often submitted for chemical analyses or technical tests such as F.S.I. or fluidity. Such samples should be fresh unless there is a special reason for testing these properties on weathered or partially weathered coal. In any event, for such dual purpose samples it is necessary to minimize loss of moisture and to prevent oxidation. When collected they should be immediately placed in appropriate containers, such as heavy duty plastic bags or in drums or cans with tight fitting lids. Sometimes such containers may be flushed with nitrogen to produce a more inert atmosphere. Storage under distilled water will also prevent deterioration. For petrological analysis only, with coals of bituminous or higher rank, loss of moisture or even mild oxidation will have little significant effect, but with lignites and subbituminous coals loss of moisture can bring about extensive physical deterioration of blocks or particles.

Finally identification data for samples should be as complete as possible, both for one's own files as well as for other laboratories where samples might be sent for analysis. The geographic location should be determined as closely as possible to mine property coordinates, to latitude and longitude, or to UTM grids on NTS maps. Underground mine locations can be determined on company coordinate maps or with reference to known points on a mine plan. If the seam is known it should be identified; formation and age should also be identified if known. Sometimes mine names are confused with seam names. If possible, both should be included in sample identification information; if only one is known it should be made clear which it is. If a seam name is unknown, or as in the case of thin and local seams it may have no name, it should be referred to the nearest known geological horizon, for example, the top or bottom of a formation or group. The type of sample should always be indicated; whether it is a grab sample, a full seam channel, etc. With column samples or with a series of mini-channels, it is important to indicate whether sampling has started at the top or bottom of the seam. Samples which have been processed in any way (for example, washed, crushed, screened, etc.) should be accompanied by such information when submitted for further analyses.

6.2 **Sample Preparation**

6.2.1 *Blocks*

If, from a stratigraphic column, it is desired to make a detailed examination of the lithology of a seam (in terms of lithotypes or microlithotypes), the procedure is to cut small blocks from the large blocks collected in the mine. This is usually done with a carborundum or diamond saw. The size of these small blocks may vary but a convenient size is about 5 x 9 centimetres (2 x 3½ inches). These are then placed in molds made of aluminum foil or stiff cardboard with the surface to be polished placed face down. At this point the block or mold should be labelled and orientation as to top and bottom indicated. The mold is then filled with a bonding material such as an epoxy resin. A number of these are available from laboratory suppliers or from plastics distributors. One such resin which has been used is Paraplex P-43 with Nuodex Cobalt and Cumene Hydroperoxide as accelerator and catalyst respectively. The mixing of these components should be done carefully and in the proper order as is indicated in the following procedure. Table 18 shows the effects of varying the proportions of constituents in the mix.

TABLE 18. RESULTS OF EXPERIMENTS SHOWING CHANGE IN SETTING TIME RELATIVE TO PARAPLEX MIXTURES

Clock Time	Appearance of Mix No. 1	Appearance of Mix No. 2
9:15 a.m.	100 ml Paraplex (P-43) 10 ml Nuodex Cobalt 5 ml Cumene Hydroperoxide Colour, dark purple	100 ml Paraplex (P-43) 5 ml Nuodex Cobalt 2.5 ml Cumene Hydroperoxide Colour, light purple
9:45 a.m.	Gummy, Jello-like	Same as initial mix
10:30 a.m.	Hard	Starting to thicken slightly
11:00 a.m.	Completely cured, ready for cutting and grinding	Gummy, Jello-like
11:30 a.m.		Hard
12:00 Noon		Completely cured, ready for cutting and grinding

Procedure for the Use of Paraplex P-43 Polyester Resin

The plastic Paraplex P-43 is a cold-setting polyester resin, requiring the use of a catalyst and an accelerator to cause it to harden into a solid mass. Certain mixtures of the resin (P-43), the catalyst (Cumene Hydroperoxide), and the accelerator (Nuodex Cobalt) will combine to form a dangerous explosive mixture unless directions are followed very carefully. The following is a list of precautions that must be observed to prevent an explosive mixture from forming.

(1) Always mix the Cumene Hydroperoxide (the catalyst) separately from the Nuodex Cobalt (the accelerator).

(2) Always use more Nuodex Cobalt than Cumene Hydroperoxide in any Paraplex mixture.

(3) Always mix the Nuodex Cobalt first with the resin, then add the proper amount of Cumene Hydroperoxide to the mixture.

Sometimes, if very friable or low rank coals are involved, it may be necessary to impregnate the blocks before much sawing is done. There are any number of resins with low viscosity which can be used for this purpose. At the Geological Survey of Canada an effective agent for this has been found to be a mixture of Ambroid glue dissolved in acetone. The block can be dipped into this mixture or the liquid can be brushed over the coal surface. The effect is enhanced if the block is placed in a low vacuum environment. Because of fumes from acetone, this whole procedure should be carried out in a fumehood.

After the mixture in the molds has hardened, the molds themselves can be removed, the blocks trimmed with a saw or grinding wheel and the identification markings transferred to the block, preferably with an engraving tool of some sort. The blocks are then ready for grinding and polishing.

The procedure for grinding and polishing is as follows:

(1) Rough grind with 240 grit silicon carbide powder or 240 grit adhesive-backed abrasive paper on rotating lap.

(2) Fine grind with 600 grit silicon carbide powder or 600 grit paper on rotating lap.

(3) Rough polish with 0.3 micrometres alumina powder and Texmet cloth on rotating lap.

(4) Fine polish with 0.05 micrometres alumina powder and silk cloth on rotating lap.

After each stage the specimen must be thoroughly rinsed in a stream of water to remove grit and grinding or polishing residue.

6.2.2 *Pellets*

Though the use of polished coal blocks is fairly common, pellets of particulate coal or grain mounts are the most frequently used specimens for microscopy. The preparation of such specimens is described in ASTM Standard D-2797-72 (ASTM, 1981). In preparing coal for such specimens it is important to remember that the end product is a relatively few grams of coal which may represent the entire thickness of a seam or a bulk sample of many hundreds of kilograms. Therefore, the preliminary crushing and reduction to subsamples must be carefully controlled. Recommended procedures are described in the ASTM Standard. In this Standard the suggested procedure is to crush the sample to either -4.75 or -2.36 millimetres. The sample is then reduced to 250 grams by riffling with a coarse riffle sampler with at least 12 divisions and riffles spaced at intervals between 12.7 and 19.1 millimetres. The 250-gram sample is then crushed again to -850 micrometres (20 mesh) by stage grinding to produce a minimum amount of fines. This material is then passed through a fine riffle sampler with at least 12 divisions and spacings between 3.2 and 6.4 millimetres, so as to obtain at least 10 grams of sample. The best control over this final crushing can be achieved using a mortar and pestle though there are varieties of mills and grinders which by careful use can produce close to the same results.

In the making of pellets two procedures are possible. In one the coal is mixed with a thermoplastic resin which requires the use of heat (up to 125 degrees Celsius) and pressure. A number of models of mounting presses are available for this purpose. The second procedure involves mixing the coal with an epoxy resin in plastic molds or in steel molds. The use of steel molds often involves the application of pressure. This second procedure does not involve heat. The effect of heat in the 'hot set' process may cause a slight increase in reflectance in the coal as was shown by Marchioni (1978) when he tested pellets of the same coal made in both 'hot set' and 'cold set' modes.

The pellets are now ready for grinding and polishing. This can be done by hand or by means of a machine such as the Buehler Automet which can polish six 25-centimetre diameter pellets at one time. An example of the procedure for the routine preparation and polishing of pellets is as follows:

Pellet Preparation and Polishing Technique
(K. Pratt, Institute of Sedimentary and Petrolem Geology, Calgary)

Step 1
When making pellets, 1-inch (2.54 cm) 'Metaserv' molds are used, in which a mixture of Buehler epoxy and coal crushed to the preferred size is allowed to cure. The mixture of coal and epoxy will result in a fairly thick paste. One should bear in mind that a paste which is too thick will cause air bubbles on the polishing surface. These air bubbles may trap particles in early grinding and polishing stages which may release during later stages, thus scratching the surface of the coal and resulting in an inferior polish. When air bubbles do occur, special care should be taken in cleaning procedures between stages.

A paste which is too thin will result in the coal particles being farther apart which will cause some inconvenience to the person studying the specimen.

Step 2
The cured pellets are then mounted in a Buehler Automet specimen holder to ensure that the surfaces to be polished are all level.

Step 3
While still in the holder, the pellets are placed on a Buehler Automet lap with a 240-grit 'Carbimet' paper disc. The Automet is turned on and pressure is applied to 245 KPa. The purpose of this stage is mainly to remove enough bonding resin to expose the grains of coal. This usually requires very little time.

Step 4
The wheel with the 240-grit disc is then replaced with one which has a 600-grit 'Carbimet' disc. The procedure of Step 3 is repeated, except this stage may require a little more time, especially after the disc has been used several times and has become worn. When removed from the lap, the pellets should have an appearance similar to brushed metal, where scratches will be hard to distinguish. A good way to look for scratches is to hold the pellets at a very sharp angle to a light. At the correct angle, the pellets will appear shiny, and any large scratches will be easy to see. If there are too many, repeat the 600 stage until a satisfactory surface is produced. Between every stage, the pellets should be placed in an ultrasonic cleaner, washed and rinsed with distilled water.

Step 5
The pellets are then placed on an Automet lap to which a 'Texmet' paper has been applied. The procedure of Step 3 is repeated except this time the paper is covered with a slurry of 0.3 micrometres Alpha alumina and water. The Automet is turned on for about 3 to 4 minutes at 245 KPa during which periodic applications of the alumina mixture are made. During the final 30 seconds to 1 minute of this stage, the lap can be flushed with distilled water. This seems to 'buff' the surface of the pellet, which will appear quite shiny at this stage.

Step 6
Finally, the pellets should be polished on a lap on which two layers of silk have been attached. This stage will require 1½ to 2 minutes and will include the application of 0.05 micrometre Alpha alumina mixed with water in the same fashion as the preceding stage. After polishing, the pellets are rinsed, placed in the ultrasonic cleaner, then rinsed with distilled water. The silk is also rinsed very well with distilled water. The pellets are then turned on the clean silk lap with no alumina, but a good supply of distilled water, at full pressure, for 20 to 30 seconds.

Step 7
The pellets are then taken out of the holder and washed with detergent and cotton balls. They are now ready for examination although a common practice is to allow their moisture content to reach equilibrium. Some coals, especially those of low rank, absorb moisture in the polishing process and this may reduce reflectance values. For this reason the ASTM (Standard D-2797, 1981) recommends that after polishing, pellets of some coals be placed in a desiccator for at least 15 hours prior to reflectivity measurement.

6.2.3 *Thin Sections*

Thin sections used to be a very common way of examining coal microscopically. They are not used much today because of the greater versatility of polished specimens, both in terms of the ranks of coal which can be examined and the variety of analyses performed. However, they are occasionally still used to study details of internal structure in some of the macerals. The preparation of coal thin sections is done in much the same way as rock sections. One surface of a small block of coal is ground to give a scratch-free and reliefless surface. This is mounted to a glass slide with a cold setting cement. Care must be taken to eliminate bubbles in the cement between the coal and the glass. After the cement has cured, excess material in the coal block may be removed by sawing, followed by coarse grinding until the thin slice of coal begins to show translucency. After this point, grinding with finer abrasive materials should be continued until the appropriate degree of translucency is achieved. For a high-volatile A bituminous coal this is obtained when the vitrinite appears reddish orange and the spores are a clear yellow. Initial stages of grinding may be carried out on a rotating abrasive lap such as a coarse-grained carborundum wheel. When the slice of coal has become quite thin (2 to 3 millimetres thick), the grinding operation should be continued on a finer grained wheel (600 grit) until the first evidence of translucency begins to appear. The grinding is then transferred to a fine-grained hone, and completed on it. A cover glass may then be mounted on the thin section. If it is desired to make a polished thin section, the specimen is not covered with a glass but instead is polished as a final step using 0.05 micrometre alumina in a water slurry.

Thin sections of peat and very low rank lignite have usually been made with a microtome. Cohen and Spackman (1972) describe a procedure for retrieving thin section material from peat cores, dehydrating the specimen, impregnating with paraffin and slicing sections with a microtome so that original association and orientation of peat constituents will be minimally disturbed.

Short descriptions for preparation of thin sections as well as polished blocks and pellets are given by Ting (1978). A method for preparing pellets or polished surfaces of peat has been described by Kalkreuth and Chornet (1982).

6.2.4 *Kerogen*

Ideally kerogen samples may be prepared both as whole rock samples or as enriched organic concentrates. The whole rock sample is used in order to study mineral matter/organic matter relationships and to measure fluorescence properties of the organic materials free from influences of fluorescing mounting media. The enriched concentrates are particularly necessary for samples poor in organic matter, because in most cases only a few organic particles can be found in the corresponding whole rock samples. With whole rock samples of this sort, analyses become very tedious and results based only on a few measurements do not have a high confidence level. Plate 12 illustrates the difference in microscopic characteristics between a whole rock sample and a concentrate prepared from the same sample. The enriched organic concentrate can be obtained by heavy liquid separation, for example, zinc bromide, bromoform, etc. (Detailed description for use of a mixture of CdI_2/KI, density = 1.95, is given in Kalkreuth, 1979, p. 284). Another method is the use of HCL and HF to demineralize a sample of its carbonates and clay minerals, thus concentrating the organic materials. For more detailed information the reader is referred to Bostick and Alpern (1977).

6.2.5 *General Remarks*

Cleanliness is probably the most important factor to remember in polishing coal pellets. Contamination at any stage by particles carried over from previous stages will almost certainly result in a poor polish. Rinsing between stages and use of the ultrasonic cleaner will aid in insuring cleanliness. For the same reason, the use of separate laps for each stage is essential. Separating the polishing operations from the grinding by some distance (at least 1 to 2 metres) is a good idea.

Another problem which can occur is the appearance of relief. This is caused by a hardness differential between the bonding material and coal, and between the various constituents of the coal. Relief causes problems during microscopic examination, especially in the measurement of reflectance. Relief can be caused by using too much water in either the polishing mixtures or in the buffing stages during polishing. If relief appears, the pellets should be reground. The causes of relief can be minimized by mixing the polishing powders with less water or by reducing the time spent buffing the pellets with water only. Time and pressure are also factors in the occurrence of relief. Too much time will exaggerate otherwise normal effects of differential grinding. Too little pressure will increase the time required, thus creating the same problem. In lower rank coals and weathered coals, relief almost always occurs and is very hard to control, but the procedure outlined above should minimize its effect given the materials and equipment available in most laboratories.

Often it is necessary to make pellets of small amounts of coaly material or of kerogen concentrates. In such instances it is necessary to alter the pellet-making procedure. If, for example, only a gram or less of the material to be analysed is available, it should be mixed with an equal amount of bonding material and this should be poured into the mold first so that the pellet surface to be polished will have a sufficient concentration of coaly material for a good analysis. Then above this initial layer of resin and coal, clear plastic should be poured to bring the pellet to sufficient height (usually about 1 inch or 2.5 centimetres) for ease of handling and polishing.

A

B

Plate 12. High reflecting organic material (OM) in whole rock (A) and in concentrate (B) from a Devonian shale, western Germany. Reflected light, oil immersion.

If the samples available for study are on the order of 0.1 grams, then another procedure may be found useful. In an already prepared pellet of clear plastic a shallow cavity can be drilled on the top surface. In this cavity the appropriate mixture of sample material and resin may be placed and allowed to set, after which it will be ready for polishing in the usual way. For samples of this size it is possible to place several in one pellet.

For samples with high contents of clay minerals, such as whole rock oil well cuttings (common in kerogen analysis), polishing with water as the medium for the polishing compounds may have disastrous results. Such samples can, however, be effectively polished with isopropyl alcohol. Special kinds of samples, such as bitumens or liquefaction residues, may cause problems if mounted in conventional epoxies, because of a tendency of some sample material to dissolve in the bonding medium. In such cases the use of dental cement may be an effective substitute as a bonding agent (Shibaoka and Ueda, 1978).

7 PETROGRAPHIC METHODS

7.1 **Reflectance of Coal Macerals**

Reflectance is defined as the percentage of an incident light beam's intensity which is reflected from a polished surface. Reflectance in oil of a coal maceral is related to refractive and absorption indices of the maceral by Beer's equation:

$$R_o = \frac{(\mu-\mu_o)^2 + \mu\kappa^2}{(\mu-\mu_o)^2 + \mu^2\kappa^2}$$

where μ and κ are the refractive and absorption indices of the maceral, respectively, and μo is the refractive index of the immersion oil. Use of this complex relationship in coal petrography, however, is avoided by comparing the maceral reflectance with that of an optical standard of known reflectance.

A maceral's reflectance is dependent on level of coalification, and hence reflectance is a widely used rank parameter. Reflectance may be measured on any of the coal macerals, although vitrinite (or a vitrinite precursor) is nearly always selected in rank studies. Vitrinite is preferred because of its tendency to occur in large homogeneous particles which are plentiful if not dominant in most coals, and because its reflectance shows good correlation with other coal rank parameters at most rank levels (Section 8.1).

As a rank parameter, determination of vitrinite reflectance is relatively rapid and precise, is applicable to coals of most ranks, and is independent of coal composition. Furthermore, reflectance is suitable for small samples such as dispersed organic matter in sedimentary rocks (Section 10.5) and may even make use of samples of oxidized coal.

Except where otherwise indicated, techniques described below conform with ASTM D2798-79. Excellent summaries of the technique of vitrinite reflectance measurement have been written by Mackowsky (Section 4.24 *in* Stach's Textbook, 1982) and Davis (1978a). This review draws heavily from these sources, and they are references for the following discussion except where otherwise indicated.

7.1.1 *Optical Characteristics of Vitrinite*

Except in extremely high rank coals, vitrinite is uniaxial negative with its optic axis perpendicular to the bedding plane (Figure 25). In any grain being observed under plane-polarized light, therefore, no matter what its orientation with respect to original bedding, there is a true maximum value of reflectance in the direction which corresponds to the bedding plane. In grains oriented parallel to bedding, of course, anisotropy disappears and the maximum value applies in all directions. There is also a minimum value, corresponding to the direction perpendicular to the maximum, for any grain, although it will nearly always be greater than the true minimum direction, which is present only when the grain surface is perpendicular to the bedding plane. Of the two values, therefore, the maximum value of reflectance of any grain has more significance, that is, it corresponds to the true maximum, while the minimum values only approach the true minimum.

A third measurable value is the random reflectance, which is the reflectance of a grain in the orientation in which it is encountered. If incident light is plane-polarized, then the random reflectance reading of a grain will be somewhere between its maximum and apparent minimum reflectances, depending on bedding orientation and position of stage. The range of possible readings is therefore between the true minimum and true maximum values for any grain. In nonpolarized light, however, the range of potential readings is less, as the reading on any grain will be the average of its maximum and apparent minimum, again depending on bedding orientation, but not depending on stage position. The lowest obtainable value on any grain will be the mean of the maximum and true minimum. Random reflectance of a grain in nonpolarized light is therefore related to the maximum and apparent minimum as follows:

$$R_om = 1/2\,(R_omax + R_o'min)$$

where R_om is random reflectance, R_omax is maximum reflectance, and $R_o'min$ is the apparent minimum.

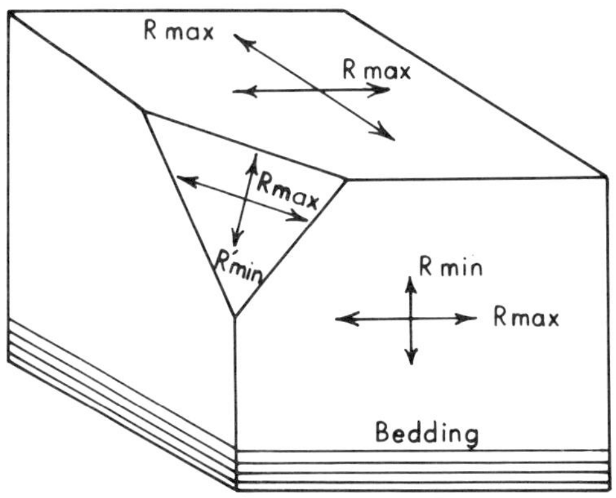

Figure 25. Anisotropy of vitrinite. The range of reflectance in polarized light depends on the orientation of grain surface with respect to bedding. $R_o$max and $R_o$min are the maximum and true minimum reflectances, respectively, while R'min is an apparent minimum direction. (Adapted from Davis, 1978).

Nonpolarized light is generally used in random reflectance determinations*. Two different values are therefore available to quantify reflectance of a vitrinite grain: (1) maximum reflectance in plane-polarized light, and (2) random reflectance in nonpolarized light. Due to natural scatter in reflectance values of vitrinite grains in a sample, it is normal to measure a large number of grains per sample and to average the values. This leads to two analytical parameters: mean maximum reflectance of vitrinite in oil ($\bar{R}_o$ max); and mean random reflectance of vitrinite in oil (R_o m). It is essential, when reporting results, to specify which parameter has been measured.

*Ting (1978) differentiates random reflectance (Re), in which polarized light is used, from average reflectance (Rav), which is equivalent to random reflectance described here (nonpolarized light).

The two parameters, when derived from the same sample, are related as follows (Ting, 1978):

$$\bar{R}_o \text{ max} = 1.066\, \bar{R}_o \text{ m}$$

Use of either parameter has advantages. Mean maximum reflectance data have a smaller standard deviation, which in turn makes the discrimination of separate coals in a blend much more straightforward. Mean maximum reflectance also has precedence. Mean random reflectance, on the other hand, is faster to determine, as each grain is measured in the orientation in which it is encountered, obviating the need for rotating the stage to locate the maximum value. Errors encountered during rotation of the stage, related to lack of centering of stage and objective, and changes in focus, are thus avoided. The method is also, therefore, more suitable for very small grains, such as dispersed organic matter in sedimentary rocks, encountered during oil and gas exploration. Furthermore, since the method omits the incident-light polarizer, more light is available for observation and measuring. Random reflectance is the method utilized by new automated systems which measure reflectance continuously along lines of parallel traverse.

Maximum reflectance is the method favoured at present by most coal petrographers, although in lower rank coals, where anisotropy is not well developed, the two are nearly interchangeable. Because it is difficult, however, to determine exactly at what rank level significant anisotropy first appears, it is probably safer to be consistent and use one method for all coals.

Reflectance is related to absorption and refractive indices by Beers equation (previously mentioned), and so is subject to variation corresponding to variation in these indices. Both indices vary with wavelength of light used, and so, by convention, petrographers have settled on a single wavelength (546 nm) for routine analysis.

7.1.2 *Instrumentation*

Earliest attempts to directly measure the reflectance of coal utilized the Berek microphotometer. This device produced a double beam of light, one part of which was reflected from the sample, the other part of which was passed through a pair of adjustable nicols. This second, or comparison, beam was adjusted using the nicols until its intensity was visually recognized as being equivalent to the first, or reflected, beam. Later this source of subjectivity was removed from the method by incorporation of a photocell.

Today a single beam method is used in coal petrography, in which the reflectance of the sample and a standard of known reflectance are compared. Photomultipliers have replaced photocells as the means of measuring reflectance.

There are three principal components of a coal reflectance measuring system: a microscope; a photometric device (photomultiplier); and a display and recording system.

The microscope (Figure 26) requires a high-powered stabilized light source. Quartz halogen lamps, rated at 12 volts and 100 watts, are most commonly used. The vertical illuminator uses a Berek-type prism which ensures maintenance of polarization direction of incident and reflected light beams, each of which uses one half of the objective aperture. This means, unfortunately, that reflected light is obliquely incident. To offset errors associated with oblique incidence in polarized light, the polarizer on the incoming light is set at 45 degrees to the plane dividing the two beams.

The objective lens should be in the range of 25x to 60x and eyepieces should be 8x or 10x magnification. Overall magnification should be in the range of 250x to 750x. (ASTM stipulates 500x to 750x). A separate viewing eyepiece is located below the photometer on some microscopes. At least one of the eyepieces should be equipped with a cross-hair or grid as a reference for the location of the area measured by the photometer. All eyepieces must be capable of being blocked off to any incoming light while readings are being taken.

The microscope stage should rotate and be equipped with a mechanical holder with 'click-stop' movements.

For routine analysis, a monochromatic filter which passes light of 546 nm wavelength should be mounted between sample and photometer (Figure 26).

A limiting aperture is also mounted between sample and photometer (Figure 26). This aperture restricts the size of the field measured by the photometer, which is generally in the 5 to 10 micrometre range. Small diameters are advantageous since the area 'seen' by the photometer must be clear of grain boundaries, scratches, and other irregularities. Small diameter apertures also reduce optic effects such as glare or diffraction which would interfere with the reflected light. Many microscopes allow simultaneous viewing of the sample and the aperture opening, which greatly aids in choosing suitable grains for measurement.

The photometers in use today are photomultipliers with high measuring current and weak dark field current, such as the EMI 6094. They require a stabilized high voltage supply, and their response should be linear in the operating range. This last qualification can and should be checked by the operator (Section 7.1.3).

Display and recording systems available today run the gamut from galvanometers or digital voltmeters which must be read visually and recorded manually to in-line computer systems which automatically read and record values. All work well, although the more sophisticated recording systems obviously save operator time.

Vitrinite reflectance also makes use of immersion oils and reflectance standards. Nondrying oils with a refractive index of 1.518 are generally used. That index is temperature-dependent, although if the reflectance of standard and sample are similar the temperature effect on both will be the same.

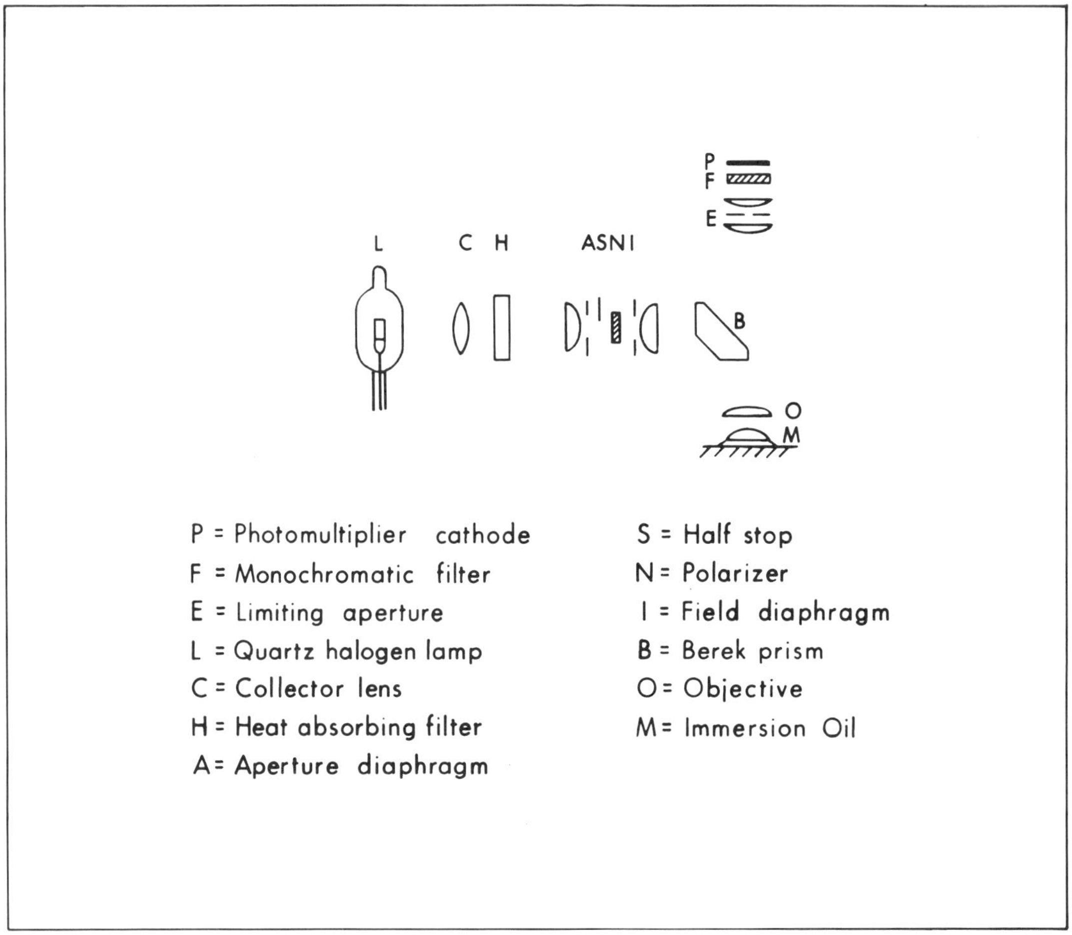

Figure 26. Schematic diagram of the components of the reflectance microscope. (Adapted from Davis, 1978).

Standards are available in a variety of natural and synthetic materials. A group of glass blocks of varying reflectance embedded in a black mounting medium is the most common variety. The requirements of a standard are that it be uniform and isotropic, chemically stable in air and water, hard, free from cleavage or other structural weakness, able to take a good polish, able to maintain a constant reflectance over time, and it should be in the reflectance range of the coals to be studied. In fact, three separate standards in the range of reflectance of coals to be studied are recommended, in order to test the photometer's linearity and to provide checks on deviations from correct values in any of the standards. Before initial use, they should be checked against other standards.

7.1.3 *Preparation and Setting Up*

Ensure that the reflectance equipment is set up on a sturdy workbench, if possible with extra vibration damping built in. The stage and objective should be carefully adjusted to ensure that the viewing system is well centred during stage rotation. Also ensure for maximum reflectance measurements that the polarizer is set at 45 degrees to the plane of incidence, and for random reflectance that the polarizer is out.

It is critical before beginning regular use of the system that the photometer is checked for linearity in the operating range. This may be accomplished by measuring reflectance of three or more standards, or by use of neutral density filters of known interference characteristics. Multiple glass standards come with data sheets indicating the reflectance of each of the standards.

The sample pellet should be well polished and free from scratches, pits, relief, and other irregularities. Coals of high-volatile bituminous and lower rank should be stored in a desiccator for at least 15 hours. A hand levelling device and plasticene are used to mount the pellet on a slide.

The photometer should be turned on for one-half to one hour to stabilize it immediately prior to use.

7.1.4 *Measurement Procedure*

The photometer readout is first adjusted to 0.00 with no light impinging on it (dark field), and then adjusted to read the reflectance of the standard. Focusing accurately on the featureless surface of a glass standard may be difficult, although there are usually a few scratches or nicks which may be used for focusing. The dark field reading (zero) should then be rechecked and the mounted sample put in place. One drop of immersion oil on the sample is sufficient.

Proper sampling of grains in the pellet requires that a grid system be established utilizing the mechanical holder. The grid should be established so that interpoint and interline distances are equal, and that the entire pellet surface is covered. Traverse along the first line continues until the eyepiece cross-hairs land on a grain of suitable vitrinite (see Section 7.1.5). A minor position adjustment is allowed to find a clear spot on that same grain. If necessary, the image may be refocused using the separate viewing eyepiece. Reflected light is then diverted to the photometer, and the eyepieces and viewing eyepiece (if separate) are blocked off against incoming light.

For random reflectance measurements the value can be read immediately. In the case of maximum reflectance, the stage is rotated slowly through 360 degrees, and the highest value is noted. Two maxima and two minima will, in fact, be encountered. Traverse of the line is continued until another suitable grain is encountered.

Check the calibration (zero and standard) every 20 to 30 minutes and at the completion of the pellet. One hundred readings on one pellet or fifty readings on each of two pellets are required per sample. (Many labs use only fifty readings total). In the case of coal blends, 500 grains total are measured, generally 250 on each of two pellets.

At the end of the day, pellets, standard, and lens should be cleaned with a xylene solution containing a wetting agent such as Kodak Photo-flo.

7.1.5 *Choosing Vitrinite for Measurement*

In coals of bituminous or higher rank, readings are either made on the total vitrinite or are restricted to that which occurs in wide monomaceralic bands (telocollinite of the ICCP, in part vitrinite A of Brown *et al.*, 1964a). The other major component of the vitrinite maceral group is desmocollinite (ICCP) or vitrinite B (Brown *et al.*, 1964a). In general the latter component has a lower reflectance than telocollinite or vitrinite A. The choice is dictated to some extent by the goals of a particular reflectance analysis. In North America the more usual practice is to measure the total vitrinite, especially if the results are to be used to predict technological behaviour. The ICCP recommends that only telocollinite be measured. On the other hand, the ASTM procedure (Standard D-2798) is not specific in this regard, indicating only that 'vitrinite' be measured, presumably meaning all vitrinite. A British draft standard (Juckes and Pitt, 1977) is more specific and requires that measurements be made on any type of vitrinite which has sufficiently large clear areas to measure. In a somewhat different approach Thompson and Benedict (1974) of Bethlehem Steel suggest a breakdown of vitrinite into *pseudovitrinite* and *reactive vitrinite* with reflectances to be determined on each. These two sets of data are used in the Bethlehem Steel coke quality prediction method. These authors also suggest that coalification or rank studies of a geological nature should be based only on reflectances determined on reactive vitrinite. This material corresponds in part to the desmocollinite of the ICCP and vitrinite B of the Australian workers. Reactive vitrinite may also include some of the vitrinite A of Brown *et al.*, 1964. Because of this variety of approaches for the selection of measurable vitrinite, it is good practice to state clearly just what was measured when reporting reflectance determinations. In low rank coals the situation is more complicated. McCartney and Teichmüller (1972) suggested that the huminite maceral ulminite B is the most suitable.

7.1.6 *Calculating and Presenting Data*

The mean of 100 readings of a sample is calculated, and generally a standard deviation value is also computed. Data are also presented in the form of a frequency histogram, divided into V-steps or 1/2-V-steps (Figure 27). A V-step represents a range of 0.1 per cent reflectance. For example, values from 0.80 to 0.89 fall into V-step 8. The frequency histogram is especially important in the case of coal blends, as it allows visual or statistical separation of the coals making up the blend. Each population may then be characterized by its mean reflectance and standard deviation. Conversely, detection of contamination of one coal sample by another should be possible.

For 100 readings on a single coal sample, the precision of the mean reflectance should be ± 0.02 per cent at the 95 per cent confidence level. Accuracy will depend on the precautions noted in this chapter, such as cleanliness of the standard, centering of the stage, linearity of the photometer, etc. Reflectance results of various laboratories using the same sample can be an enlightening demonstration of the potential for error. Table 19 indicates results of a 'round-robin' analysis within the Canadian Coal Petrographers Group and Table 20 summarizes a similar comparison among British petrographers (Juckes and Pitt, 1977). Discrepancies in excess of one V-step are apparently common between laboratories.

TABLE 19. VARIATION IN REFLECTANCE RESULTS ($\bar{R}_o max$) OBTAINED BY 11 CANADIAN LABS

	1	2	3	4	5	6	7	8	9	10	11
Sample E	1.21	1.19	1.25	1.14	1.18	1.16	1.17	1.05	1.06	1.15	1.16
Sample J	1.29	1.23	1.26	1.19	1.23	1.26	1.19	1.16	1.06	1.18	1.19

TABLE 20. RANGE OF REFLECTANCE RESULTS ($\bar{R}_o max$) OBTAINED BY 11 BRITISH LABS (Juckes and Pitt, 1977)

	Coal A	Coal B	Coal C
Range	0.86 - 0.94	1.14 - 1.24	1.45 - 1.56

7.2 **Random Point Count Analysis — Macerals and Microlithotypes**

Random point count analysis has applications whenever it is desirable to characterize the maceral and microlithotype composition of a coal. This method is used routinely in evaluating coal for all technological and most geological purposes. A variety of methods can and have been used in analysis and the following discussion is limited to a review of some of the more accepted methods and methodologies.

7.2.1 *Methods*

Random point count analysis requires: (1) crushing of the coal to the prescribed size; (2) removal of one or more representative samples; (3) formation of a coal pellet by mixing of the coal with a binder (generally an epoxy or a thermoplastic); (4) grinding and polishing of the sample; and (5) counting of a large number of points in order to characterize the sample. Methods of obtaining representative samples and sample preparation for point count analysis are discussed in Chapter 6.

ISPG COAL SECTION TECHNOLOGY
REFLECTANCE MEASUREMENTS
RANK DETERMINATION

****

GRAIN PELLET	MAXIMUM REFLECTANCE
STANDARD	1.02
SAMPLE	MOOSEBAR CREEK 719/82
COMMENTS	
DATE	09.46--03/11/1982
OPERATOR	WK

****

STATISTICAL REPORT

PARAMETER — MAXIMUM

PROPERTIES OF THE MEAN

MEAN VALUE	1.040 +/- 0.0058
CONFIDENCE (95%, 99%)	0.0113 -- 0.0149

PROPERTIES OF OBSERVATIONS

NUMBER OF POINTS	50
STANDARD DEVIATION	0.041 +/- 0.0041
VARIANCE	0.0017

MAXIMUM REFLECTANCE VALUES

0.95	0.96	0.97	0.98	0.99	1.00	1.00	1.00	1.00	1.00
1.00	1.00	1.01	1.01	1.02	1.02	1.02	1.02	1.02	1.03
1.03	1.03	1.03	1.04	1.04	1.04	1.04	1.05	1.05	1.05
1.05	1.05	1.06	1.06	1.06	1.06	1.06	1.06	1.07	1.07
1.08	1.08	1.09	1.09	1.09	1.09	1.10	1.11	1.12	1.13

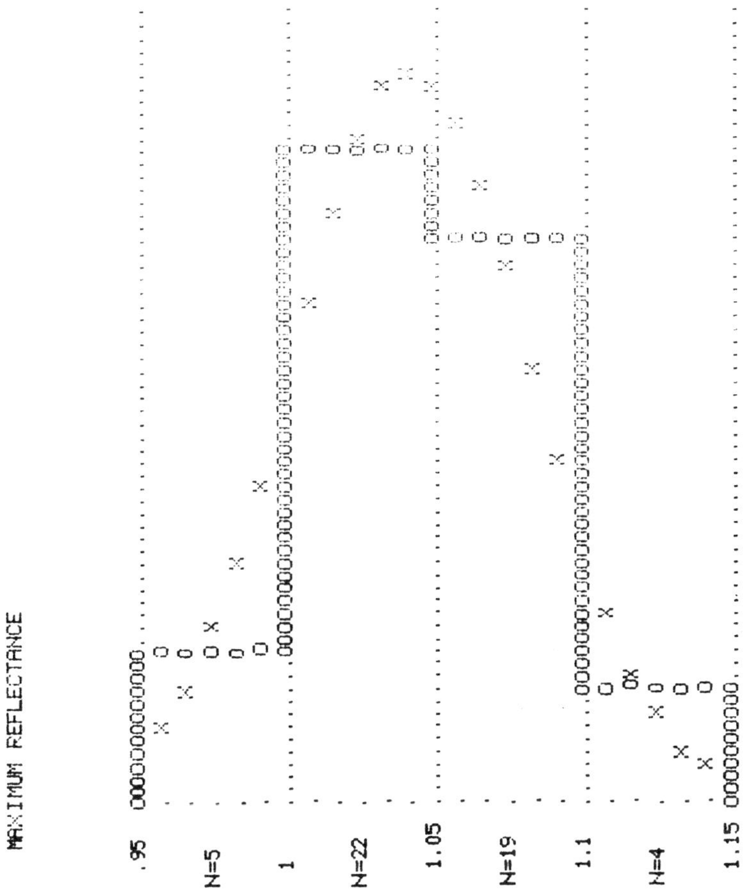

Figure 27. Typical example of reflectance frequency histogram. This diagram represents analysis of a Gething Formation (northeastern British Columbia) high-volatile coal. $R_o$max = 1.04% and standard deviation = 0.041%.

7.2.2 *Point Counting*

For point count analysis a microscope, equipped with a vertical illuminator (or if thin sections are used, transmitted light) and a mechanical stage, is required. Any ore or metallurgical microscope may be used, as long as the resolution is in the order of 1 to 2 microns and the magnification is about 400x (ASTM D-2799). It is highly recommended that an oil immersion objective be used because of the much better resolution. The microscope eyepiece requires a cross-hair or graticule for maceral analysis. The maceral counted is that maceral that lies directly on the cross-hair, or on selected intersections on the graticule.

If microlithotypes are to be quantified, a 20-point graticule is used, in which the distance between the outermost intersections is 50 microns (Figure 28a). In the analysis the area falling under the whole graticule is considered a point. The microlithotype composition of that area depends on the macerals occurring under each intersection of the graticule. For example, if in a given field the maceral distribution under the graticule is as follows: vitrinite 15 intersections, liptinite 5 intersections, then according to Table 21 (also see Figure 28c), the composition is that of clarite and one point is counted for clarite. In a microlithotype analysis, for a point to be considered for counting, at least 10 intersections of the graticule must fall on macerals or minerals (see Mackowsky, 1982b). For microlithotype analysis it is recommended that a magnification of 250x be used.

Microlithotype and maceral analysis may be combined in any number of imaginative ways (see Mackowsky, 1982b).

The mechanical stage for point count analysis must be capable of movement in two perpendicular directions. The specimen needs to be advanced about 0.5 to 1.00 millimetres after each point for maceral analysis or 0.3 to 0.5 millimetres for microlithotype analysis. If a grain size other than the usual 20 mesh

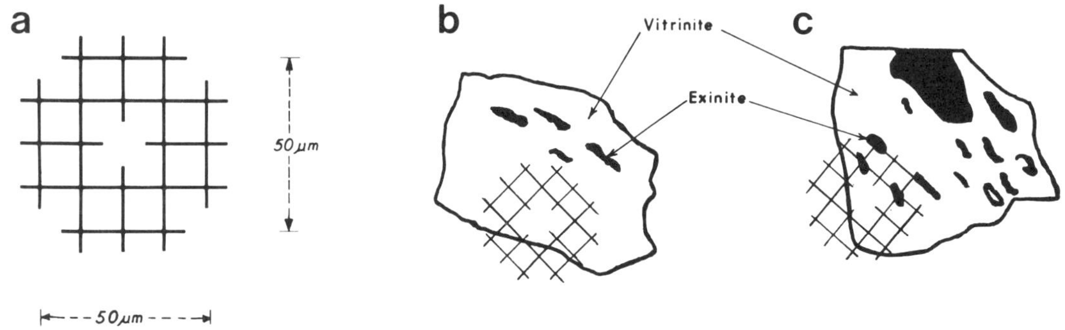

Figure 28. 20-point graticule used for microlithotype analysis. An encounter such as shown in this sketch would be counted as vitrite. An encouter such as shown in this sketch would be counted as clarite. Refer to the text for a full explanation.

(850 micrometres) is used, the stepping distance should be about one-half the maximum particle diameter. A stage with interchangeable mechanical stop clicks is preferred such that the distance between steps is readily apparent and controllable during operation. After the sample is traversed completely, and the specimen is moved at right angles by a space equal to the stepping distance, the procedure is repeated until a statistically significant number of points have been counted. As each point is counted it is recorded either manually or with counters that are commercially available. The mounting medium is not counted unless the analysis is performed by counting occurrences per unit area, a method seldom used in coal petrology.

The quickest and most convenient method of point count analysis is with an electro-mechanical stage and point counter such as a Swift. The stage moves automatically in preset increments as the points are counted, enabling recording of the data and movement of the stage without stopping to record or manually move the stage.

TABLE 21. MICROLITHOTYPE ANALYSIS USING A 20-POINT GRATICULE (MODIFIED FROM MACKOWSKY, 1982). REFER TO SECTION 4.2 FOR A DEFINITION OF THE MICROLITHOTYPES

Microlithotype	Position of the graticule intersections with respect to macerals
Vitrite	All intersections on vitrinite
Liptite	All intersections on exinite
Inertite	All intersections on inertinite
Clarite	All intersections on virtinite and exinite with at least 1 intersection on each
Durite	All intersections on inertinite and exinite, with at least 1 insection on each
Vitrinertite	All intersections on vitrinite and inertinite, with at least 1 intersection on each
Trimacerite*	At least one intersection on each of the three maceral groups

*Trimacerite can be subdivided into:
Duroclarite (V>I+E), Clarodurite (I>V+E), Vitrinertoliptite (E>V+I).

Once a statistically significant number of points have been counted, the abundance of each constituent is determined. Point count analysis determines the area of each constituent, but, because it is assumed that the components are randomly distributed, the area and the volume are proportional (ASTM D-2799). Mineral matter is commonly difficult to recognize during maceral analysis and thus should be calculated from the ash content determined from proximate analysis and the sulphur content, if known. The weight per cent of mineral matter can be calculated using Parr's formula:

$$\text{Mineral matter} = 1.08\ (\text{per cent ash}) + 0.55\ (\text{per cent sulphur})$$

The mineral matter content on a weight basis is then converted to volume per cent by first dividing the calculated mineral matter on a weight basis by two (because the specific gravity of mineral matter is about twice that of the organic fraction) and recalculating the maceral and mineral matter content (see also Ting, 1978).

7.2.3 *Statistical Evaluation*

Variation in maceral or microlithotype analysis of coal may result from variability in the coal (sampling error), or inaccuracy in identification (operator error). Whenever possible the sampling error should be determined by examination of duplicate samples from the parent population. This error can be determined by routine statistical analysis (see Davis, 1973).

A number of different methods for statistical evaluation of a point count analysis exist (see Soloman, 1963, for a comprehensive evaluation of counting and sampling error). The ASTM (D-2799) suggests that a minimum of 1 000 points on each of two samples should be counted (ignoring the mounting medium). If the results from the two pellets agree within 2 per cent mean variance, the average of the results are acceptable. The mean variance is equal to the sum of the differences between each component in each sample set divided by the number of components counted. For example, if vitrinite (v), liptinite (l), and inertinite (i) macerals are counted, the mean variance is calculated by:

$$\Delta V = (\Delta V_v + \Delta V_l + \Delta V_i)/3$$

where ΔV = difference in volume per cent of each maceral between the two samples

If the two sets differ by greater than 2 per cent mean variance, a third pellet is counted and all three sets are then averaged or the data from the two sets with the lowest mean variance are averaged and reported (ASTM 2799).

TABLE 22. MACERAL ANALYSIS FROM POINT COUNTING OF A LOW-VOLATILE BITUMINOUS COAL; CANADIAN COAL PETROGRAPHY GROUP (COMPILED BY CAMERON)

	Range %	Mean %	Standard Deviation
Vitrinite	46-64	53.6	8.4
Semifusinite	17-41	31.9	9.8
Fusinite	2.5-23.5	8.9	5.9
Other inerts	0-16	5.3	4.7
Liptinite	0-.5	0.1	0.2

Operator variability may be an additional source of error, because maceral identification, to a great degree, is subjective. Replicate sample analysis in any given lab may show little variation (depending on operator expertise); however, the variation between labs may be great. The results from the analysis of one sample by members of the Canadian Coal Petrography Group are shown in Table 22 (compiled by Cameron). The large variation obtained between labs attests to the necessity of establishing strict standards.

7.3 Fluorescence Microscopy

Fluorescence microscopy has been used in coal petrology for more than 40 years (Schochardt, 1936) but only in recent years have the equipment and the methods been developed to such a degree of precision that fluorescence microscopy can now be applied successfully in coal petrology and in maturation studies of organic materials (kerogen) in sedimentary rocks (Gijsel, 1963, 1975, 1981; Jacob, 1964, 1969, 1973, 1974; Ottenjann *et al.*, 1974; Teichmüller and Wolf, 1976; Ottenjann, 1980, 1982; Robert, 1981).

The basis of all investigations is the observation that certain components in coals and dispersed organic material, in particular the macerals of the liptinite group, show auto-fluorescence when irradiated with blue light or ultra-violet (UV) light.

$$\text{coal} \xrightarrow{\text{excitation}} \text{luminescence} \begin{cases} \nearrow \text{auto-fluorescence (ends after excitation)} \\ \searrow \text{phosphorescence (continues} > 10^{-6}\text{ sec.)} \end{cases}$$

As can be seen from Table 23, both fluorescence colour and fluorescence intensities of the macerals change with increase in rank in a characteristic manner:

Liptinites — Display high fluorescence intensities in peat and low rank coals. With increase in rank the fluorescence intensities decrease. At the same time the fluorescence colours shift to longer wavelengths. The fluorescence colours for the liptinite macerals are, however, not uniform, but differ from one maceral to the other. On Plate 13 the changes in the fluorescence colours with rank for some macerals are demonstrated. The fluorescence colours change from green-yellowish in the peat stage to orange-brown in high-volatile A bituminous coals. In medium-volatile bituminous coals at a reflectance level of 1.3 per cent, the fluorescence of liptinite macerals disappears.

Huminite/vitrinite — Some of the huminites and low to medium rank vitrinites also fluoresce. The fluorescence colours change from yellow in lignites to brown in bituminous coals. The fluorescence in vitrinite is probably caused by a relatively high proportion of hydrogen-rich materials from lipoid substances which are absorbed or chemically incorporated in the vitrinite.

Inertinites — Do not fluoresce.

TABLE 23. FLUORESCENCE COLOURS AND INTENSITIES FOR THE DIFFERENT MACERAL GROUPS AT DIFFERENT COAL RANKS USING BLUE-LIGHT EXCITATION (FROM INT. HANDBOOK OF COAL PETROGRAPHY, 1975)

	SOFT BROWN COAL (LIGNITE)	HARD BROWN COAL (LIGNITE)	LOW-RANK BITUMINOUS COAL	HIGH-RANK BITUMINOUS COAL
LIPTINITE	STRONG; GREEN, YELLOW, ORANGE, AND BROWN	STRONG TO MODERATE; GREENISH YELLOW, YELLOW-ORANGE, AND BROWN	STRONG TO WEAK, YELLOW, ORANGE, AND BROWN	NO FLUORESCENCE
HUMINITE OR VITRINITE	STRONG TO WEAK, YELLOW AND BROWN; OR NO FLUORESCENCE	VERY WEAK, BROWN; OR NO FLUORESCENCE	VERY WEAK, BROWN; OR NO FLUORESCENCE	NO FLUORESCENCE
INERTINITE	NO FLUORESCENCE	NO FLUORESCENCE	NO FLUORESCENCE	NO FLUORESCENCE

GSC

The differences in the fluorescence behaviour of the macerals are not well understood. Both fluorescence intensities and fluorescence colours seem to be related to presence of hydrogen in unsaturated bonds (Cook, 1980). In the past 10 to 15 years methods have been developed to describe both fluorescence intensity and fluorescence colours quantitatively. Changes in fluorescence intensity are measured by the so-called monochromatic fluorescence microscopy, whereas changes in fluorescence colours are measured by spectral fluorescence microscopy. Both methods, along with a brief description of sample preparation, are described below.

Plate 13. Macerals of the liptinite group from various rank levels after blue-light excitation. Reflected light, oil immersion.

Figure 1. *Pollen grain* in peat. Peat from Villeroy-Lotbiniere district, Quebec, Pleistocene in age.

Figure 2. *Suberinite* (Su) associated with phlobaphinite (P) in greenish-yellowish peat matrix. Peat from Villeroy-Lotbiniere district, Quebec, Pleistocene in age.

Figure 3. *Pollen grain* (P) in greenish-yellowish peat matrix. Peat from Villeroy-Lotbiniere district, Quebec, Pleistocene in age.

Figure 4. *Sporinite* (S) in a subbituminous coal. Drumheller area, Alberta, Late Cretaceous in age.

Figure 5. *(Mega)-sporinite* (S) in a high-volatile A bituminous coal. Sydney Coalfield, Nova Scotia, Carboniferous in age.

Figure 6. *Sporinite* (S) at the transition from high-volatile to medium-volatile bituminous coal. Crowsnest Coalfield, British Columbia, Early Cretaceous in age.

Figure 7. *Resinite* of "crystalline" appearance in a lignite. Hat Creek deposit, British Columbia, Tertiary in age.

Figure 8. *Resinite* (R) as cell-filling in a lignite from Greece, Tertiary in age.

Figure 9. *Resinite* (R) in a high-volatile B bituminous coal. Sydney Coalfield, Nova Scotia, Carboniferous in age.

Figure 10. *Resinite* (R) in a high-volatile A bituminous coal. Peace River Coalfield, British Columbia, Early Cretaceous in age.

Figure 11. *Resinite* (R) at the transition from high-volatile to medium-volatile bituminous coal, Crowsnest Coalfield, British Columbia, Early Cretaceous in age.

Figure 12. *Cutinite* (C) in a subbituminous coal. Drumheller area, Alberta, Late Cretaceous in age.

Figure 13. *Cutinite* (C) in a high-volatile A bituminous coal, Crowsnest Coalfield, British Columbia, Early Cretaceous in age.

Figure 14. *Cutinite* (C) at the transition from high-volatile to medium-volatile bituminous coal, Peace River Coalfield, British Columbia, Early Cretaceous in age.

Figure 15. Bottryococcus-type *alginite* (A) in a high-volatile bituminous coal from an oil shale zone in the Pictou Coalfield, Nova Scotia, Carboniferous in age.

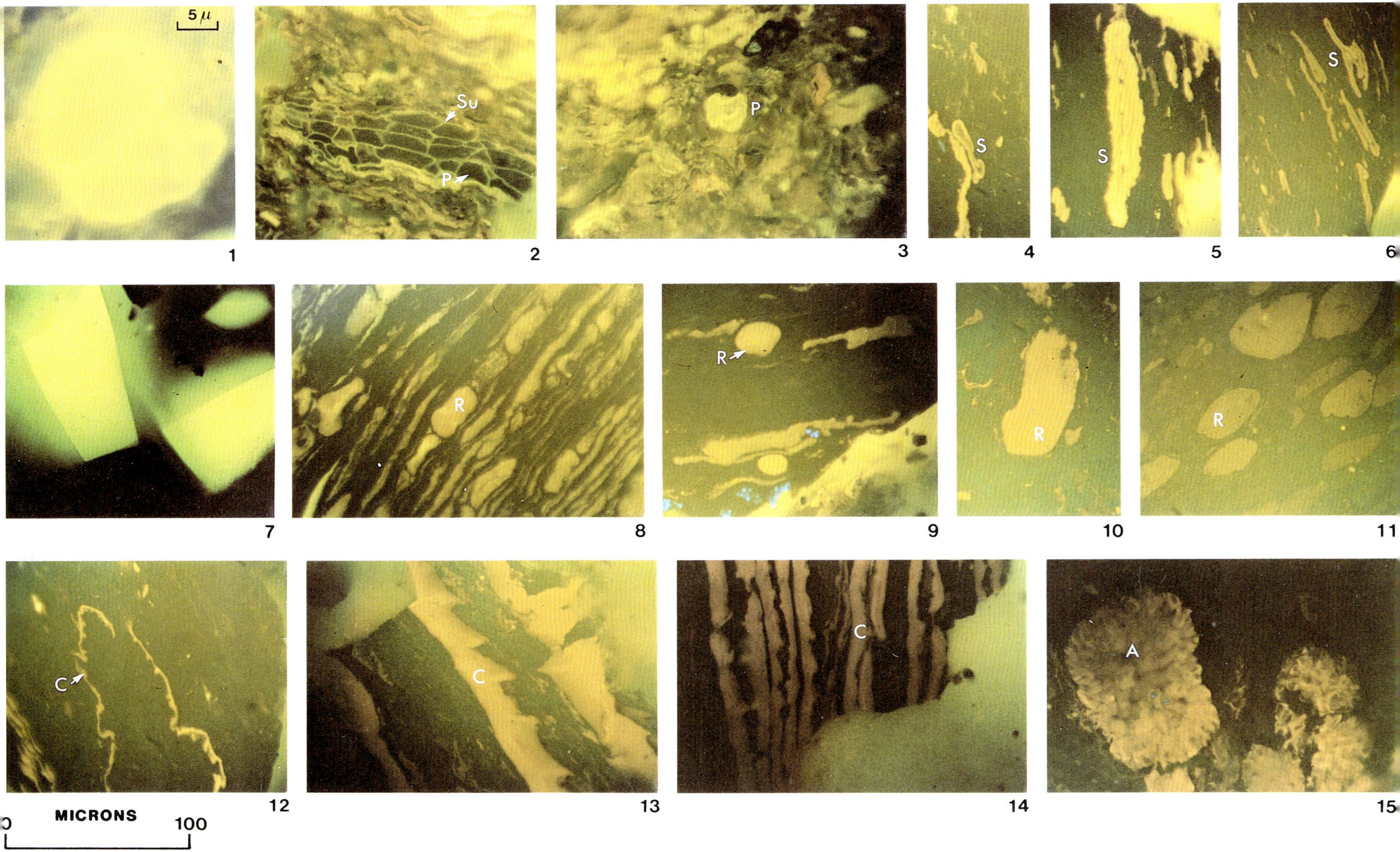
5 µ
Su
P
P
S
S
S
R
R
R
R
C
C
C
A
1
2
3
4
5
6
7
8
9
10
11
12
13
14
15
0
MICRONS
100

7.3.1 *Quantitative Fluorescence Photometry*

7.3.1.1 **The Fluorescence Microscope**

On Figure 29 is shown a schematic diagram of a fluorescence microscope, which can be used both for monochromatic and spectral fluorescence photometry.

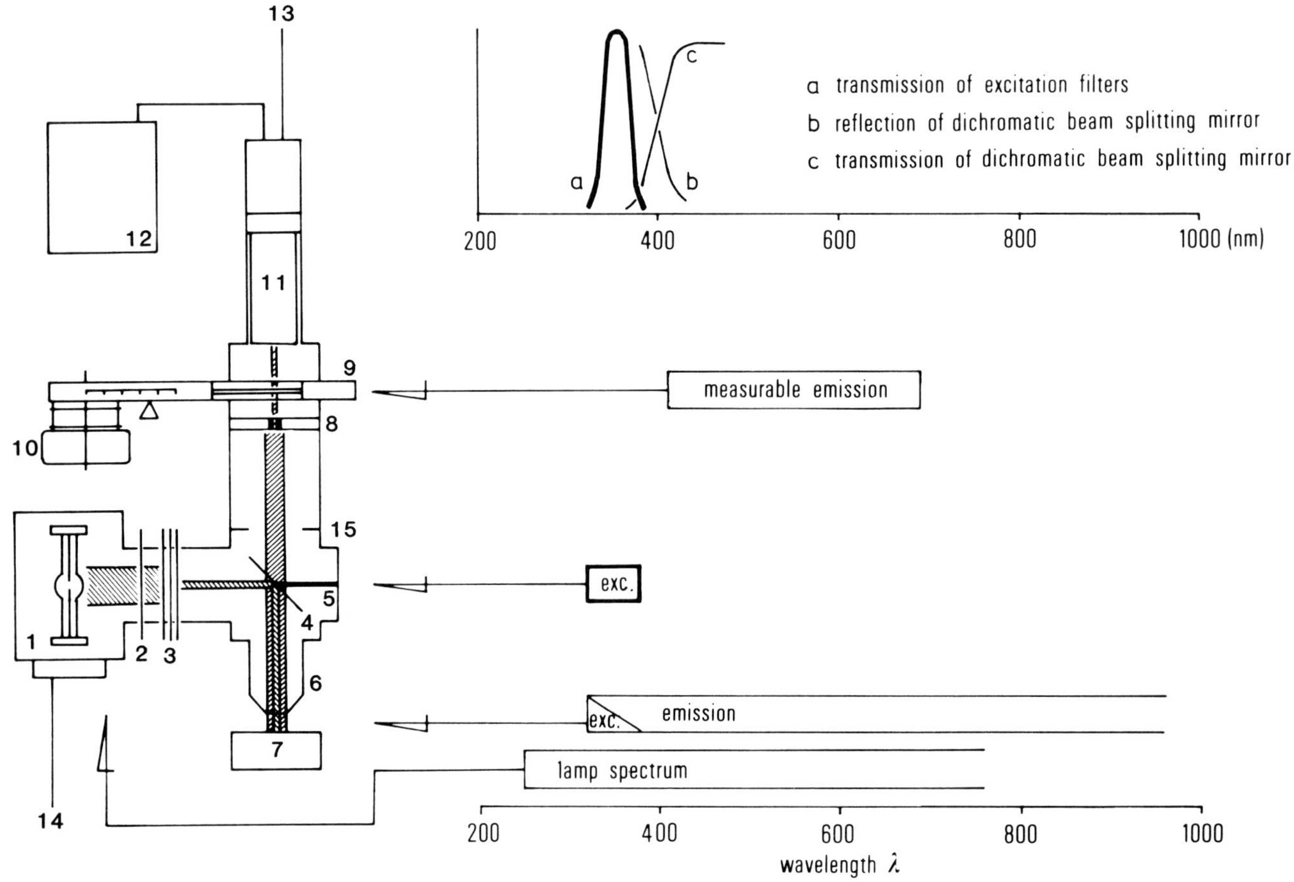

Figure 29. Schematic diagram illustrating apparatus for spectral fluorescence photometry in reflected light, modified after International Handbook of Coal Petrography (1975). 1 - Lamp housing with 100 W high pressure mercury lamp; 2 - heat absorbing filter; 3 - excitation filter(s); 4 - beam-splitting mirror; 5 - reflected light fluorescence illuminator; 6 - objective; 7 - sample; 8 - measurement aperture; 9 - variable interference filter; 10 - variable motor drive; 11 - photomultiplier; 12 - interface, computer, printer, plotter; 13 - multiplier balance voltage; 14 - lamp power supply; 15 - barrier filter.

The following description of some special requirements is based on the procedure outlined in the International Handbook of Coal Petrography (1975).

Excitation Lamp (1)*
A high-powered stabilized mercury vapor lamp is used to provide the specific lines of the mercury spectrum. By use of excitation filters, suitable mercury lines such as 365/366 nm can be projected to the sample.

---

*Number in brackets refers to part numbers on Figure 29.

Excitation Filter (3)

Blue-light filter. The blue-light filter is mainly used for observation purposes. Fluorescence intensities of many organic substances are higher when excited with blue light as compared to UV-light excitation. Blue-light excitation can be used in monochromatic fluorescence photometry where the fluorescence intensities are recorded at 546 nm. The blue-light filter is not suitable for spectral fluorescence photometry because the use of the barrier filter K530, which has to go with blue-light excitation, cuts off all fluorescence spectra below 510 nm.

UV-light filter. Enables the recording of complete spectra as the barrier filter K430, which is used with UV-light excitation, has a transmission from 420 nm on. Disadvantage of UV-light excitation is the lower fluorescence intensity as compared to the use of blue-light excitation.

Objectives (6)

All objectives with negligible internal fluorescence and large apertures are suitable. Dry objectives are preferable over oil immersion objectives because use of oil tends to dissolve bituminous materials out of the organic materials when excited and interferes with the measurement.

Reflected light fluorescence illuminator (5)

Highest intensities are obtained by use of a special incident fluorescence illuminator (for example, Ploem-Opak from Leitz) which has built-in beam splitting mirror and barrier filter that provides a desired separation of excitation light and emitted light, see Figure 30.

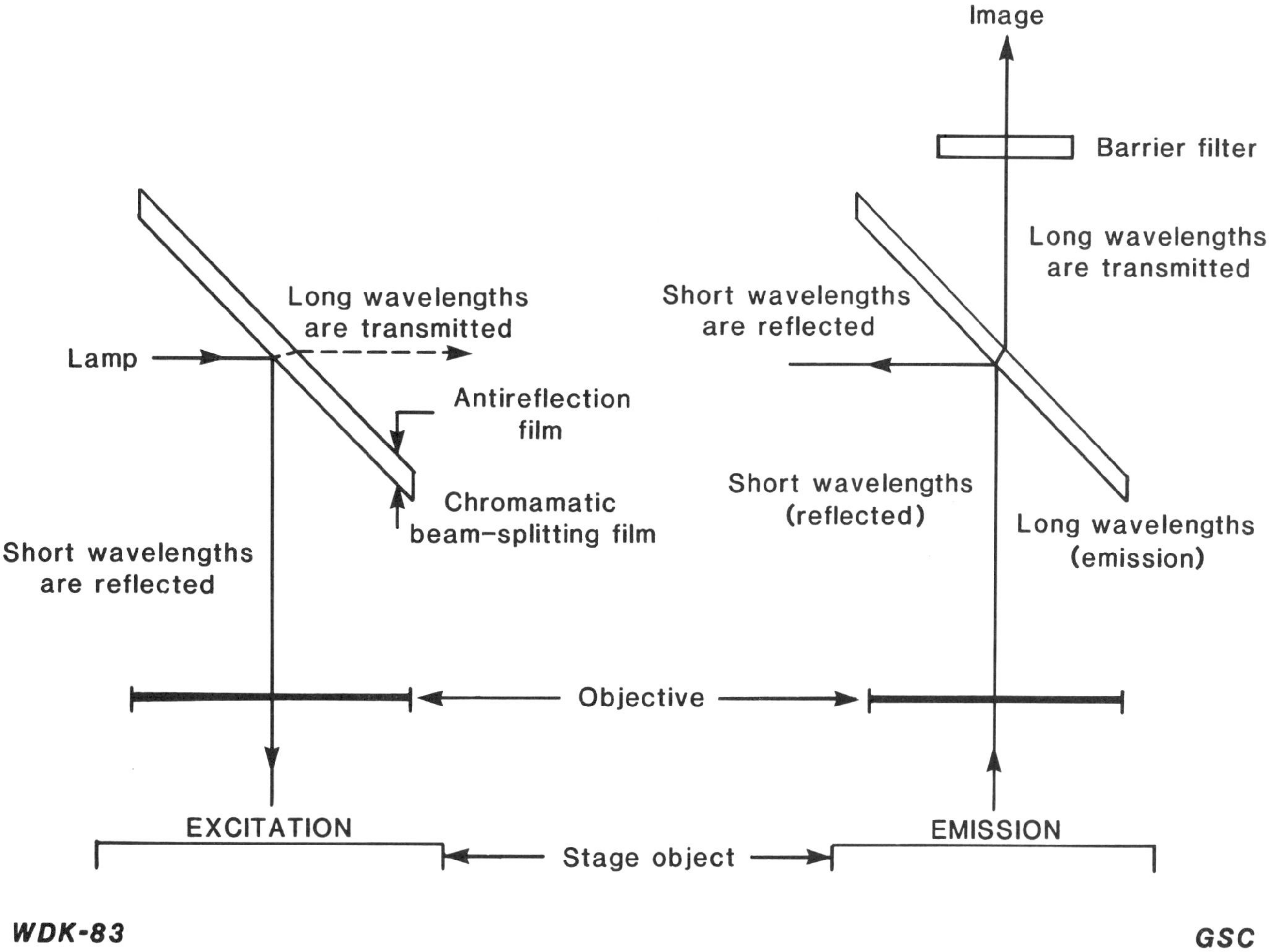

Figure 30. Function of beam-splitting mirror (modified after Cook, 1980).

Barrier filter (15)
Prevents any scattered UV light from entering the eye and damaging it. There is a range of barrier filters available, most commonly used are the K430 in connection with UV-light excitation and the K530 in connection with the blue-light excitation.

Photomultiplier (11)
Must have adequate sensitivity to red light, such as the EMI 9558A-S 20, and maximum general sensitivity.

Variable interference filter (10)
This filter should cover the range from 400 to 700 nm. In connection with synchronous motor drive, it permits rapid measurements of spectral fluorescence intensities. At 546 nm it is also used for monochromatic fluorescence measurements.

7.3.1.2 **Sample Preparation**

Fluorescence properties of organic materials can be studied both in transmitted and incident light. Incident excitation is preferable because it provides higher fluorescence intensities. In principle polished blocks or pellets for fluorescent light observation are prepared in the same way as for normal white light observation. Cold-setting bonding resins are preferred to avoid alterations of low rank materials in the course of preparations. The mounting medium should fluoresce as weakly as possible and polymerize quickly to prevent dissolution of organic materials.

7.3.1.3 **Monochromatic Fluorescence Photometry**

In this technique the sample is excited with blue light and the fluorescence intensity is measured at a specific wavelength, such as 546 nm. The fluorescence intensity is recorded in relation to a standard whose fluorescence intensity has been set to 100 scale units on the digital voltmeter. A uranyl glass is used as standard, which is the only stable fluorescence standard available at the present. This standard gives reproduceable fluorescence measurements within one laboratory. To compare data with other laboratories, the uranyl glass standards have to be calibrated against each other. This method is in limited use at the moment because of the standardization problems mentioned previously and also because it can only be applied to very low rank materials up to 0.3 per cent reflectance (lignite stage). The advantage of this method is that it is easy to combine monochromatic fluorescence photometry with reflectance photometry in a single working procedure (Jacob, 1973). An application of monochromatic fluorescence photometry on bitumens is shown on Figure 44.

7.3.1.4 **Spectral Fluorescence Photometry**

In this technique the organic material is excited with UV light (365/366 nm) and the emitted fluorescence intensities are recorded over the range from 400 to 700 nm. In order to get comparable results, the measured spectrum has to be corrected for:

(a) *Background*

The background data are obtained by measuring a nonfluorescing particle and are then subtracted from the sample spectrum. The background noise may consist of dark currents and fluorescence of the objective and beam-splitting mirror used, and, if measurements are done under oil, the fluorescence properties of the immersion oil used.

(b) *Microscope Properties* (spectral sensitivity of the microscope)

The background corrected curve has to be further corrected due to various degrees of transmittance of the interference filter and lenses used and the spectral sensitivity of the photomultiplier. The correction is obtained by measuring the spectrum of a lamp with defined colour temperature under equal conditions (suitable light source, for example, a 12 volt, 100 watt tungsten filament lamp with a colour temperature of 3,400 degrees K). The correction factor for the microscope is formed by the quotient of the known spectrum of this lamp according to Planck's formula and the spectral intensities measured. The final curve is then calculated by multiplying the background corrected data with the correction factor. An example of the correction procedure for a fluorescence spectrum is shown on Figure 31. Also illustrated is the corresponding uncorrected spectrum, the background curve, and the final spectral fluorescence spectrum.

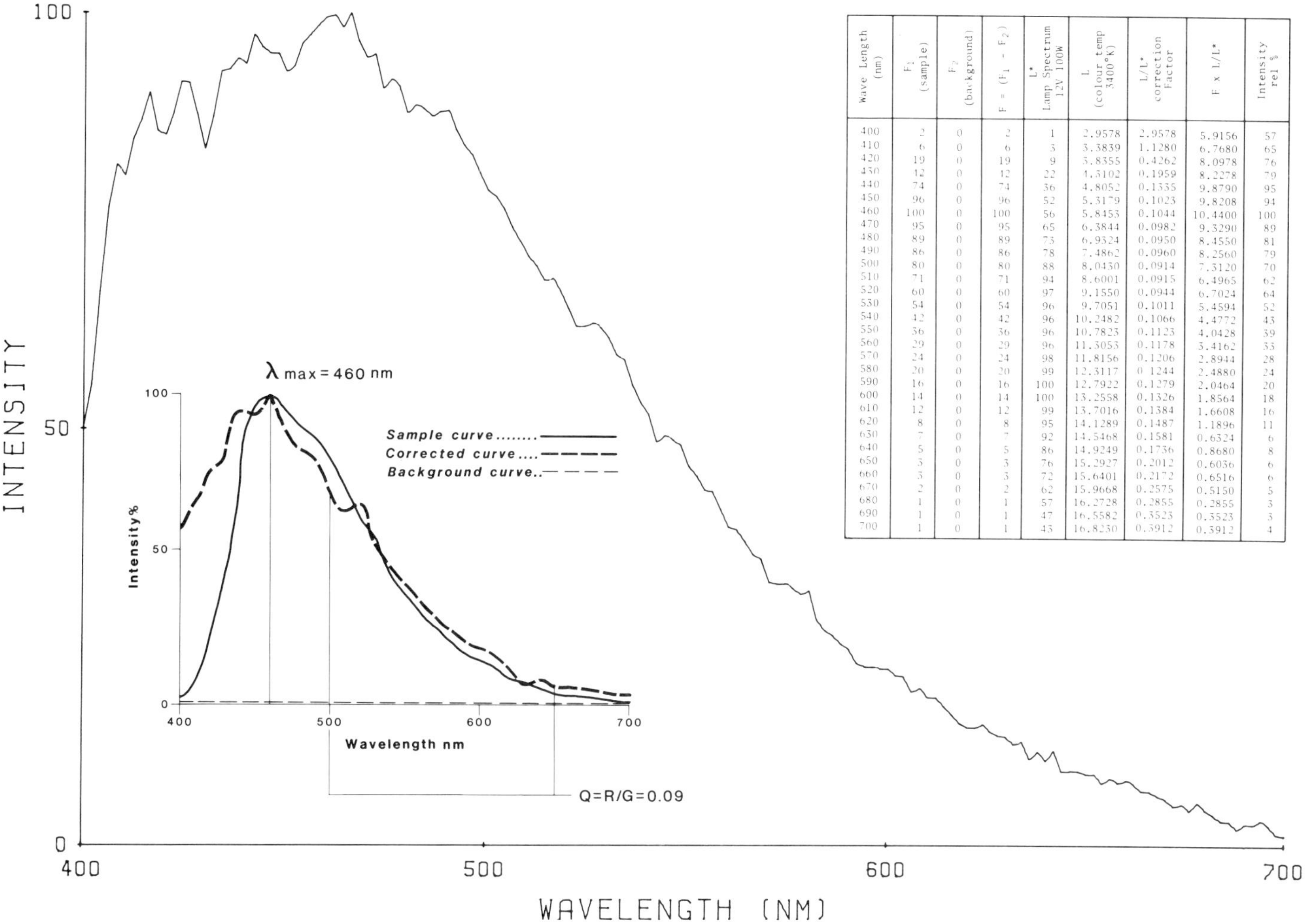

Wave Length (nm)	F_1 (sample)	F_2 (background)	$F = (F_1 - F_2)$	L* Lamp Spectrum 12V 100W	L (colour temp 3400°K)	L/L* correction Factor	F x L/L*	Intensity rel %
400	2	0	2	1	2.9578	2.9578	5.9156	57
410	6	0	6	3	3.3839	1.1280	6.7680	65
420	19	0	19	9	3.8355	0.4262	8.0978	76
430	12	0	12	22	4.3102	0.1959	8.2278	79
440	74	0	74	36	4.8052	0.1335	9.8790	95
450	96	0	96	52	5.3179	0.1023	9.8208	94
460	100	0	100	56	5.8453	0.1044	10.4400	100
470	95	0	95	65	6.3844	0.0982	9.3290	89
480	89	0	89	73	6.9324	0.0950	8.4550	81
490	86	0	86	78	7.4862	0.0960	8.2560	79
500	80	0	80	88	8.0430	0.0914	7.3120	70
510	71	0	71	94	8.6001	0.0915	6.4965	62
520	60	0	60	97	9.1550	0.0944	6.7024	64
530	54	0	54	96	9.7051	0.1011	5.4594	52
540	42	0	42	96	10.2482	0.1066	4.4772	43
550	36	0	36	96	10.7823	0.1123	4.0428	39
560	29	0	29	96	11.3053	0.1178	3.4162	33
570	24	0	24	98	11.8156	0.1206	2.8944	28
580	20	0	20	99	12.3117	0 1244	2.4880	24
590	16	0	16	100	12.7922	0.1279	2.0464	20
600	14	0	14	100	13.2558	0.1326	1.8564	18
610	12	0	12	99	13.7016	0.1384	1.6608	16
620	8	0	8	95	14.1289	0.1487	1.1896	11
630	7	0	7	92	14.5468	0.1581	0.6324	6
640	5	0	5	86	14.9249	0.1736	0.8680	8
650	3	0	3	76	15.2927	0.2012	0.6036	6
660	3	0	3	72	15.6401	0.2172	0.6516	6
670	2	0	2	62	15.9668	0.2575	0.5150	5
680	1	0	1	57	16.2728	0.2855	0.2855	3
690	1	0	1	47	16.5582	0.3523	0.3523	3
700	1	0	1	43	16.8230	0.3912	0.3912	4

FILE : RESINITE HAT CREEK B.C. MICRO : CONSTANTS 2 R/G : 0.11

DATE : 05.24--07/02/1983 RUNS : 1 MAX@ : 462.52

Figure 31. Example for procedure of correction of fluorescence spectra (Table) and illustration of sample curve, background curve, and final corrected curve for the maceral resinite in a lignite from Hat Creek, British Columbia. A photomicrograph of this resinite is shown on Plate 13.

To characterize the fluorescence spectra, the following parameters are in use:

(a) 2 max = the wavelength where the fluorescence spectrum reaches its maximum

(b) $Q = \text{red/green quotient} = \dfrac{\text{relative intensity at 650 nm}}{\text{relative intensity at 500 nm}}$

An example of changes in maximum wavelength at different rank levels is illustrated by the maceral sporinite on Figure 32.

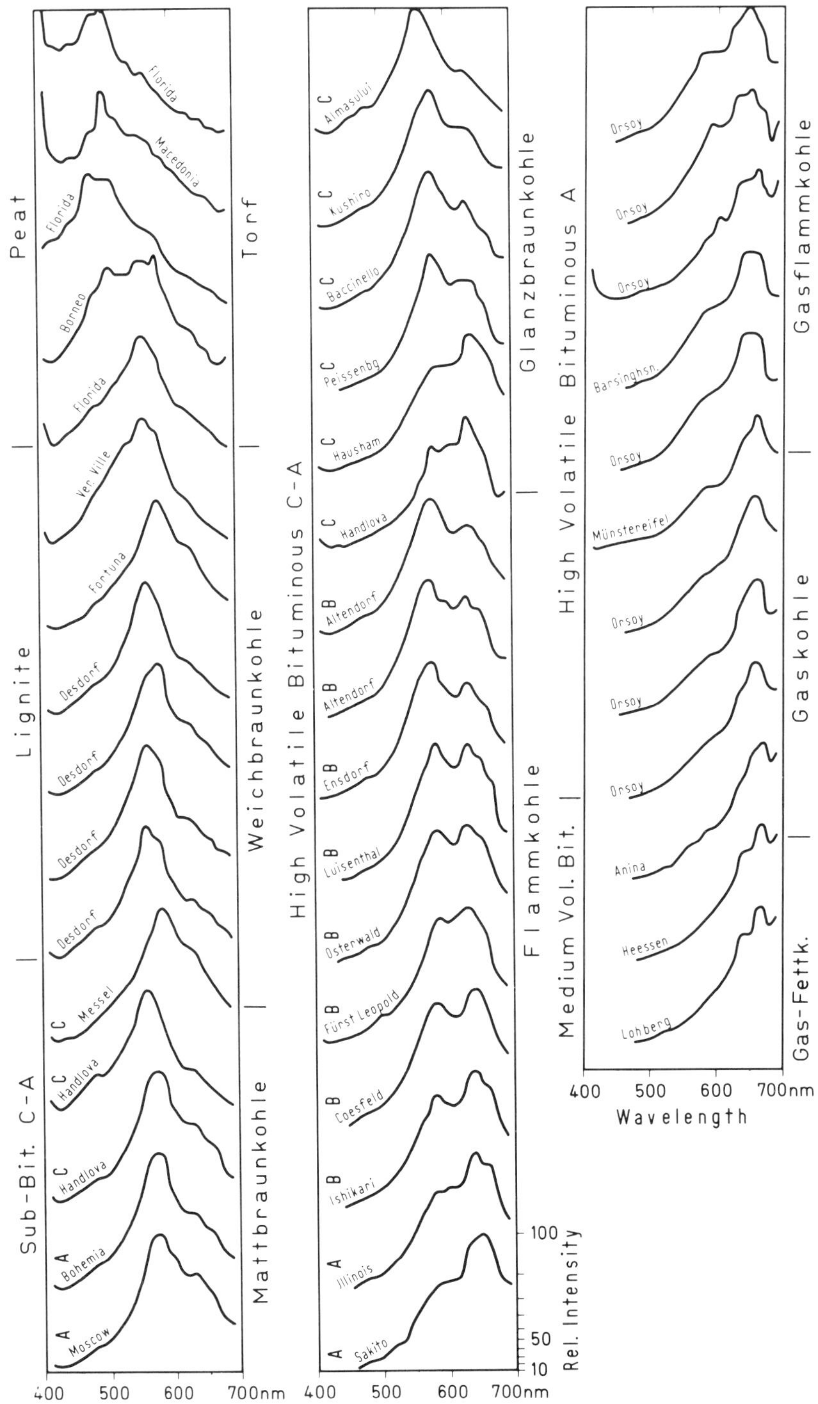

Figure 32. Changes of fluorescence spectra of sporinite with increase of rank (after Ottenjann *et al.*, 1974).

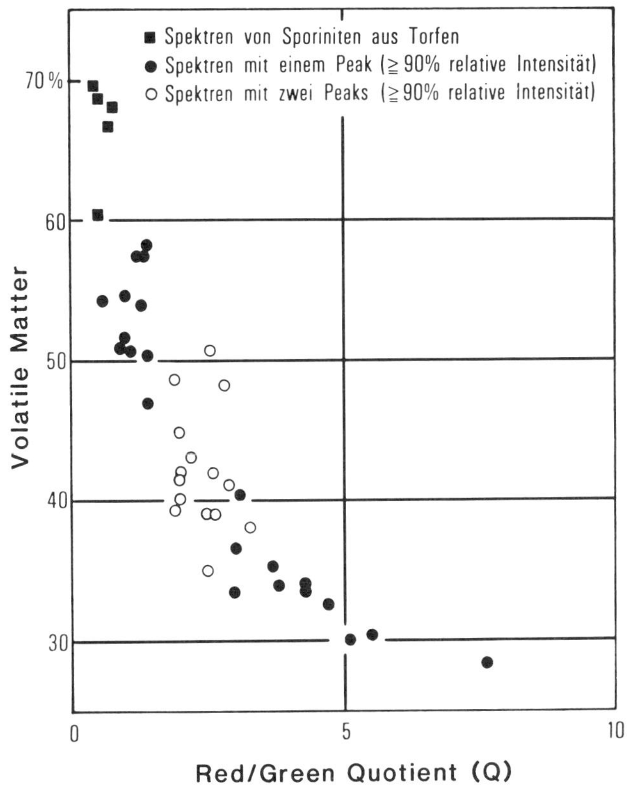

Figure 33. Relation between volatile matter (maf) and red/green quotient of sporinites at different rank levels (modified after Ottenjann *et al.*, 1974).

The maximum wavelength shifts from values of about 500 nm at the stage of peat and lignite to values of over 650 nm in medium-volatile bituminous coals. Along with the shift to longer wavelengths with increase in rank goes an increase of the red/green quotient (Figure 33).

7.3.1.5 **Alteration**

Fluorescence colours and intensities of many organic materials may change during excitation with blue or UV-light. This effect is called alteration or fading, although the latter term suggests that fluorescence properties fade away, which is not the case. In many substances, in fact, an increase in fluorescence intensity can be observed. The effect of negative or positive alteration can be used to characterize organic materials and is commonly measured by excitation over a period of 30 minutes. An example for the alteration behaviour of bituminous groundmass in an oil shale at different rank levels is shown on Figure 34.

Low rank samples are characterized by a decrease of fluorescence intensity within the first 4 minutes, followed by a rapid increase in fluorescence intensities. At higher rank levels only weak changes in fluorescence intensity are noticeable.

7.3.1.6 **Computerization**

In order to undertake all necessary corrections in acceptable time, especially when spectral fluorescence photometry is applied, a computer program is essential to evaluate the raw spectral fluorescence data. A computer in connection with a monitor, printer, and plotter is desirable to produce the data and curves in the desired form (see Figure 35).

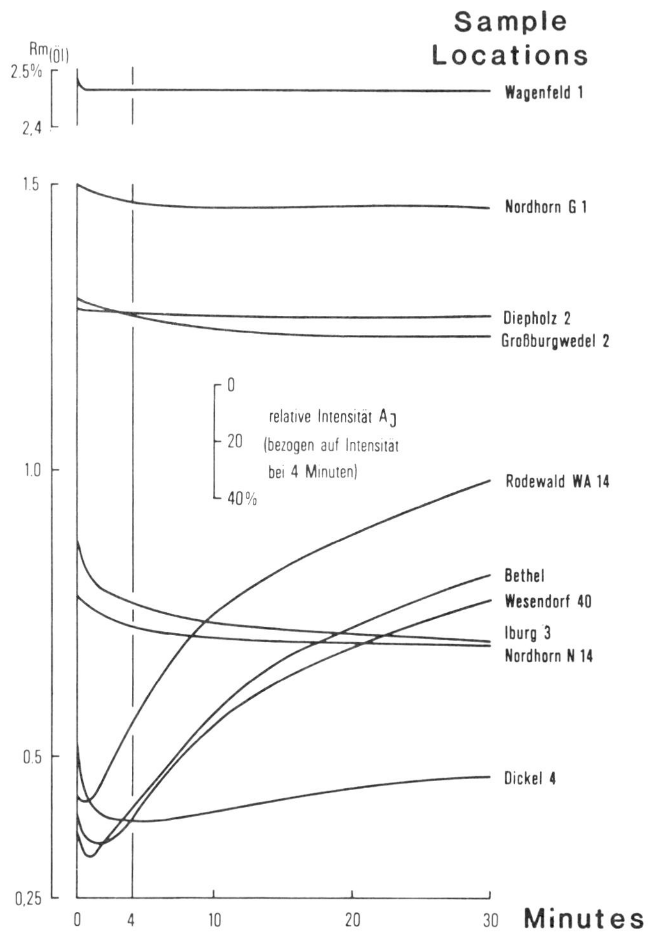

Figure 34. Alteration of bituminous groundmass at different rank levels (modified after Teichmüller and Ottenjann, 1977).

Figure 35. Fluorescence microscope, computer (Apple II), monitor, printer, and plotter used at ISPG for spectral fluorescence photometry.

7.3.2 *Applications of Fluorescence Microscopy*

In the past few years the use of fluorescence microscopy has become such a wide field that here emphasis is given only to the main applications in coal petrology and in maturation studies of kerogen. In addition, some new approaches such as the determination of fluorescence properties of extracts (soluble organic materials), and the relationship between fluorescence properties of vitrinite and technological properties are discussed briefly.

7.3.2.1 **Application in Coal Petrology**

(a) Identification of liptinite macerals, is facilitated by fluorescence microscopy in particular in low rank coals, where both shape and internal structure become more visible in fluorescent light as compared to normal reflected light (see Plate 4, Section 4.1.1.2). Some liptinite macerals like fluorinite, bituminite, and exsudatinite can only by identified in fluorescent light (Teichmüller, 1974b).

(b) Fluorescence microscopy enables differentiation of liptinite macerals from mineral matter. In particular structureless liptinites that cannot be distinguished from mineral matter in normal reflected light are visible in fluorescent light are visible in fluorescent light (see Plate 4, Section 4.1.1.2).

(c) Fluorescence properties of liptinite macerals can be used as an additional rank parameter, especially in low rank coals where fluorescence intensities of the liptinites are relatively high and and the reflectance of huminite/vitrinite is low. In low rank coal vitrinite reflection shows a relatively wide scatter due to original differences in plant material.

7.3.2.2 **Application in Kerogen Studies**

Kerogen is that part of the organic matter in sedimentary rocks which is insoluble in organic solvents such as chloroform. Its concentration, composition, and state of maturation are essential criteria for the formation of oil and gas.

(a) As in the case of coals, some of the liptinitic components can only be identified and/or distinguished from the mineral matrix in fluorescent light (see Plate 18). The combined use of normal reflected light and fluorescent light gives a more accurate determination of the type of organic material.

(b) The determination of rank by fluorescence properties of liptinite macerals has proven to be very useful in sequences characterized by the absence of vitrinite particles as observed, for example, in many oil shales. In samples characterized by the occurrence of more than one vitrinite population (recycled vitrinites or contaminations) a precise identification of the 'true' vitrinite population is very difficult. The relationships of the fluorescence parameters of the liptinite macerals sporinite, alginite, and cutinite to the associated vitrinite reflectances of coals in a range from lignite to high-volatile bituminous coals are shown on Figure 36.

7.3.2.3 **Spectral Fluorometric Analysis of Soluble Organic Matter (Extract)**

This new approach has been developed by Hagemann and Hollerbach (1980, 1981) as a rapid method of source rock evaluation for example determination of the stage of maturation. Instead of measuring the fluorescence properties of individual liptinite macerals, they looked at the fluorescence properties of the soluble organic material (extract), commonly obtained by Soxhlet extraction. They found that similar to the macerals of the liptinite group, the properties of the extracts change in a characteristic manner with increase in maturation. An example for the application of this method is demonstrated on Figure 37.

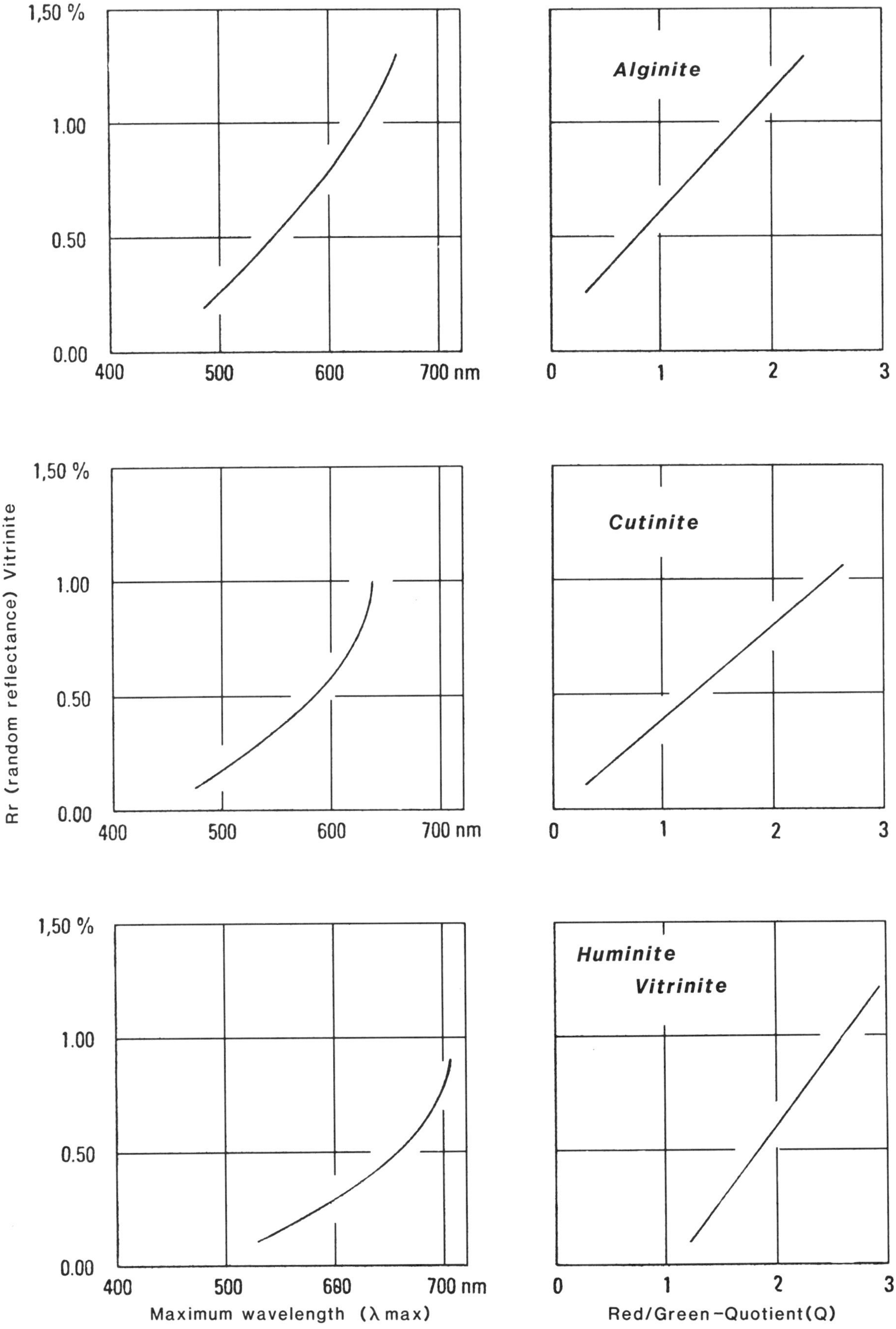

Figure 36. Relationship between vitrinite reflectance and fluorescence properties of alginite, sporinite, and cutinite (modified after Ottenjann, 1982).

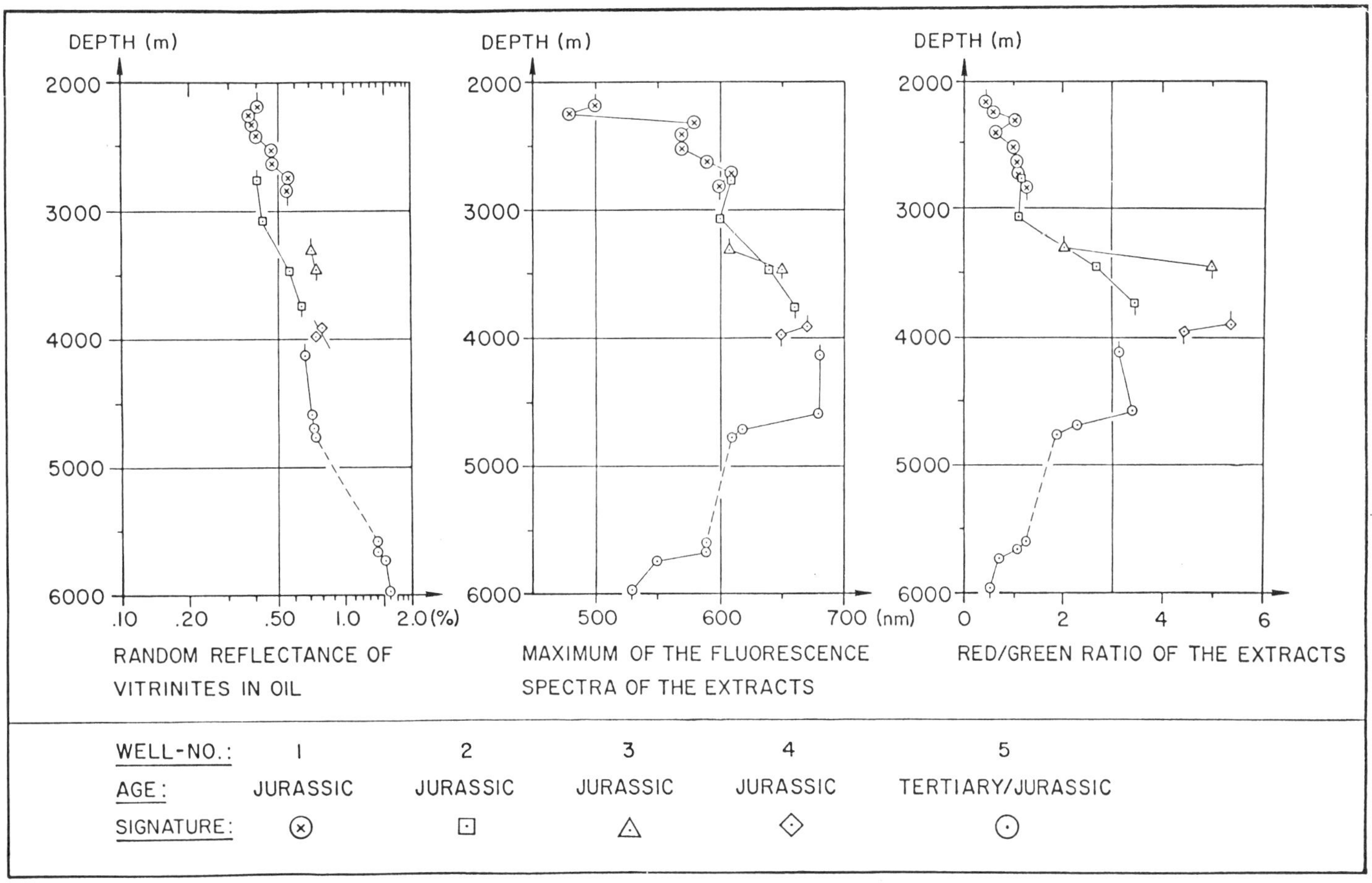

Figure 37. Variations of rank parameter with increasing depth (after Hagemann and Hollerbach, 1983).

Vitrinite reflectances ranging from 0.37 to 1.61 per cent from five boreholes representing a depth from 2 000 to 6 000 metres, are compared with the fluorescence properties of the extracts. Up to a depth of 4 600 metres maximum wavelength and red/green quotient increase regularly with increase in maturation as expressed by vitrinite reflectances. From there on an inverse trend of the fluorescence properties can be observed although vitrinite reflectances still indicate an increase in maturation. This discrepancy is interpreted here as an indication of migrated mature oil in borehole No. 5. Crude oils of different maturation levels (hydrocarbon content varying from 30 to 90 per cent) show an opposite trend with increase in maturation, such as a shift to lower wavelengths and lower red/green quotients (Figure 38).

7.3.2.4 Fluorescence Properties of Vitrinites and Their Relation to Technological Properties of Coals

It has been demonstrated recently (Radke *et al.*, 1980) that not only huminite/vitrinite of low rank coals fluoresce, but also that the amount of fluorescing vitrinite increases with rank to a stage of high-volatile bituminous A coal (0.9 per cent R_om (random reflectance); see Figure 39).

At higher rank levels the amount of fluorescing vitrinite decreases, accompanied by a decrease of visible fluorescence of the liptinite macerals. The fluorescence observed in the vitrinite is believed to be caused by bituminous substances formed in the course of coalification. In strong fluorescing vitrinites the reflectance values are lowered considerably by the absorption of these bituminous substances (Kalkreuth, 1982). In these coals vitrinite reflectance is not an accurate indicator of rank. Determination of the fluorescence properties may give, in this situation, a better indication of the 'true' rank of the sample.

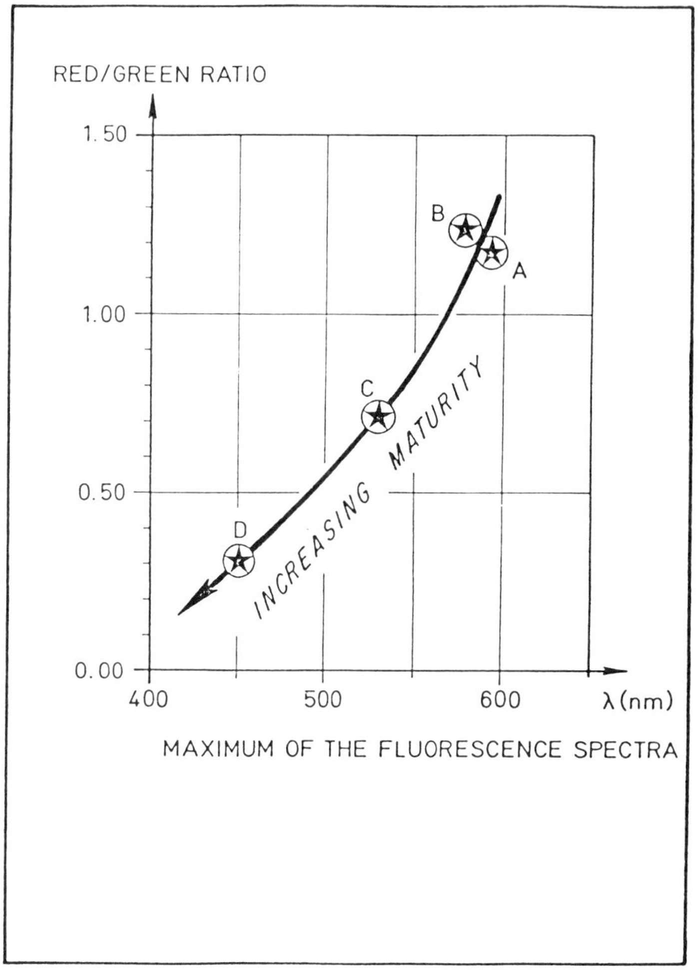

Figure 38. Variation of fluorescence parameters of crude oils (after Hagemann and Hollerbach, 1983).

Fluorescence properties of vitrinites of Carboniferous coals from the Ruhr and Saar region of western Germany are shown on Figures 40 and 41. They show that fluorescence intensities increase with rank from 0.65 to 0.95 per cent R_om (high-volatile C bituminous to high-volatile A bituminous). At higher rank levels the fluorescence intensities decrease due to increase in aromatization of the bituminous materials. The decrease in fluorescence intensity is associated by a shift toward longer wavelength and an increase in the red/green quotient.

Preliminary results by Ottenjann *et al.* (1981) indicate that some of the technological properties of Carboniferous coals in western Germany can directly be related to fluorescence properties of the vitrinites. Figure 41 shows the relation between softening point of the coals and fluorescence intensity of the vitrinites. It can be seen that coals having low softening points are characterized by high fluorescence intensities. Free swelling index cannot be related to fluorescence intensity but shows correlation to the red/green quotient derived from spectral fluorescence measurements (Figure 42).

The relations are not fully understood yet and need further investigation, but it seems that study of the fluorescence behaviour of vitrinites can give worthwhile additional information on rank where the reflectance is lowered by the presence of bituminous substances, and at the same time can give a better understanding of the technological behaviour of coals.

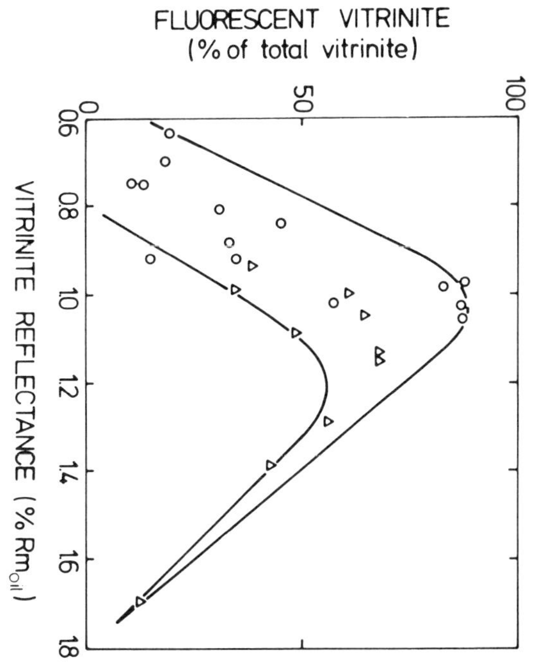

Figure 39. Proportion of fluorescent vitrinite (% of total vitrinite) versus mean vitrinite reflectance for coal samples of two wells in western Germany (well Kohlhaus and well Uedem 1, circles and triangles, respectively). After Radke *et al.* (1980).

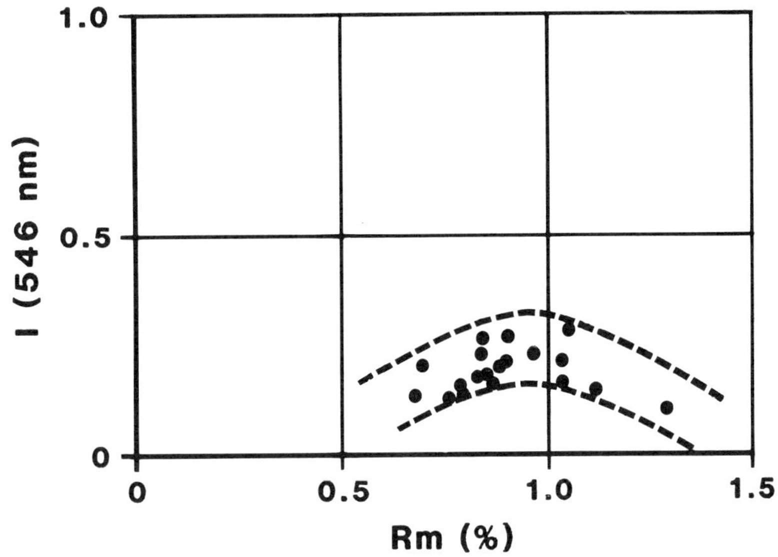

Figure 40. Relation between rank (% Rm vitrinite reflectance) and fluorescence intensities of vitrinites at 546 nm (after Ottenjann *et al.*, 1981).

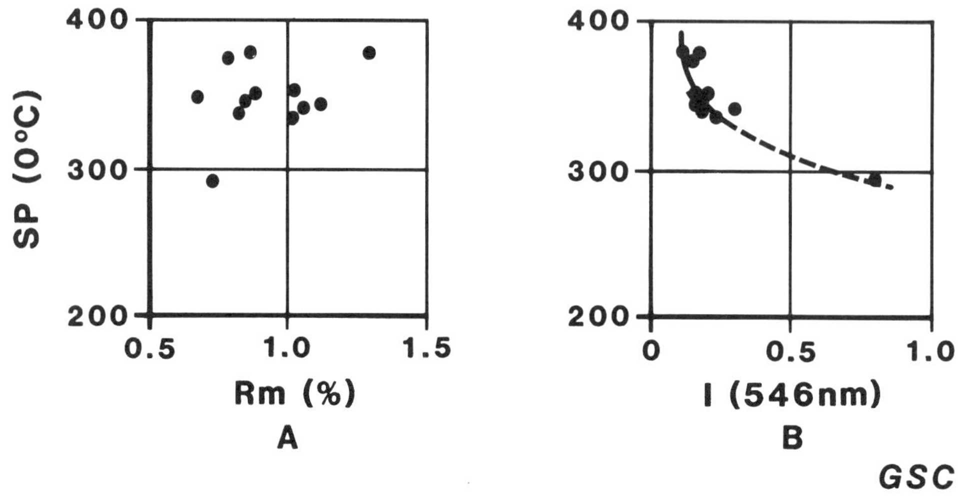

Figure 41. Relation between softening point and vitrinite reflectance (A) and fluorescence intensities of vitrinite (B). After Ottenjann *et al.* (1981).

7.3.2.5 **Characterization of Solid Bitumens**

Bitumens are generally formed from natural crude oils. Most commonly they are found in carbonate rocks where they adopt the form of the cavities in which they occur. Bitumen of varying degrees of alteration can be found in one sample indicating several phases of bitumen generation. A genetic classification of naturally occurring bitumens is given on Figure 43.

The solid bitumens can be distinguished by means of reflectance and fluorescence properties, by micro-solubility and micro-flowing point (Jacob, 1975). As can be seen from Figure 44, the optical properties vary widely. Reflectances range from less than 0.01 to 1.5 per cent. Fluorescence intensity can be applied on the low reflecting bitumens (<0.30 per cent $R_o$max) to distinguish one from the other.

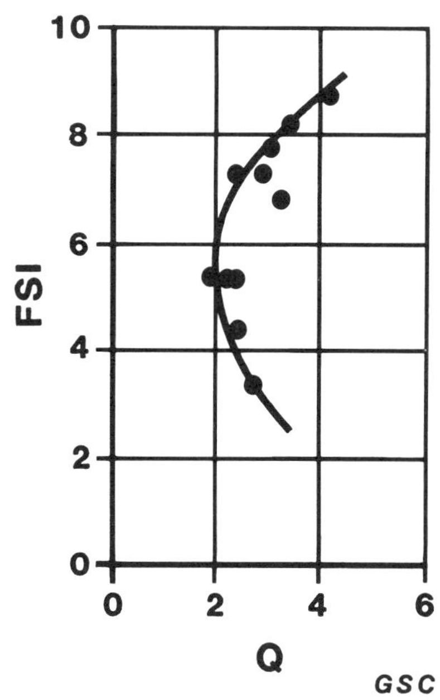

Figure 42. Relation between Swelling Index and the red/green quotient of Carboniferous coals (after Ottenjann *et al.*, 1981).

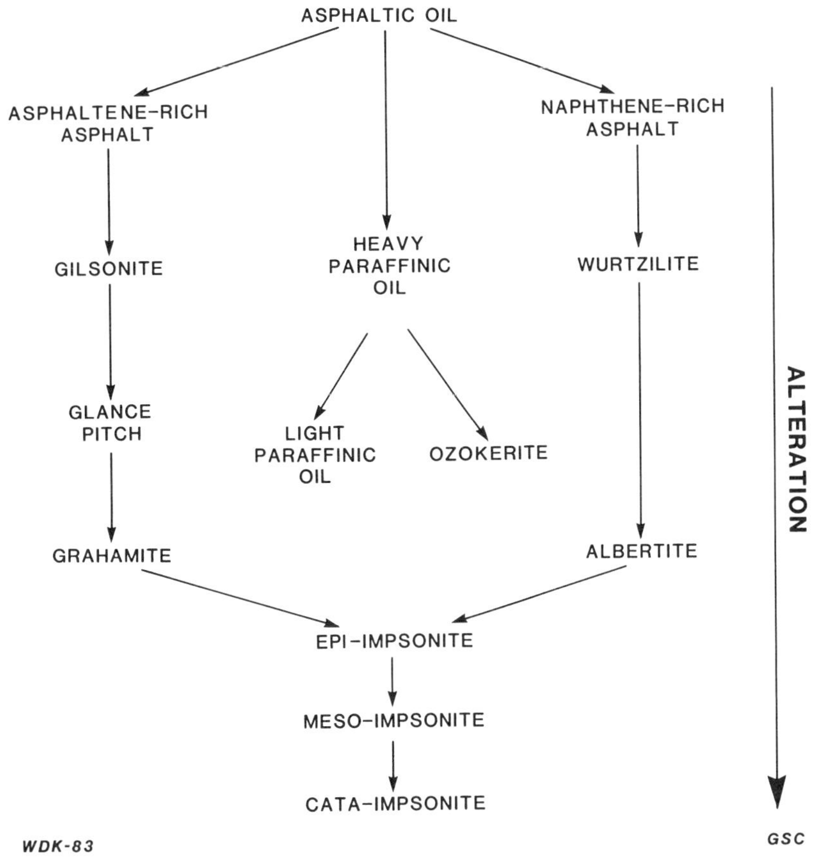

Figure 43. Genetic classification of solid bitumens (after Jacob, 1976).

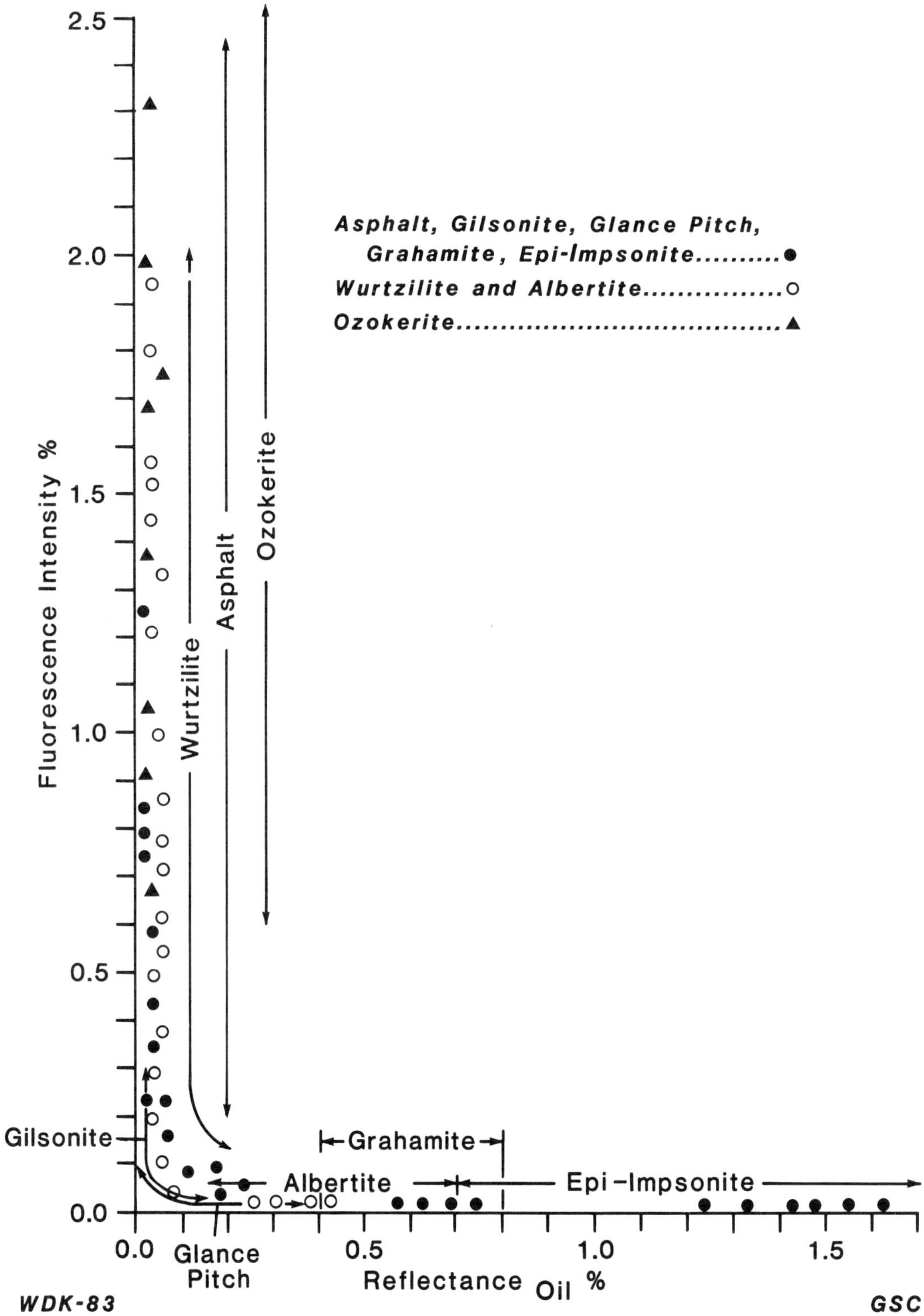

Figure 44. Bitumen fluorescence intensities versus random reflectance (modified from Jacob, 1976).

8 RELATIONSHIP OF PETROGRAPHIC PROPERTIES TO OTHER COAL PROPERTIES

Coal petrography provides another means, in addition to chemical analysis, of determining rank and other compositional data on coals. Thus a broader spectrum of analytical data is made available for a better understanding of the basic constitution of coal and its varied technological behaviour.

It is necessary at this point to consider the relationship between coal petrographic parameters and the results of chemical analyses and physical tests used to characterize coals, in order to predict the variations in coal properties implied by variations in coal petrographic results. This chapter will first consider the applicability of vitrinite reflectance as a rank parameter, and show comparisons with the ASTM rank classification. Next, the relationship between coal petrography and some of the chemical properties arrived at during proximate and ultimate analyses, namely volatile matter, and ultimate carbon and hydrogen, will be discussed. Next, the relationship between coal petrography and results of rheological tests such as free-swelling index, maximum dilatation, and Gieseler fluidity will be considered. Lastly, some of the methods of classifying coals which utilize one or more coal petrographic parameters will be outlined.

8.1 **Vitrinite Reflectance and Coal Rank**

The most fundamental concept relating to both the coalification history and utilization potential of a coal is its rank (Chapter 3). The ASTM rank classification is based in part on volatile matter, which, as will be demonstrated in the following, is dependent on both rank and coal composition. This suggests a weakness of the ASTM rank system: two coals of different composition may be assigned to the same rank value, even though their levels of coalification may differ considerably.

Table 24 gives recommended boundaries to the ASTM rank categories based on vitrinite reflectance (McCartney and Teichmüller, 1972; Davis, 1978a). In the bituminous coal range the boundaries based on $R_o$max are very similar to those based on R_om (compare columns 1 and 2, Table 24). Figure 45 shows an empirical relationship between vitrinite volatile matter content and vitrinite reflectance ($\bar{R}_o$max), with ASTM rank category boundaries indicated. Figure 10 (Chapter 2) shows ASTM rank categories as a function of vitrinite reflectance ($\bar{R}_o$m) and other rank parameters, including volatile matter, ultimate carbon, bed moisture, and calorific value. Figure 10 also shows the applicability of various rank parameters over the complete range of coalification, demonstrating that vitrinite reflectance is the most widely applicable parameter in use today.

McCartney and Teichmüller (1972) observed that vitrinite reflectance is the best single criterion of rank in bituminous coals with less than 30 per cent volatile matter. They also stated that it is highly likely that reflectance is also the best rank parameter for anthracites, and that reflectance is nearly comparable to moisture content as a rank indicator in high-volatile bituminous coals. They suggested that the same is not true for lower rank coals, but subsequent studies by Castano and Sparks (1974), Bustin *et al.* (1977), and Marchioni (in press) have demonstrated the utility of reflectance in even low rank lignitic coals, provided that care is taken in the selection of the component measured.

TABLE 24. SUGGESTED LIMITS OF ASTM RANK CLASSES IN TERMS OF VITRINITE REFLECTANCE (from Davis, 1978a; McCartney and Teichmüller, 1972).

Rank	Maximum Reflectance (% $\bar{R}_o$max)	Random Reflectance (% $\bar{R}_o$m)
Subbituminous High-volatile bituminous	$<$0.47	
	C 0.47 - 0.57	
	B 0.57 - 0.71	0.50 - 1.12
	A 0.71 - 1.10	
Medium-volatile bituminous	1.10 - 1.50	1.12 - 1.51
Low-volatile bituminous	1.50 - 2.05	1.51 - 1.92
Semi-anthracite	2.05 - 3.00 (approx.)	1.92 - 2.50
Anthracite	$>$3.00	$>$2.50

Vitrinite reflectance has also been correlated with other thermal maturation indicators used in studies of diagenesis and low-grade metamorphism (see, for example, Heroux *et al.*, 1979). These include liptinite fluorescence, organic geochemical parameters used in oil and gas exploration (for example, kerogen, bitumen, and gas chemistry), and mineralogical criteria such as illite crystallinity and presence of specific clay minerals, zeolities, or low-grade metamorphic minerals. Changes in fluorescence characteristics of liptinite with increasing rank are shown in Section 7.3. The range of coal rank corresponding to oil and gas generation is discussed in Section 10.5. Mineralogical criteria are most applicable in relatively high rank coals. For example, kaolinite is destroyed at a maturity level corresponding to approximately 1.3 per cent $\bar{R}_o$max, laumontite is stable over a range corresponding to approximately 0.5 to 2.5 per cent $\bar{R}_o$max, and prehnite and pumpellyite occur mainly in rocks which have achieved a level corresponding to 2.5 per cent $\bar{R}_o$max.

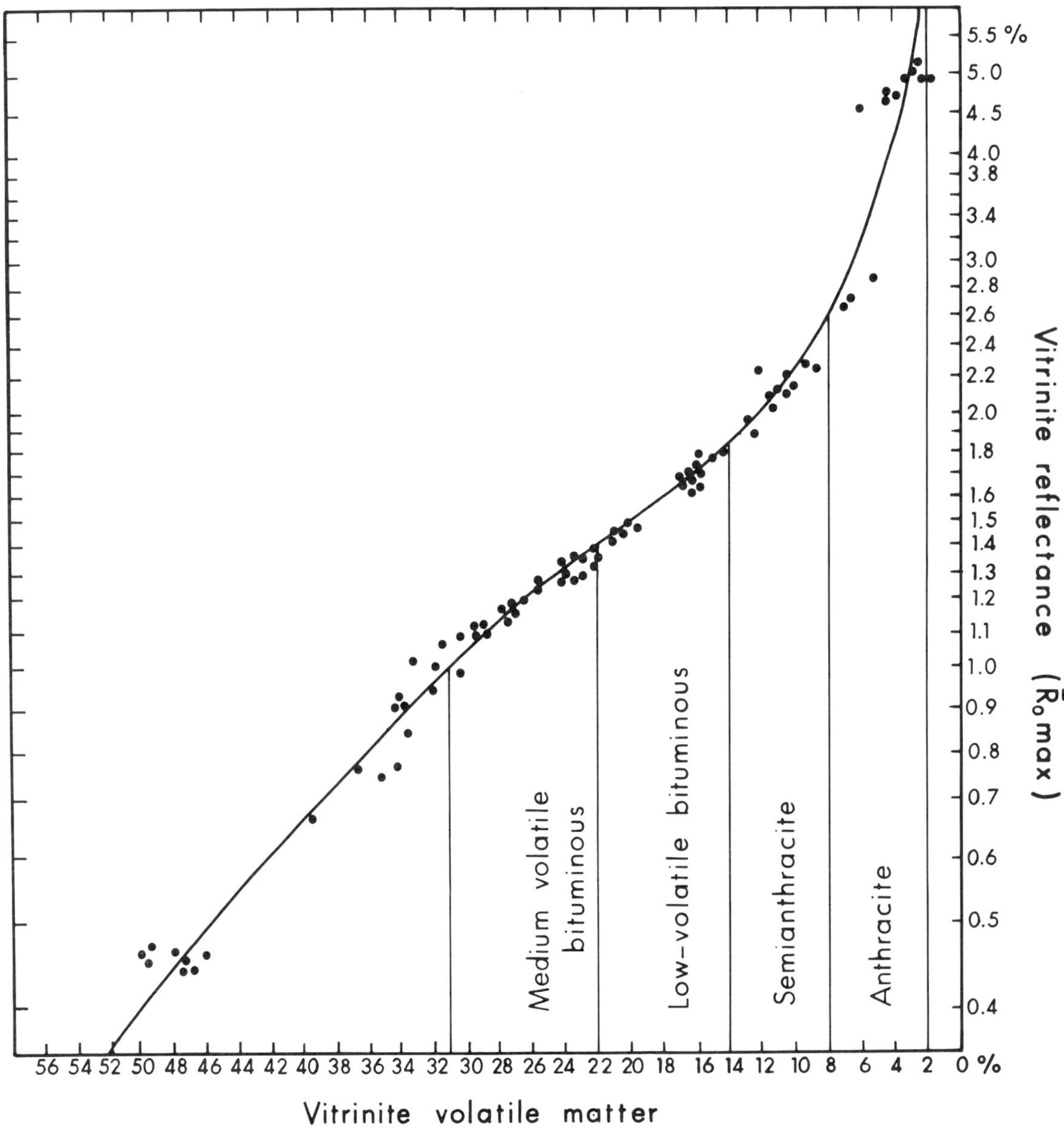

Figure 45. Relationship of volatile matter and reflectance of vitrinite from European coals. ASTM rank categories based on volatile matter are also plotted (from Kotter, 1960; Cameron, 1975).

8.2 Relationship of Coal Petrography to Chemical Parameters

The volatile matter content of a coal is dependent on both rank and maceral content. Figure 46 indicates the general relationship of volatile matter to vitrinite reflectance. Volatile matter can be seen to decrease with increasing vitrinite reflectance, with the rate of change being greatest at low reflectance, especially below 1 per cent, and least at high reflectances, especially above 2 per cent. Correlation is best at higher reflectance levels, as is indicated by the width of the curve.

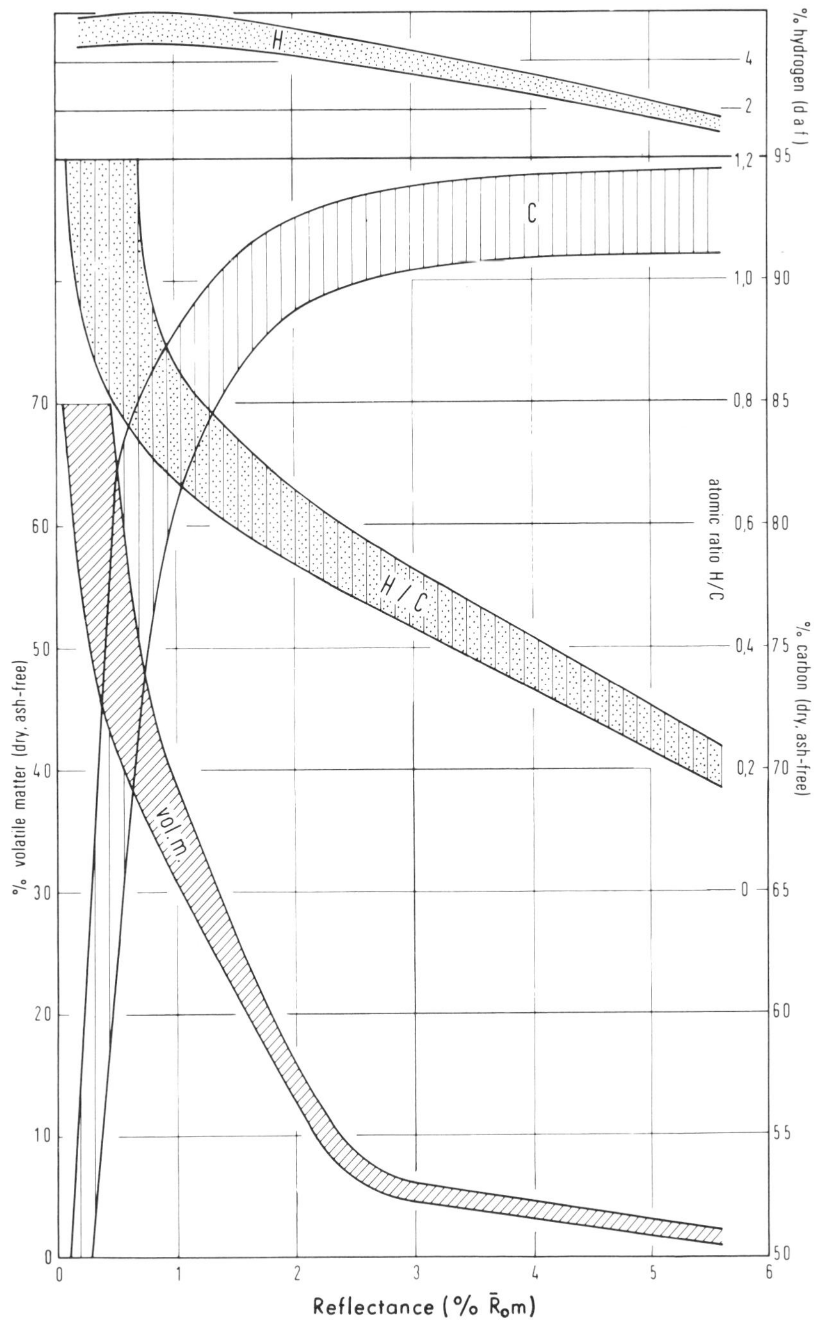

Figure 46. Relationship between vitrinite reflectance ($\bar{R}_o$m) and volatile matter, hydrogen and carbon contents, and the ratio hydrogen/carbon (from Teichmüller and Teichmüller, 1982).

The relative proportions of the various maceral groups also influence volatile contents. In Section 4.1 the differences in properties of the various maceral groups were explained. At any given coal rank level inertinite macerals are lower in volatile contents than vitrinites, which in turn are lower in volatiles than liptinites. The contrast is greater in lower rank coals. Therefore, increase in inertinite content of a coal seam tends to lead to a decrease in volatile content, other factors being equal.

Figure 47 shows sets of empirical curves based on Australian coals which demonstrate the combined effects of variations in reflectance and maceral content on volatile matter of coal. Maceral content is quantified as vitrinite plus liptinite. As expected, volatile matter increases with decreasing reflectance and increasing vitrinite-plus-liptinite content (Strauss *et al.*, 1976).

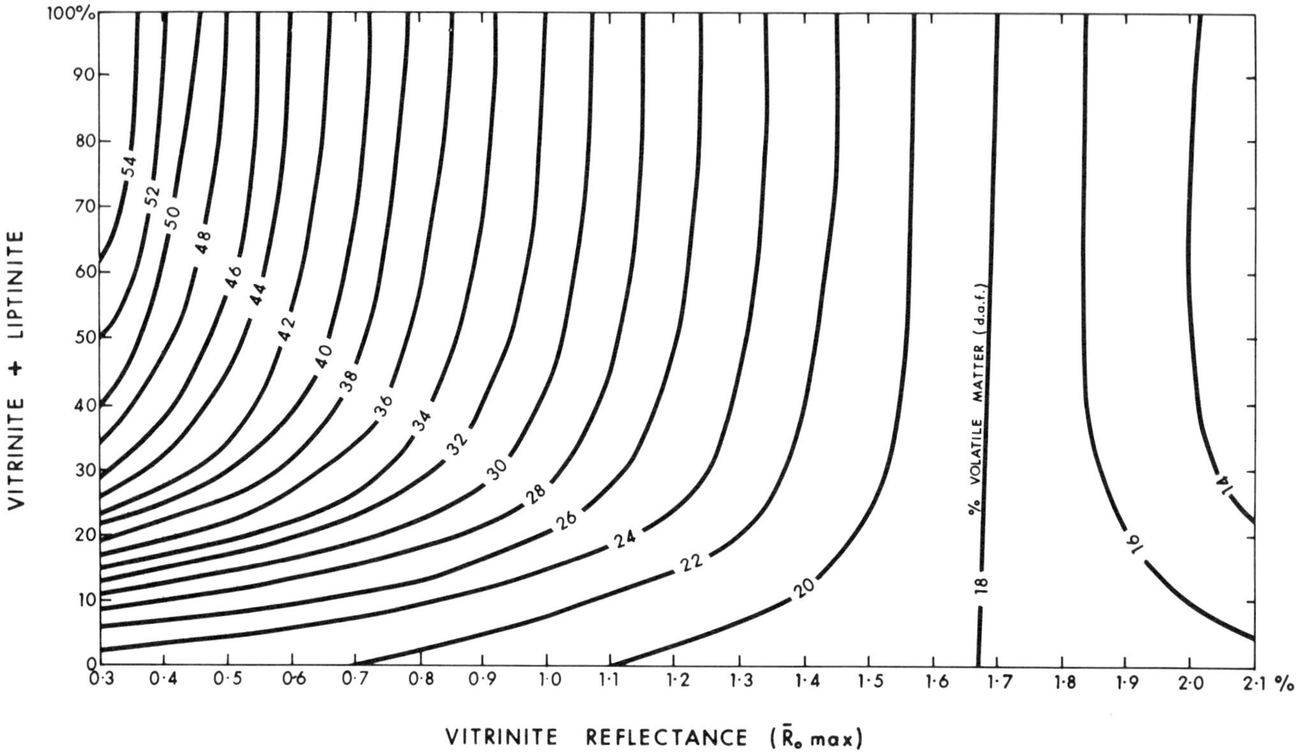

Figure 47. Volatile matter contents of Australian coals as a function of vitrinite reflectance and maceral composition (from Strauss *et al.*, 1976).

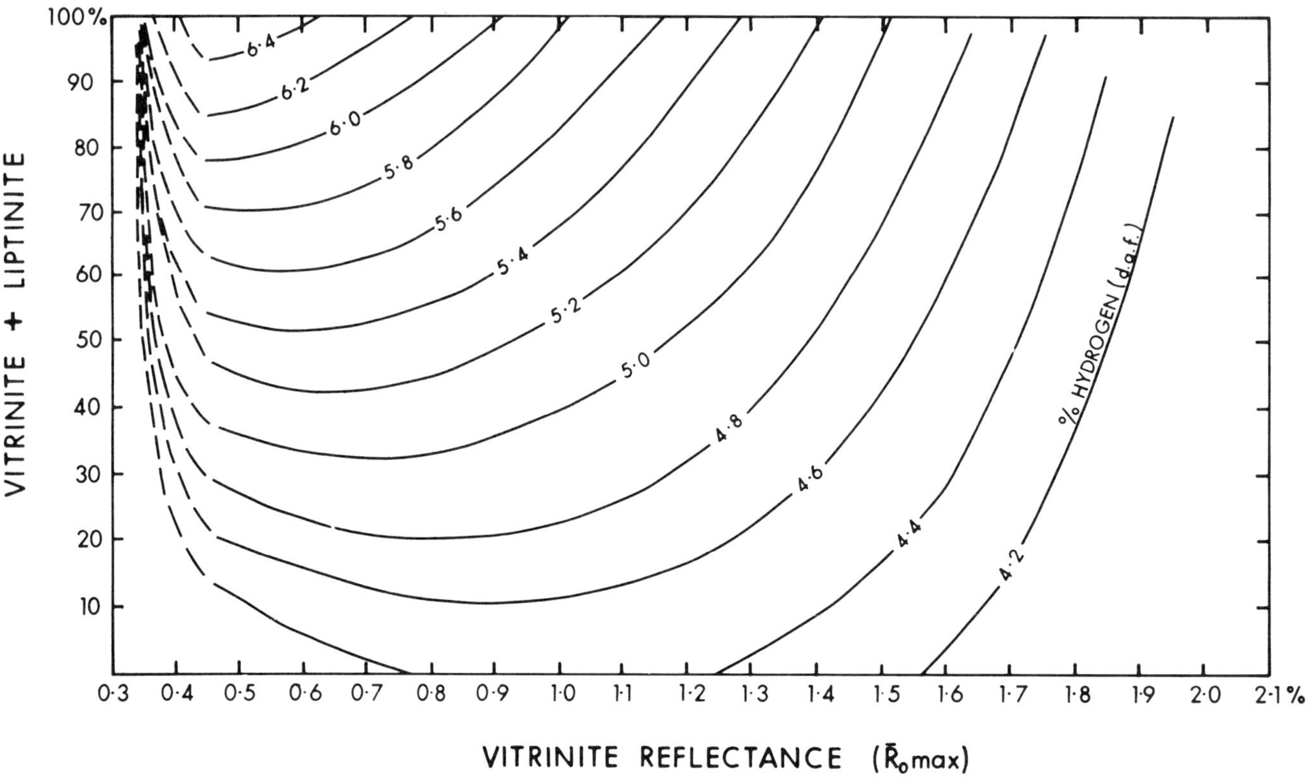

Figure 48. Hydrogen content of Australian coals as a function of vitrinite reflectance and maceral composition (from Strauss *et al.*, 1976).

Ultimate analytical results are also dependent on rank and composition of coal. Ultimate carbon content increases with increasing vitrinite reflectance, while hydrogen content and the ratio H/C both decrease with increase in vitrinite reflectance (Figure 46).

Maceral composition of coals will also influence C and H contents, as each of the maceral groups has a distinctive chemistry (Section 4.1 and Section 2.5.2). In short, inertinite has higher C and lower H contents than the other maceral groups, while exinite has lower C and higher H contents than the other groups.

On Figure 48 the combined effects of composition and reflectance on hydrogen content of Australian coals are displayed. Through subbituminous and bituminous rank ranges, hydrogen content increases with decreasing vitrinite reflectance and inert maceral content. The relationship breaks down at low rank levels, especially below 0.45 per cent $R_o$max (Strauss *et al.*, 1976).

8.3 **Relationship of Coal Petrography to Rheological Properties**

The behaviour of a coal in the various tests of its carbonization potential can be related to rank and composition of the coal. Detailed petrographic analysis has allowed for prediction of rheological test results and utilization potential based on petrographic data.

Rheological tests have been devised to quantify the agglomerating performance of a potential coking coal which takes place during heating and devolatilization. The commonest of these tests involve rapid heating expansion (free swelling index or FSI), slow heating expansion (dilatation), and apparent fluidity (see Pearson, 1980). FSI determination is carried out by rapidly heating to 820 degrees Celsius 1 gram of -60 mesh (-250 μm) powdered coal (ASTM D720-67). The resulting coke button is compared with a series of standard profiles, and assigned the number from zero to nine which corresponds with the most similar profile. Dilatation is determined by heating -60 mesh (-250 μm) powdered coal in a confined tube from 330 to 600 degrees Celsius at a rate of 3 degrees Celsius per minute. The change in the length of the coal pencil in the tube is recorded as a function of temperature. Parameters recorded include the softening temperature, maximum contraction, temperature of maximum contraction, maximum dilatation, and temperature of maximum dilatation. The Gieseler fluidity test utilizes the Gieseler plastometer, in which 5 grams of -40 mesh (-425 μm) powdered coal is placed in a crucible with a 10 kilogram load (ASTM D2639-74). A constant torque is applied to a mechanical stirrer as the sample is heated from 300 to 550 degrees Celsius at a rate of 3 degrees Celsius per minute. Parameters recorded include the softening temperature, maximum fluidity (in dial divisions per minute - ddm), temperature of maximum fluidity, and temperature of solidification.

Figure 49 demonstrates empirical variations in free swelling index as a function of maceral content and vitrinite reflectance, based on analysis of western Canadian coking coals (Pearson, 1980). Petrographic composition is quantified as percentage of inerts, which includes all inert macerals, two-thirds of the semifusinite, and all of the mineral matter based on the Parr formula (assuming constant ash and sulphur contents of 9.0 and 0.6 per cent, respectively). Figure 49 shows that for any given composition, highest swelling values are obtained at vitrinite reflectance values between 1.2 and 1.4 per cent. Further, for any given reflectance value, swelling increases with increasing content of reactive macerals. Highest values are obtained on coals with reactive contents greater than 70 per cent.

Marchioni (1981) demonstrated an application of the dependency of FSI on maceral composition, by pointing out that successive samples subjected to the FSI test during development of an adit do not necessarily provide a good index of coal oxidation. Variations in swelling results are related to composition variations as well as to the extent of oxidation. Petrographic examination of samples is therefore a useful supplement to this method of determining oxidation.

Shibaoka and Bennett (1976) turned the relationship between maceral composition and swelling properties around, by using swelling values for a single seam to estimate the vitrinite content in different areas.

Figure 50 indicates the relationship between petrographic parameters and maximum dilatation utilizing the Ruhr dilatometer, as determined empirically on western Canadian coking coals (Pearson, 1980). Maximum dilatation values are obtained on reactive-rich coals in the reflectance range 0.8 to 1.2 per cent. A similar diagram relating to maximum Gieseler fluidity (Figure 51) indicates that coals of the same approximate rank and composition also display the highest fluidity values (Pearson, 1980).

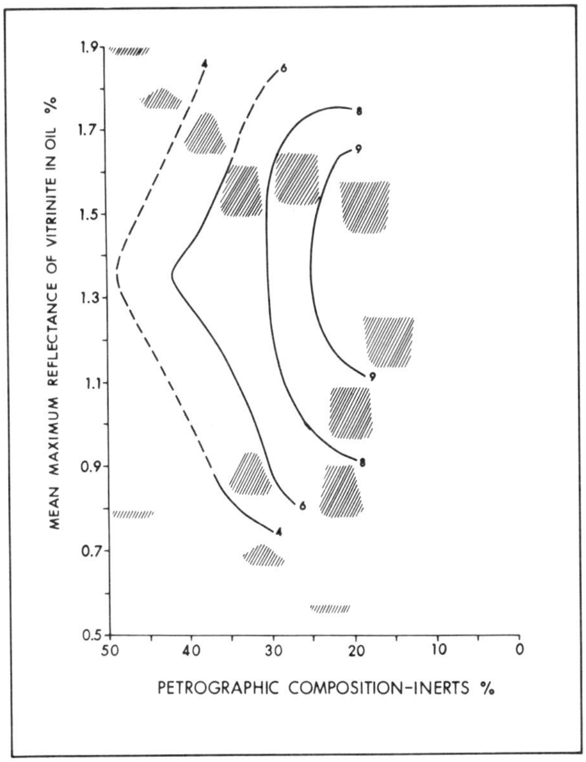

Figure 49. Free swelling index of western Canadian coals as a function of vitrinite reflectance and inert content (from Pearson, 1980).

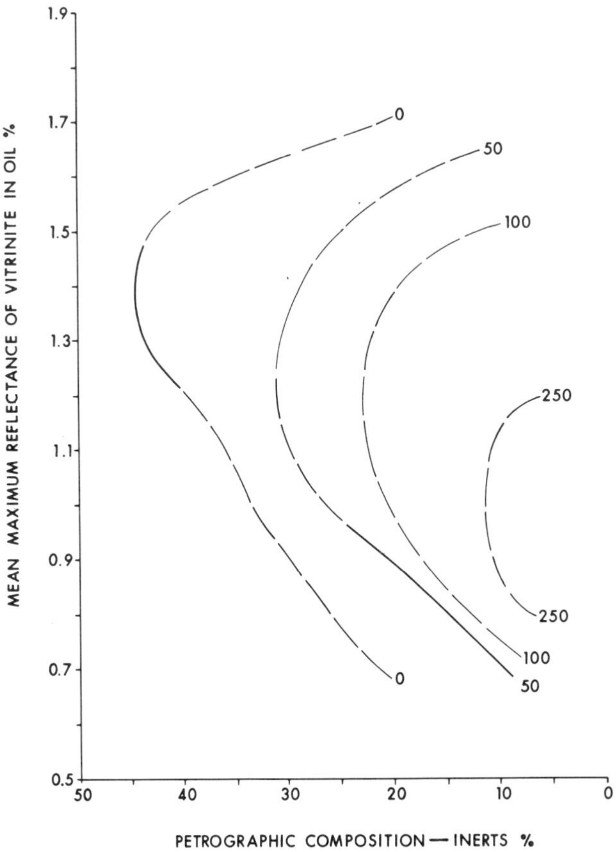

Figure 50. Maximum dilatation of western Canadian coals as a function of vitrinite reflectance and inert content (from Pearson, 1980).

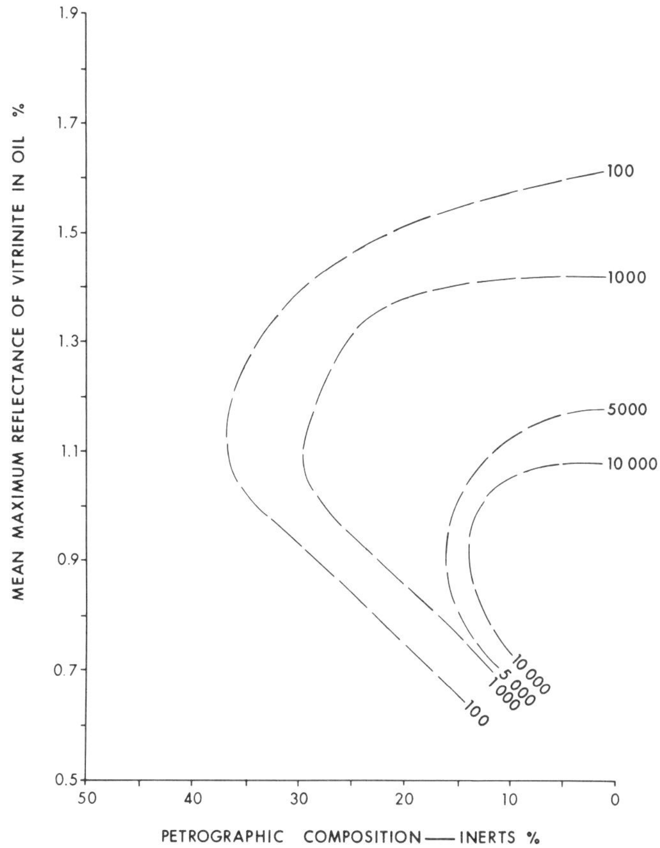

Figure 51. Maximum Gieseler fluidity of western Canadian coals as a function of vitrinite reflectance and inert content (from Pearson, 1980).

8.4 Application of Petrography to Classification

Classification schemes which utilize vitrinite reflectance as a rank parameter, in combination with either petrographic composition or a non-petrographic parameter, are becoming more common. The nature of any classification system generally depends on its proposed application, as no system is universally applicable. Nonetheless, the desirability of using a system which is based on independent parameters has been pointed out by several authors (for example, Eremin *et al.*, 1976). Systems based on vitrinite reflectance and maceral or microlithotype composition satisfy this requirement.

Bennett and Taylor (1970), for example, proposed such a scheme in which the composition parameter was the total of vitrinite-rich microlithotypes. Strauss *et al.* (1976) established a system based on vitrinite reflectance and the sum of vitrinite-plus-liptinite (Figures 47 and 48). The major application of their system is as a prediction tool for exploration, helping to provide preliminary assessments of coal quality without costly sampling and testing. Pearson (1980) established a six-group classification, corresponding to the major blending groups in the Japanese coke-making industry, based on vitrinite reflectance and inert content (Figure 52). The system was designed to show how western Canadian coals entering world markets compare with international coals, and to provide preliminary quality assessments (Figures 49, 50, and 51). Predicted maximum fluidity of coals of the Fernie Basin, southeast British Columbia, based on field samples, is shown on Figure 53.

As was mentioned previously, some classification systems utilize reflectance in combination with a non-petrographic property. For example, Marshall (1976) established a four-digit system based on vitrinite reflectance and maximum dilatation (Figure 54) for application to the coke-making industry. The first two digits correspond to the reflectance V-step (see Section 7.1.6), while the second two digits correspond to the first two integers of the maximum dilatation. For example, class 11.17 represents a coal with a reflectance between 1.10 and 1.19 per cent and a maximum dilatation (Ruhr dilatometer) of between 160 and 180 per cent. Coals with a dilatation of less than 0 are assigned a value of 00 with a qualification relating to whether the coal dilated, contracted only, or did not soften. Besides being easily standardized, the system has the advantage that changes in a coal's properties related to oxidation may show up in its classification.

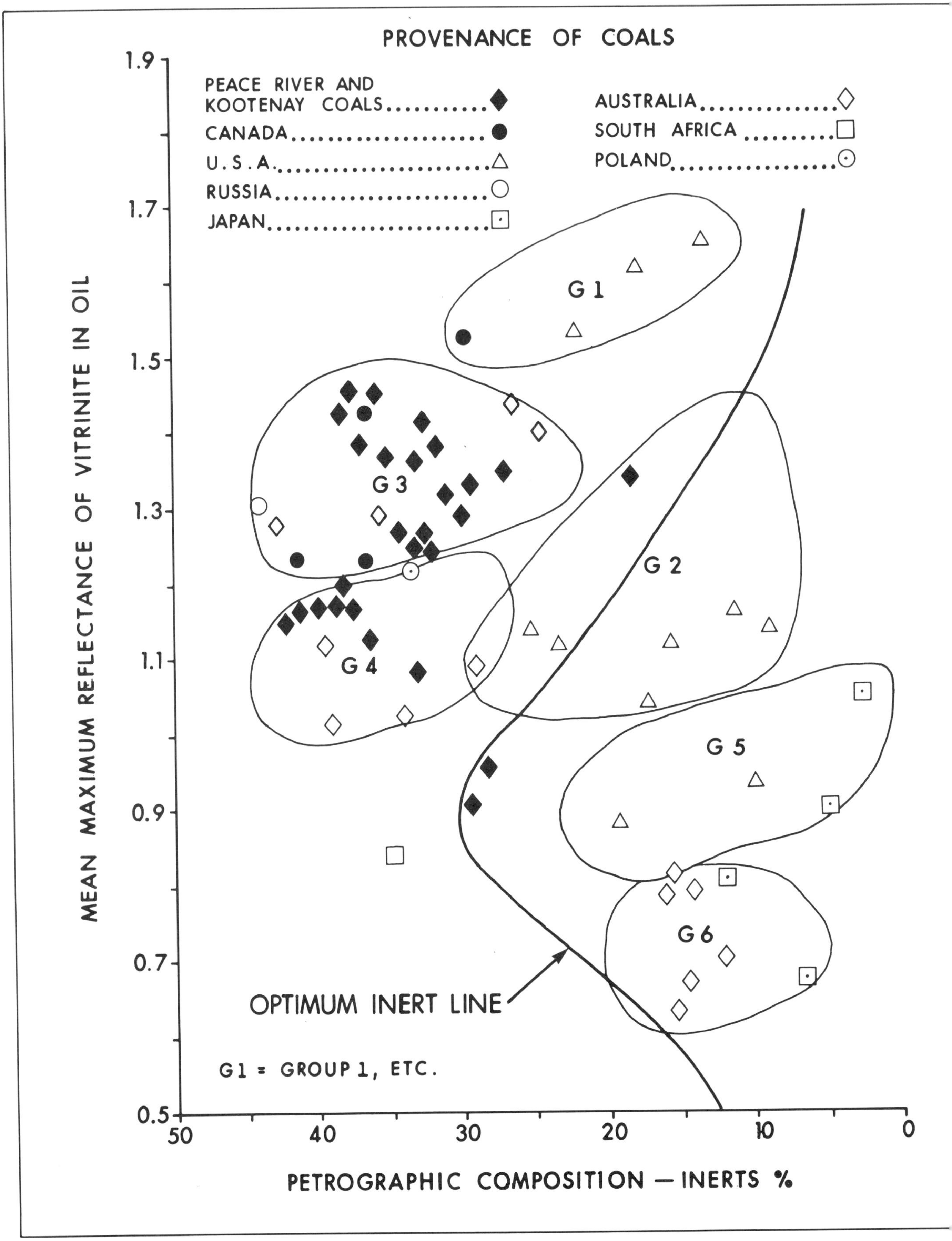

Figure 52. Coking coal groups and international coals plotted as a function of vitrinite reflectance and inert contents (from Pearson, 1980). Note location of optimum inert line with respect to these parameters (Section 10.3.3).

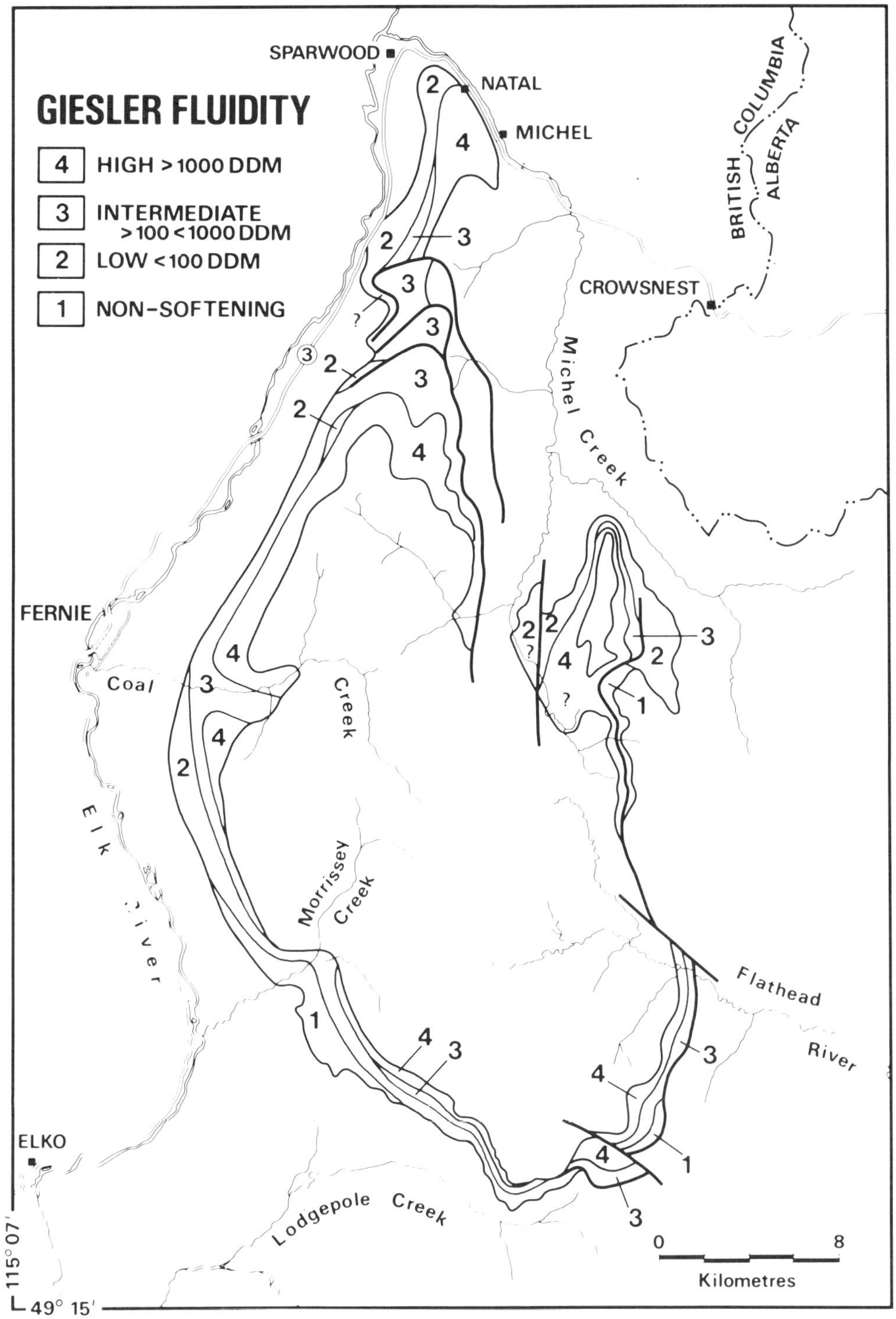

Figure 53. Predicted maximum Gieseler fluidity of coals of the Fernie Basin, southeast British Columbia, based on samples collected in the field. Predictions made using Figure 51 (from Pearson and Grieve, in preparation).

Of similar application is the system used by the Japanese steel industry, based on vitrinite reflectance and maximum Gieseler fluidity (Okuyama *et al.*, 1970). The chart (Figure 55) is divided into four blending quadrants corresponding to high and low reflectance and fluidity values, and the field for blended coal for coke making has been plotted. Coals from all four quadrants are generally required to produce a blend with the required characteristics (mean reflectance of about 1.2 per cent and fluidity between 60 and 1 000 ddm). Reactive-rich coals generally fall on the indicated line.

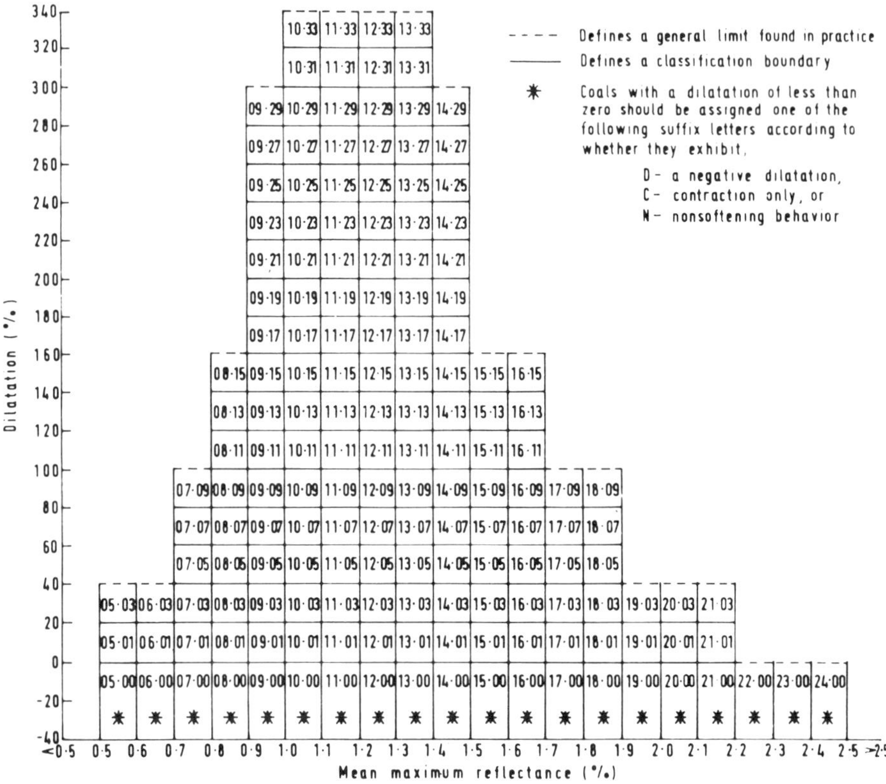

Figure 54. Coking coal classification based on vitrinite reflectance and dilatation (Marshall, 1976).

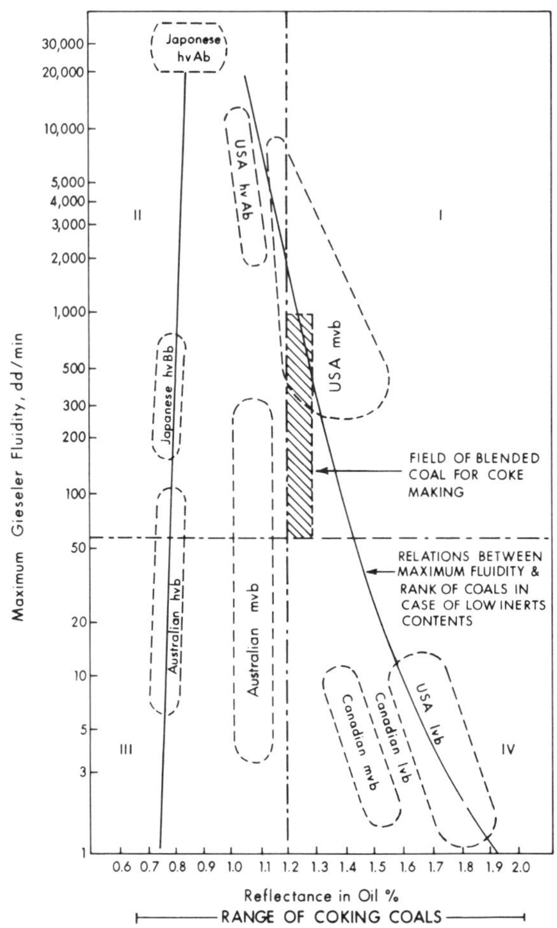

Figure 55. Coking coal classification based on vitrinite reflectance and Gieseler fluidity (Okuyama *et al.*, 1970).

9 GEOLOGICAL APPLICATIONS

9.1 Seam Correlation

Precise identification of coal seams is a critical aspect of coal exploration and mine planning. There are many methods of seam correlation, of which coal petrography is one of the less frequently used. The demonstrated usefulness of coal petrography in certain cases, however, suggests it should not be overlooked, even if only as a complement or accessory to other correlation methods. Its use is certainly preferable to use of rheological, proximate, and ultimate analytical parameters, which are dependent in the first place on petrography (Chapter 8).

9.1.1 *Macroscopic Criteria*

The use of macroscopic criteria in seam identification is fairly common. These include general seam geometry (that is, thickness and/or number of benches and partings), unusual seam features (such as coal balls, siderite concretions, and high pyrite, clay or carbonate content), and characteristics of interbedded strata (for example, marine or plant fossils in roof, distinctive seatearths, and channel sandstones or conglomerates in intervening strata). On a finer scale, the sequence of coal lithotypes and thin partings may provide a distinctive seam 'fingerprint' which can be identified at different localities. Five column samples of four seams in the southern part of the Illinois Basin were characterized according to lithotype proportions shown in Table 25 (Cameron, 1978). The generalized macroscopic petrology for each of the five columns was then plotted (Figure 56). Comparison of the samples shows that the profiles for the two samples from the same seam are quite similar, while those for the other three are different from each other and from the first two. This indicates that seams can be distinguished from one another by this form of description, and also that, in at least some cases, there appears to be a considerable amount of lateral persistence (at least 19 kilometres in this case) in petrologic characteristics.

TABLE 25. CLASSIFICATION OF COAL TYPES USED ON FIGURE 56 (from Cameron, 1978).

Coal Type		Proportions of Lithotypes
Bright	(B)	>75% vitrain plus bright clarain
Semibright	(SB)	50 - 75% vitrain plus bright clarain
Semidull	(SD)	25 - 50% vitrain plus bright clarain
Dull	(D)	<25% vitrain plus bright clarain
Parting	(P)	

Detailed macroscopic laboratory descriptions of selected column samples can be combined with in-place seam descriptions to produce an even more precise 'fingerprint.' Identification of laterally persistent index or marker layers is especially useful in this regard. For example, the series of columns on Figure 57 represents a stratigraphic cross-section of the Harbour seam, Sydney Coalfield, Nova Scotia (Cameron, 1971). Laboratory analysis of column samples identified a number of thin dull layers, some of which were found during field analyses to be extremely continuous. The 'I' band, for example, can be traced across the entire coalfield (Figure 57).

Presence of a tonstein within a coal seam provides probably the most reliable and successful method of seam correlation to date, as individual tonsteins can often be traced laterally over great distances (Figure 58). Most successful applications of the method have been in western Europe. Work by Grieve (unpublished) and Meriaux (1972) has also established the existence of tonsteins in the Kootenay Group of southeast British Columbia. Macroscopic identification of a particular tonstein may rely on its colour, thickness, and relative position within a seam, although all these properties are variable. Positive identification may therefore require petrographic, X-ray, or chemical analysis.

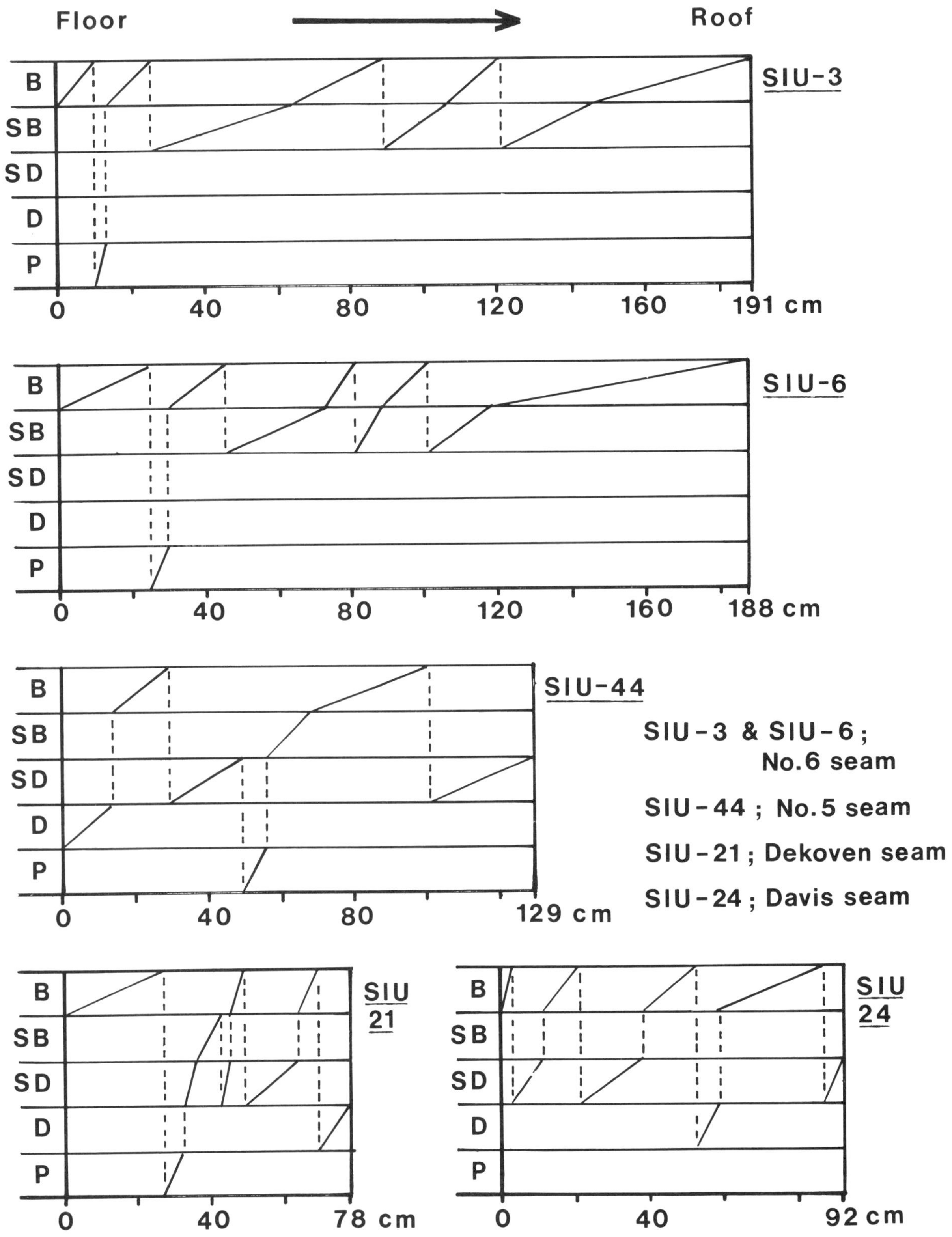

Figure 56. Distribution of lithotypes in intervals of column samples from the southern Illinois Basin (from Cameron, 1978). For explanation of abbreviations see Table 25.

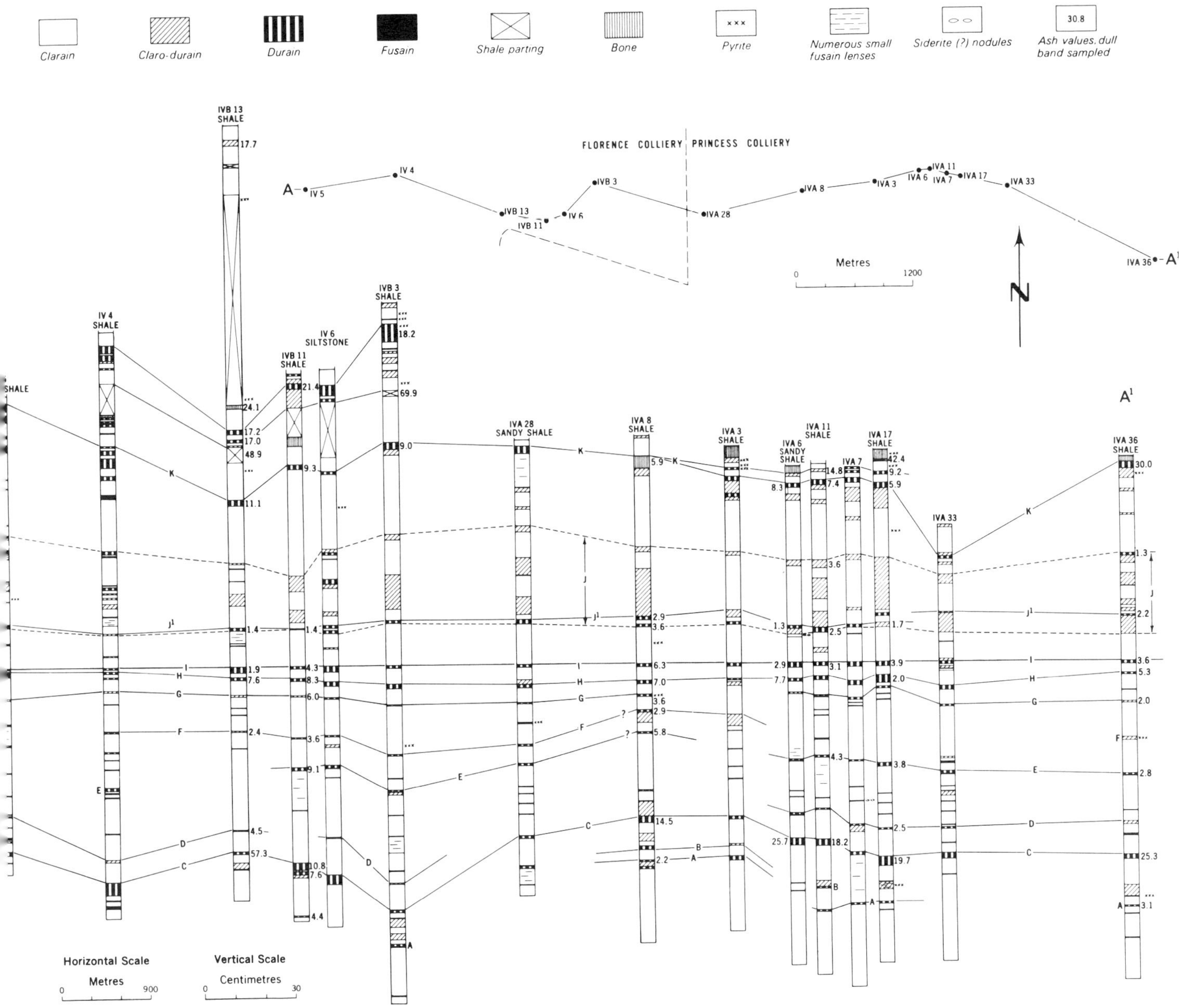

Figure 57. Stratigraphic cross-section of lithotypes of the Harbour seam in the Princess and Florence collieries, Sydney Coalfield, Nova Scotia (from Cameron, 1971).

In summary, any macroscopic characteristic of a seam or its related strata which differentiate it from other seams may serve to correlate it. So-called marker horizons provide quick and inexpensive exploration and mining tools.

9.1.2 *Microscopic Criteria*

Microscopic identification of seams may concentrate on mineral matter, vitrinite reflectance, maceral or microlithotype analysis, or some combination of these.

Mineral analysis may include recognition and identification of tonstein layers, or recognition of primary minerals such as siderite, pyrite, and quartz.

Vitrinite reflectance has been successfully applied as a correlation tool in certain cases. Within an individual multi-seam colliery, assuming reasonable coalification gradients and inter-seam thicknesses, it may be possible to characterize individual seams by their reflectance. A successful application by Canmore in the

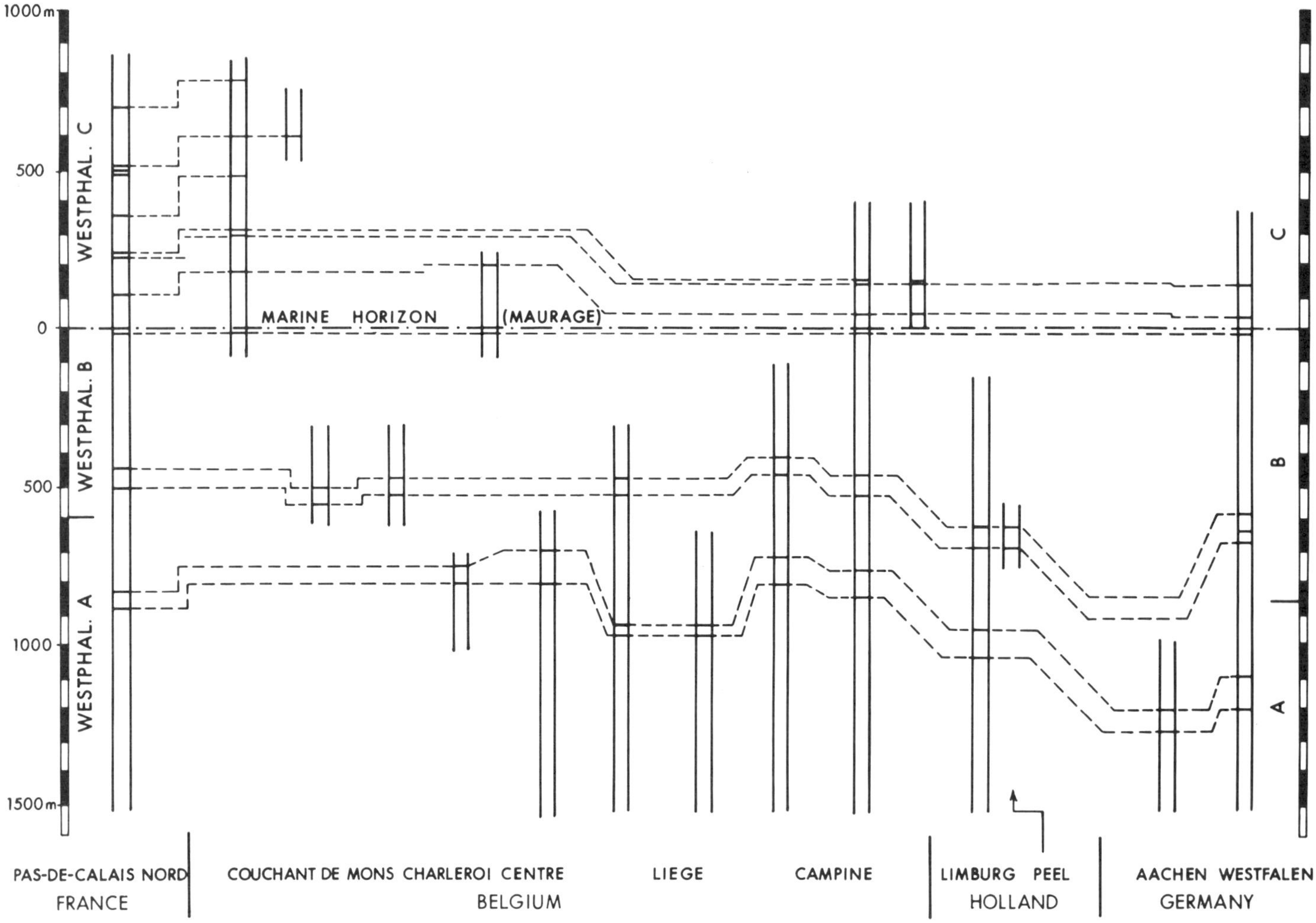

Figure 58. Lateral extent of tonstein bands in western Europe (ICCP, 1963).

Alberta Rocky Mountains was noted by Hacquebard and Donaldson (1974). In areas where lateral variations in seam rank occur, such a method may not be applicable.

Whole-seam maceral analysis as applied to correlation may make use of the presence of certain normally rare maceral types, or of the tendency for individual seams or groups of seams to be of relatively similar composition at various locations. In southeast British Columbia, for example, it was found that the seams in the lowest 250 metres of the coal-bearing Mist Mountain Formation (Kootenay Group) can, in general terms, be distinguished from those higher in the section on the basis of petrographic composition (Figure 59; Cameron and Kalkreuth, 1982). This corresponds with the trend recognized by Cameron (1972) that the lower portion of the section in this area is richer in inertinite macerals and poorer in vitrinite than the upper portion (Figure 60). Identification of individual seams was not possible here, but the method apparently has had more success in Australia, where microlithotype analysis appears to have been capable of providing fairly accurate differentiation of seams (Figure 61; Taylor and Warne, 1960). The latter study also demonstrated the potential usefulness of combining microlithotype analysis with vitrinite reflectance in characterizing seams. Data for the same seams (Figure 62) show an even more pronounced separation, and this has allowed for successful applications of this correlation method.

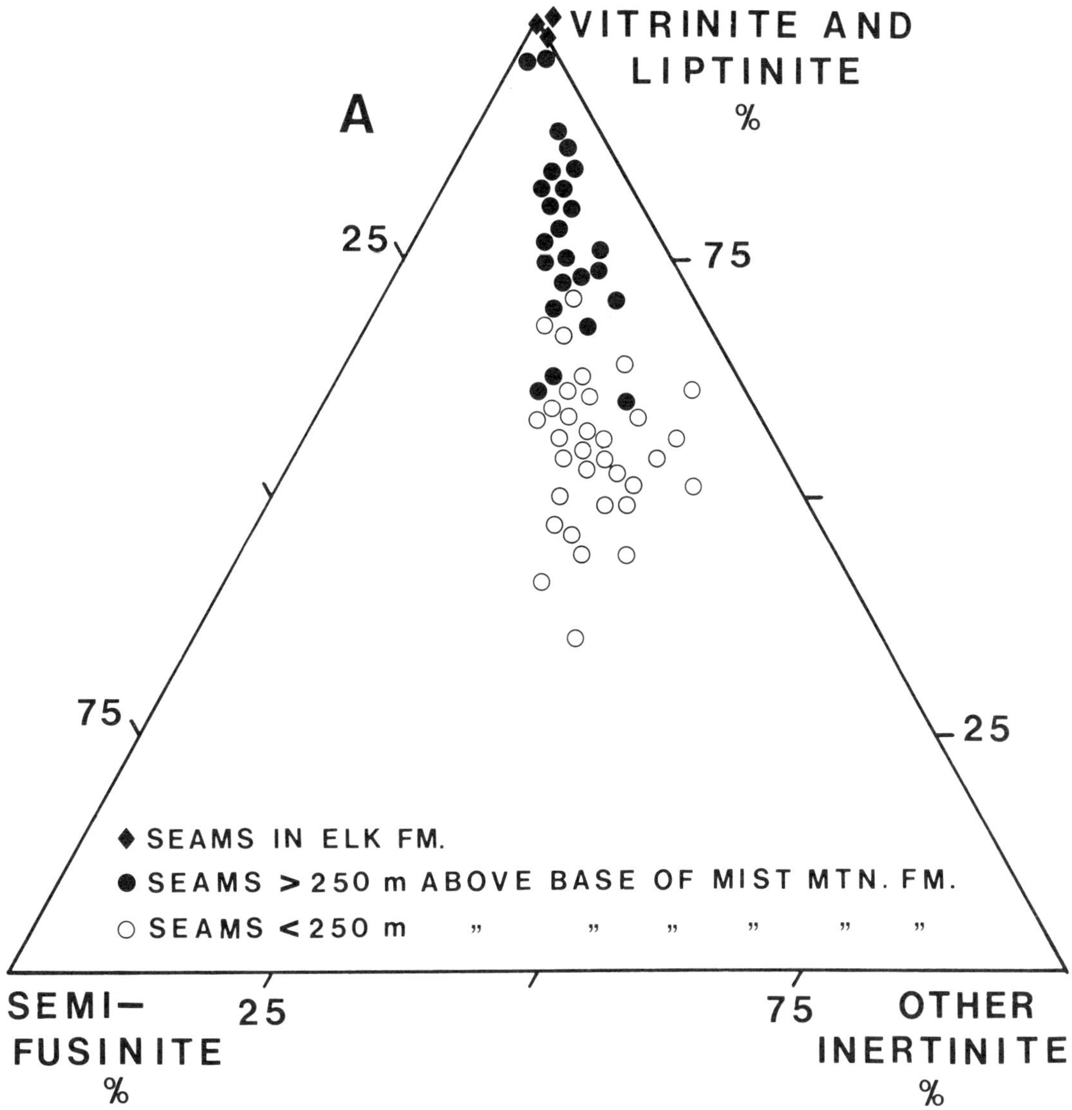

Figure 59. Maceral composition of Kootenay Group coals from southeast British Columbia, based on relative position within the group. Elk Formation overlies Mist Mountain Formation and is characterized by thin alginite-rich coals. (Cameron and Kalkreuth, 1982).

Petrographic analysis of ply-by-ply samples offers a more refined approach than whole-seam analysis, in that the depositional history of a coal seam is a unique feature which should be better reflected in its profile than in its overall composition. Vitrinite content profiles of the Bulli seam, Sydney Basin, New South Wales (Figure 63), show a definite cyclicity of compositional banding which is repeated throughout at least two sections (Shibaoka and Smyth, 1975). Profiles of seam No. 10 (Balmer) at three locations in the Michel area of southeast British Columbia (Figure 64) show that individual intervals and the sequence of intervals are traceable laterally (Cameron and Babu, 1968). Microlithotype analysis of the Tracy seam, Sydney Coalfield, Nova Scotia, at two locations (Figure 65) shows that individual intervals are continuous (Hacquebard, 1952). In this case, the petrographic analysis confirmed correlation made on stratigraphic evidence.

Work on the Harbour seam in Cape Breton (Cameron, 1971) utilized a combination of macroscopic and microscopic techniques. Marker durain bands, such as the 'I' band, were analysed microscopically to confirm their identification (Figure 57).

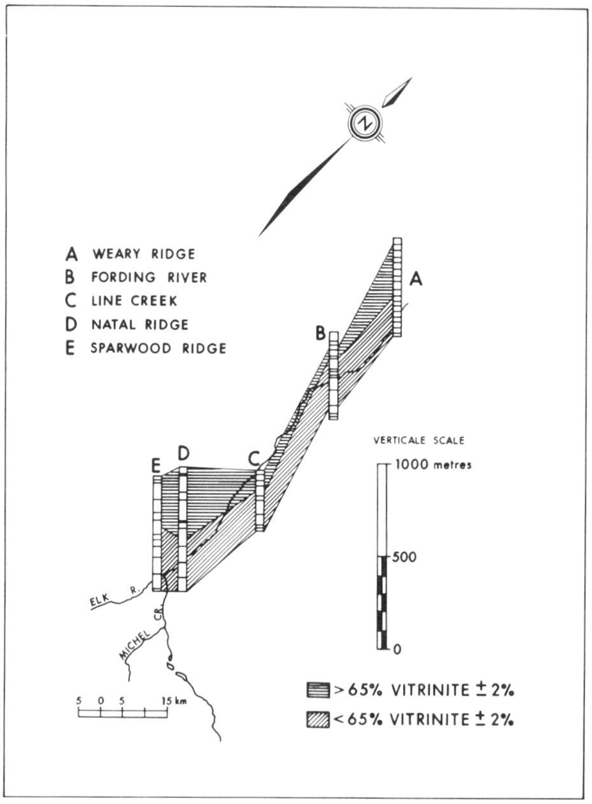

Figure 60. Panel diagram showing vertical distribution of Kootenay Group seams with high and low vitrinite content (from Cameron, 1972).

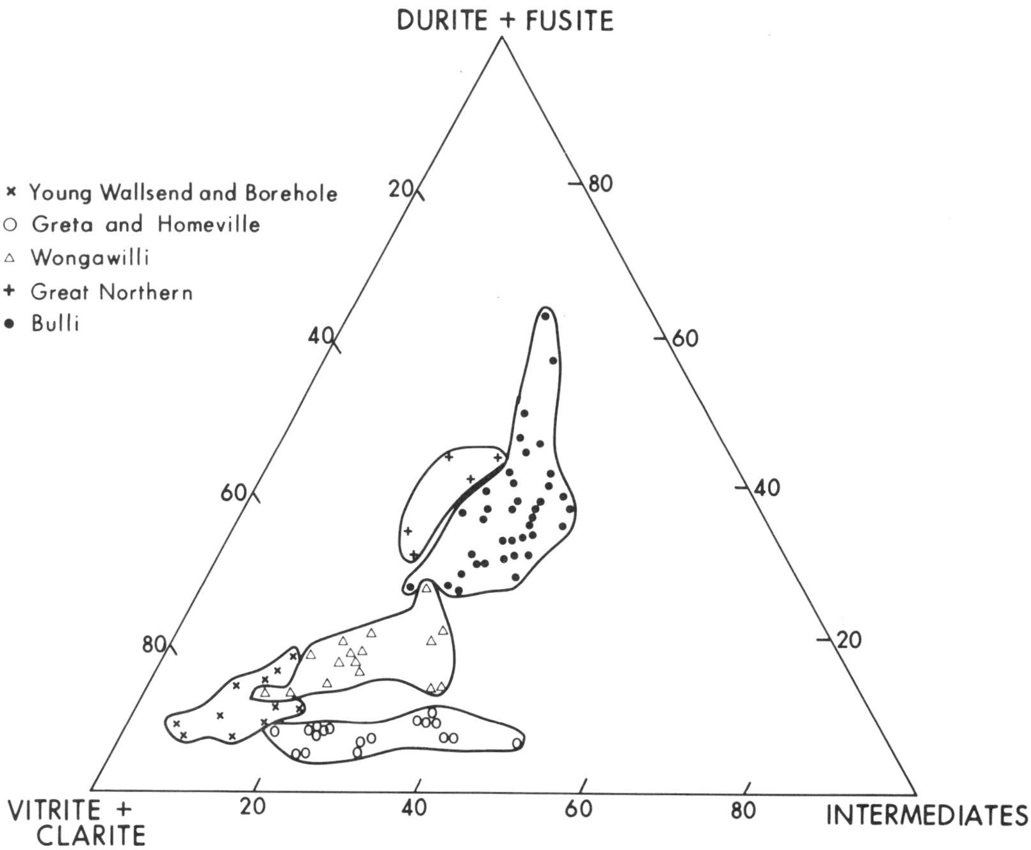

Figure 61. Microlithotype composition of several Australian coal seams from various locations (from Taylor and Warne, 1960).

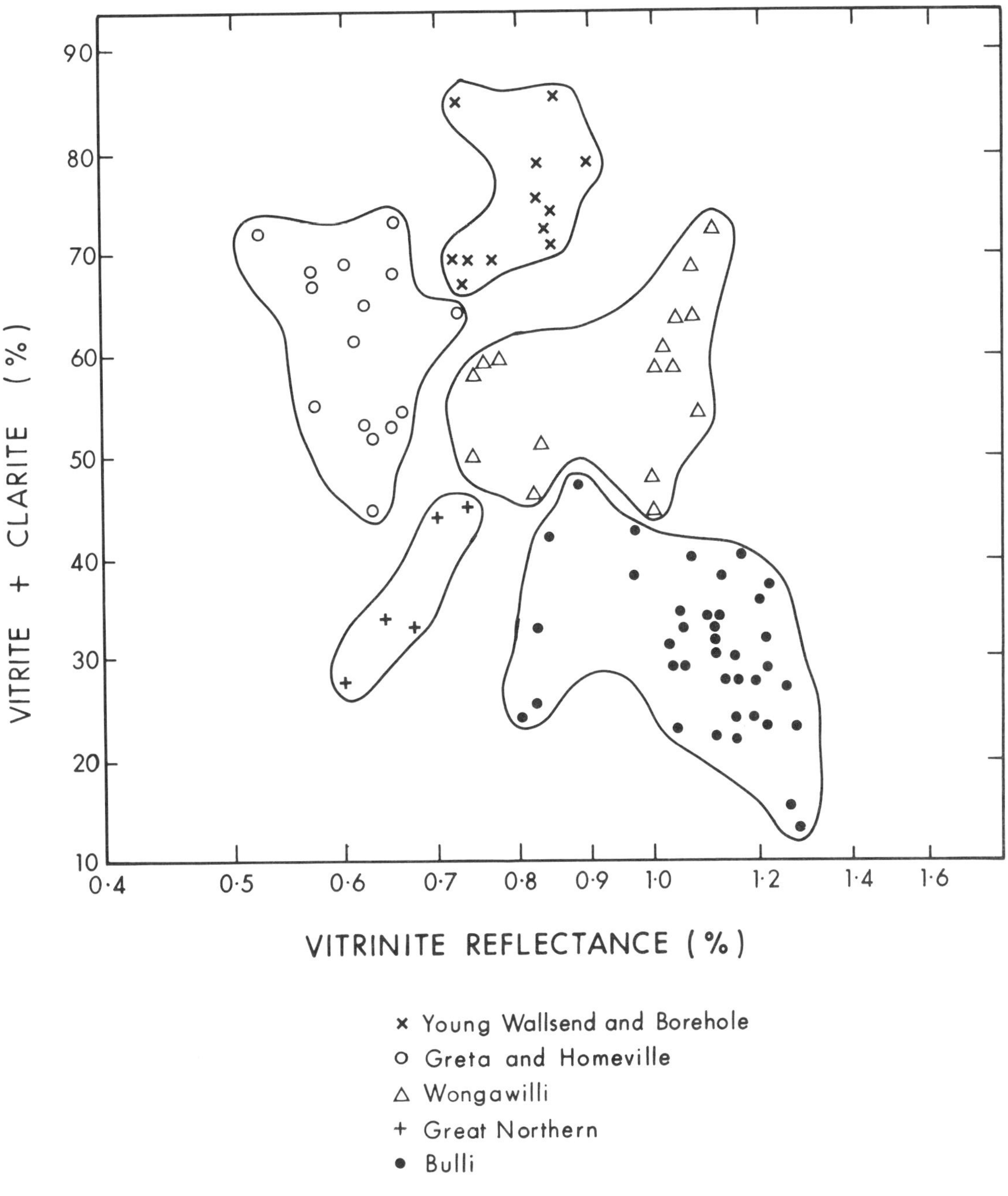

Figure 62. Vitrinite reflectance and microlithotype composition of several Australian coal seams from various locations (from Taylor and Warne, 1960).

9.1.3 *Formation Correlation*

Identification of a particular geologic formation may sometimes be made on petrologic criteria. In southeast British Columbia, for example, the coal-bearing part of the Kootenay Group is subdivided into the Mist Mountain and Elk Formations by the presence in the latter of vitrinite and liptinite-rich coals that occur in thin lenticular seams (Figure 59). Macroscopic identification of the Elk coal is readily made, especially when it occurs as 'needle coal' (Gibson, 1977), which appears as clusters of algal bodies similar in appearance to conifer needles. Microscopically the Elk coals are readily identified by the presence of masses and ovoid bodies of alginite (Kalkreuth, 1982).

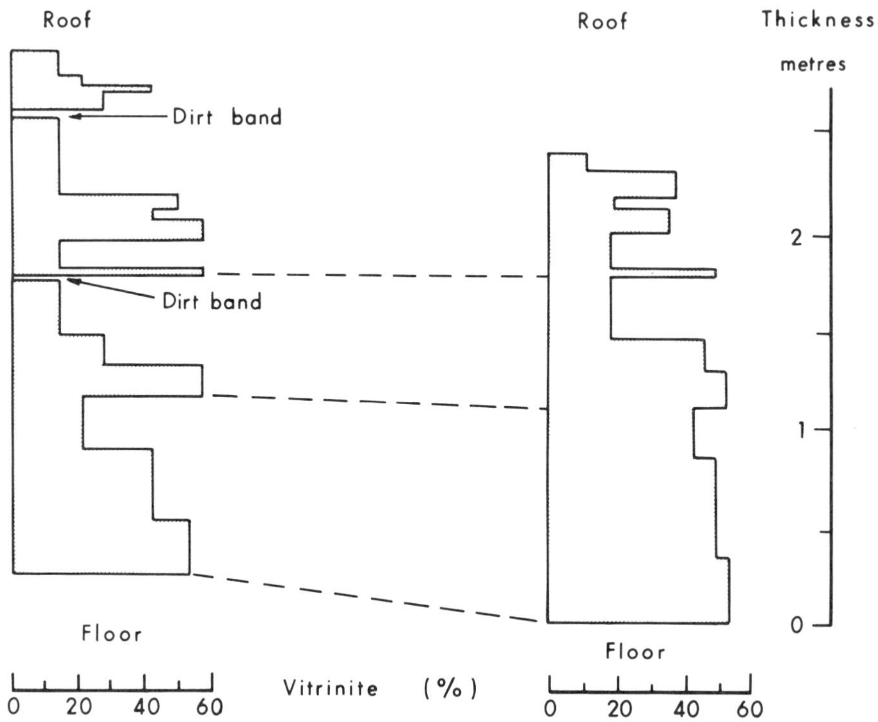

Figure 63. Vitrinite content profiles of the Bulli seam, Sydney Basin, New South Wales (from Shibaoki and Smyth, 1975).

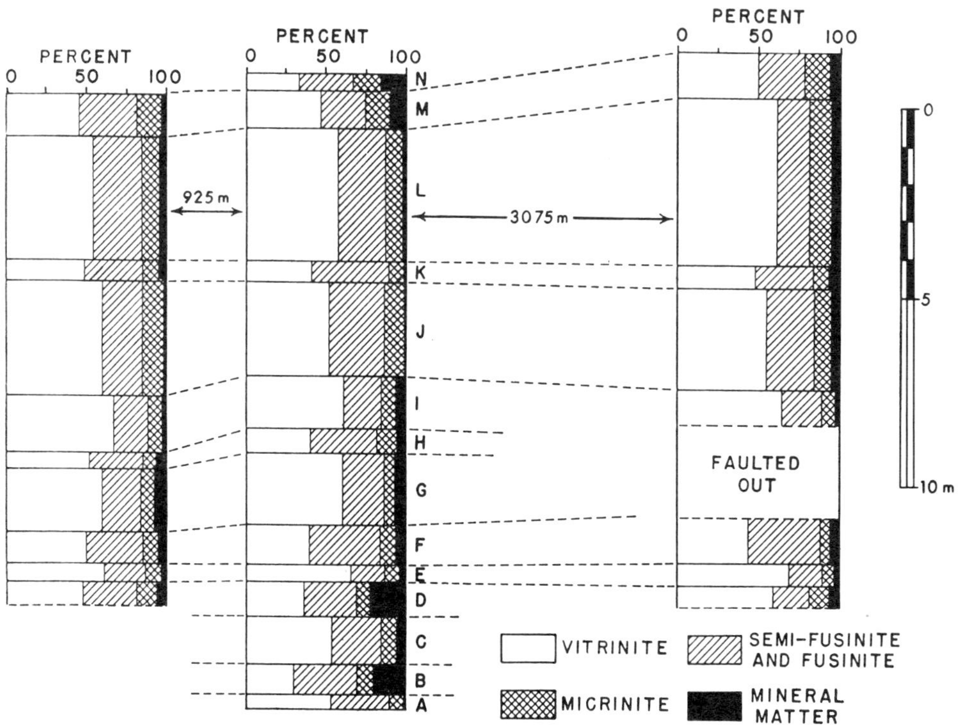

Figure 64. Maceral content profiles of the Balmer seam, Fernie Basin, southeast British Columbia (from Cameron and Babu, 1968).

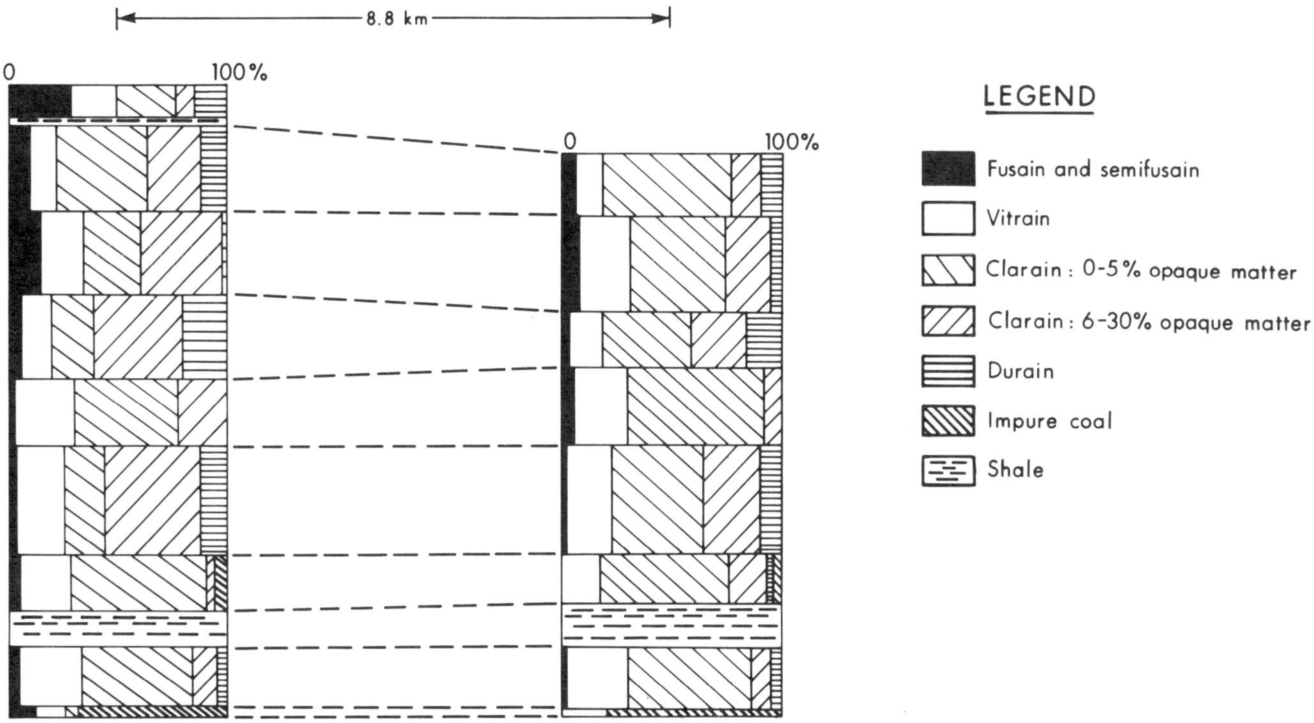

Figure 65. Composition profiles of the Tracy seam, Sydney Coalfield, Nova Scotia (from Hacquebard, 1952). Note that terms in legend actually refer to microlithotypes.

Two coal-bearing units in northeast British Columbia, the Minnes Group and the Gething Formation, are distinguished by the presence of alginite-rich sapropelic coals in formations within the former (Kalkreuth, 1982). The compositional similarity of Minnes Group and Elk Formation coals, in combination with their similar stratigraphic position (underlying the Cadomin conglomerate) and age, suggests contemporaneous similarity in depositional environments between northeast and southeast British Columbia.

9.2 **Interpretation of Geological Structure**

Petrographic data, mainly vitrinite reflectance, when plotted on geologic maps and cross-sections or borehole and measured section profiles, can aid greatly in interpretation of the structure. Basic to this application is elucidation of the relative timing of coalification and structural events.

Such an approach to structural analysis has significance for both exploration and mining, for, besides allowing interpretation of major structures, it allows prediction of changes in coal quality by projection of structurally related rank variations.

9.2.1 *Timing of Coalification and Structural Events*

The existence of rank gradients related to basin geometry and varying geothermal gradients will be covered in Section 9.4. Rank variation may also occur in connection with structural deformation, reflecting the relative timing of coalification and major structural events. In other words, coalification may have been largely completed prior to the deformation (pre-tectonic coalification), may have been mainly confined to time after the deformation (post-tectonic), or may have occurred concurrently with or, the equivalent situation, both before and after the deformation (syntectonic, pre-tectonic or post-tectonic).

In the case of folding, if coalification was largely pre-folding, then isorank lines will be parallel with the strata (Figure 66). That is, a given seam will have constant rank, assuming no rank variation related to other factors. An example from the Ruhr Basin is given on Figure 67 (Teichmüller, 1982). In the case where nearly all coalification occurred post-folding, a considerably different picture emerges. Isorank lines will tend to be horizontal in cross-section (Figure 66), and thus a given seam will increase in rank toward the core of a syncline, and a stratigraphically high seam in the core of a syncline may have as high a rank as a stratigraphically lower seam on the limbs. Such is the case in the Sydney Coalfield of Nova Scotia (Figure 68; Hacquebard and Donaldson, 1970). If coalification was both pre-folding and post-folding, an intermediate situation occurs (Figure 66), that is, the rank of a given seam will increase toward the core of a syncline but at a lower rate than in the post-folding situation. In other words, isorank lines are folded, but not to the same extent as the strata. This is the case in the Fernie Basin of southeastern British Columbia, where the rank of a particular seam increases down-dip, but isorank lines dip at a shallower angle than the bedding (compare Figures 66 and 69; Pearson and Grieve, in press).

It is possible to establish the post-folding coalification gradient of an area by calculating the reflectance increase of a seam with depth (Hacquebard and Donaldson, 1970). Conversely, the pre-folding coalification gradient can be determined by calculating the down-section rank increase at constant elevation. In any given situation, the post-folding and pre-folding coalification gradients add up to the total coalification gradient, which is defined in Section 9.4.2. Examples from two locations in the Fernie Basin in southeast British Columbia are illustrated on Figure 70. The relative proportions of pre-tectonic and post-tectonic coalification gradients are not the same at these sites. In the Fernie Basin in general, the proportion of post-tectonic coalification gradient varies, with the highest ratios coinciding with areas of highest rank (Pearson and Grieve, in press).

Timing of coalification can also be determined with respect to major faults. Normal and reverse faulting are opposite cases. If coalification was mainly confined to time prior to the faulting, there will be either a gap in rank values in the case of a normal fault (Figure 71a) or a repetition of rank values in the case of a thrust fault

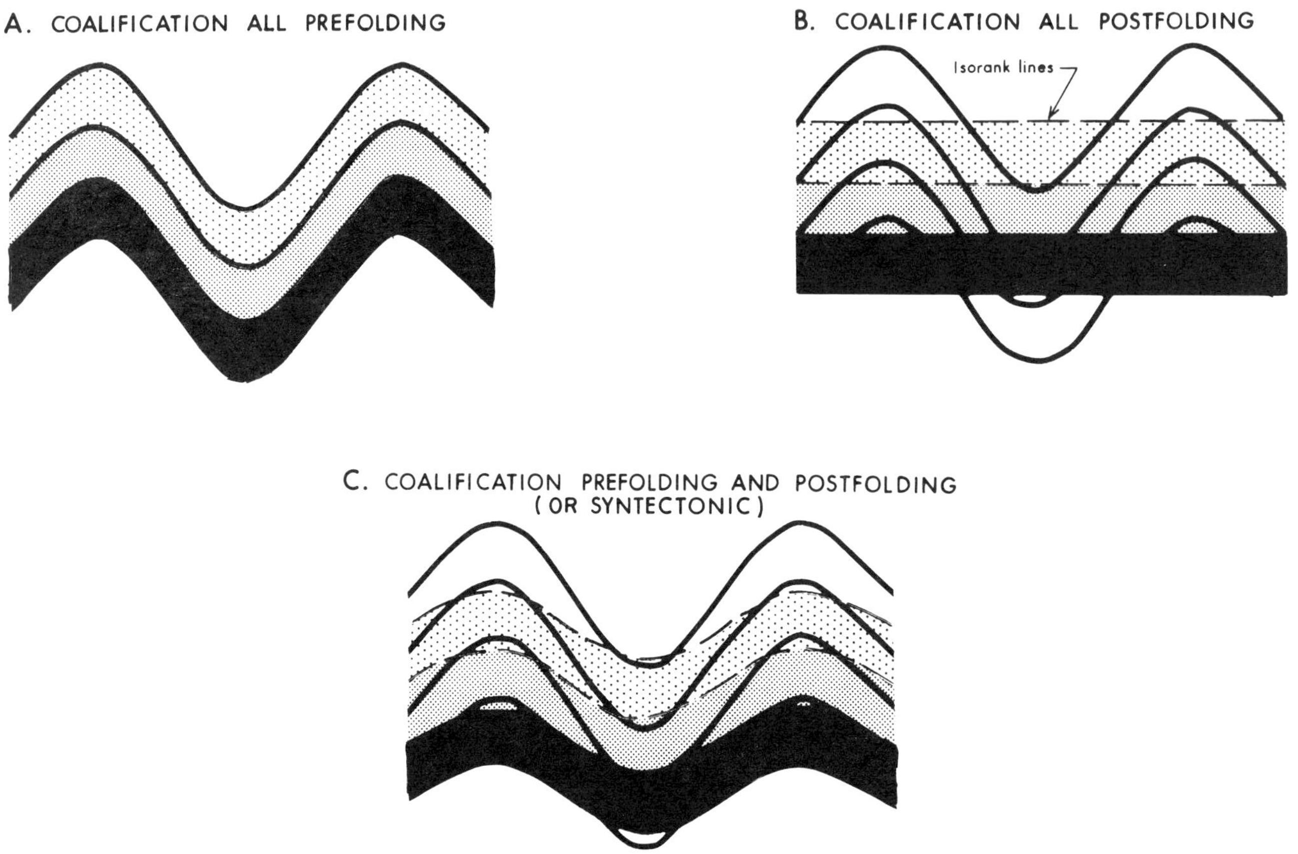

Figure 66. Idealized relationships between folded coal seams and isorank lines in cross-section in cases of: (a) pre-folding; (b) post-folding; and (c) pre- and post-folding (or syntectonic coalification) (from Pearson and Grieve, in press).

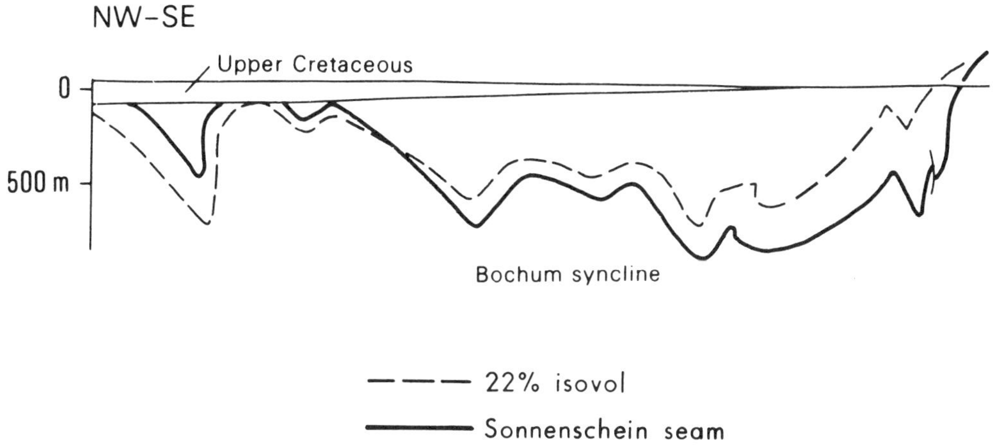

Figure 67. Cross-section through the Bochum megasyncline (Ruhr Basin) showing the pre-tectonic 22% isovol approximately parallel to the Sonnenschein seam (from Teichmüller, 1982).

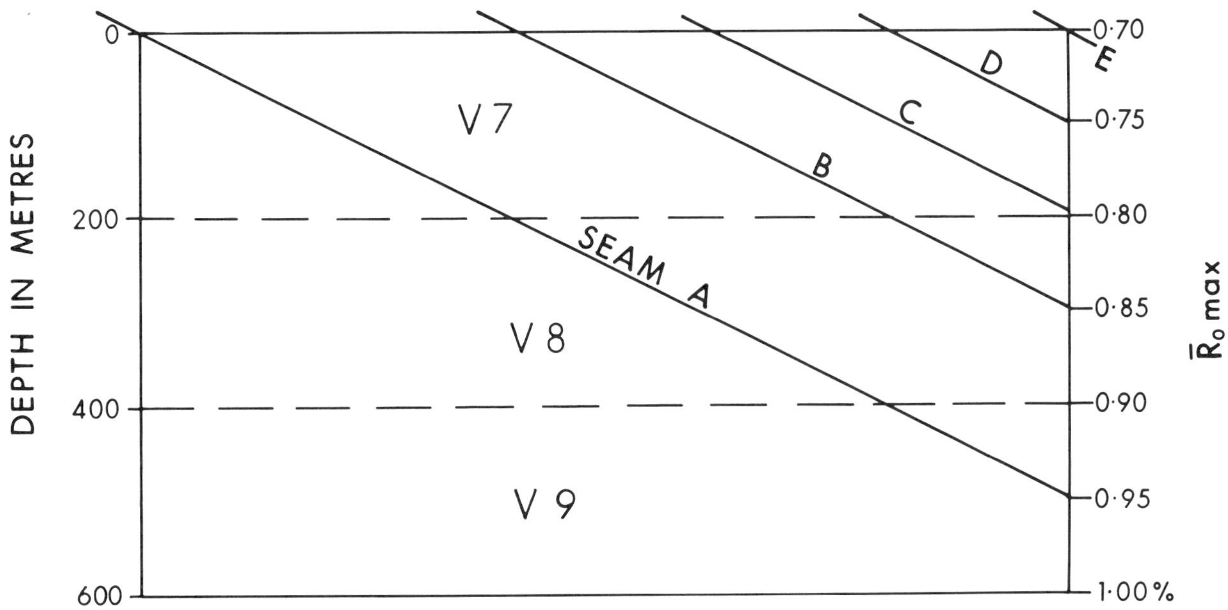

Figure 68. Cross-section of inclined coal seams in Sydney Coalfield, Nova Scotia, with horizontal (post-tectonic) isoreflectance lines (from Hacquebard and Donaldson, 1970). See Section 7.1.6 for explanation of V-types.

(Figure 71b). A given seam will have the same rank value in both hangingwall and footwall. If, on the other hand, coalification was largely restricted to post-faulting time, then there will not be a detectable break in rank values across the fault. However, in the case of a normal fault a given seam will have a higher rank in the hangingwall than it does in the footwall (Figure 71c), while the opposite applies in the case of a thrust fault (Figure 71d). In all instances, the movement must be of a certain minimum displacement, dependent on the coalification gradient, to allow detection of the rank changes across the fault.

In some cases, coalification can be demonstrated to have occurred between two distinct stages of disturbance. In the Fernie Basin, for example, a significant portion of coalification occurred after the folding (Figure 69), but all the coalification was complete prior to normal faulting (Pearson and Grieve, in press).

Timing of coalification with respect to structural deformation is not necessarily constant throughout an area. In the Ostsauerland anticline in Germany (Figure 72), for example, coalification was pre-folding in the western part of the area and syn-folding and post-folding in the eastern part (Kalkreuth, 1979).

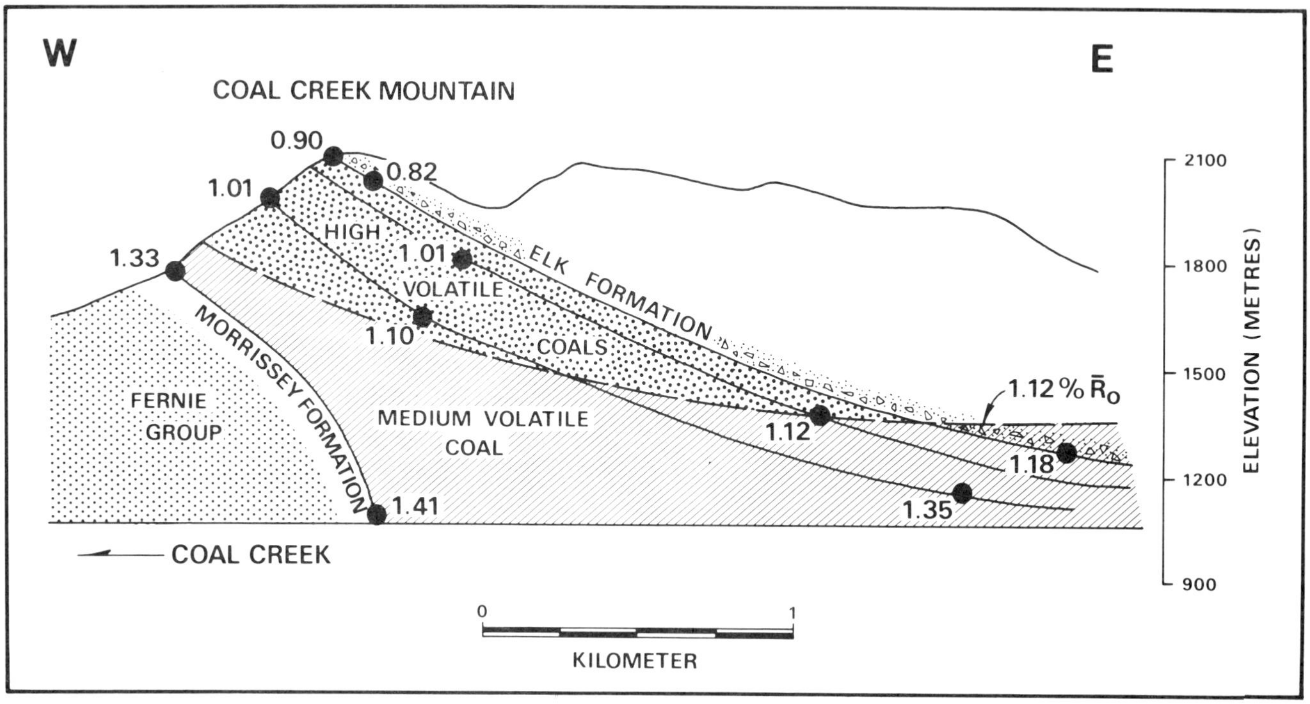

Figure 69. Cross-section of Coal Creek Mountain, Fernie Basin, southeast British Columbia, with reflectance values on four coal seams in the Mist Mountain Formation (Kootenay Group). Note how the 1.12% isoreflectance line dips in the same direction as the strata but at a shallower angle (pre- and post-folding coalification) (from Pearson and Grieve, in press).

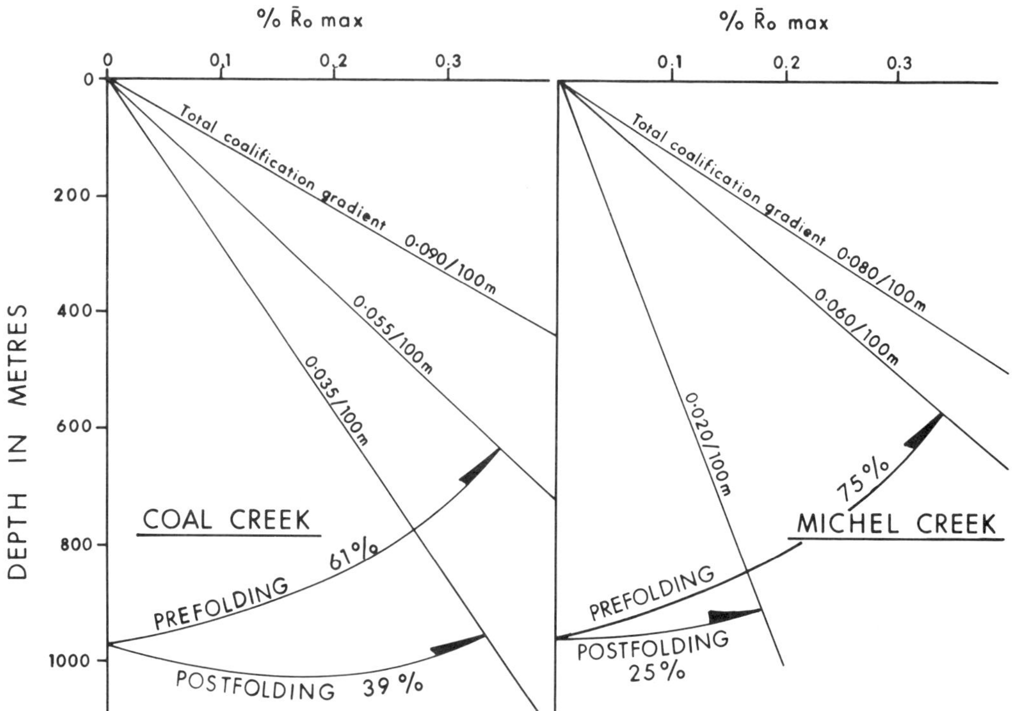

Figure 70. Comparison of pre- and post-folding coalification gradients and total coalification gradient for 10-seam (Balmer) at Michel and B-seam at Coal Creek in the Fernie Basin, southeast British Columbia (from Pearson and Grieve, in press).

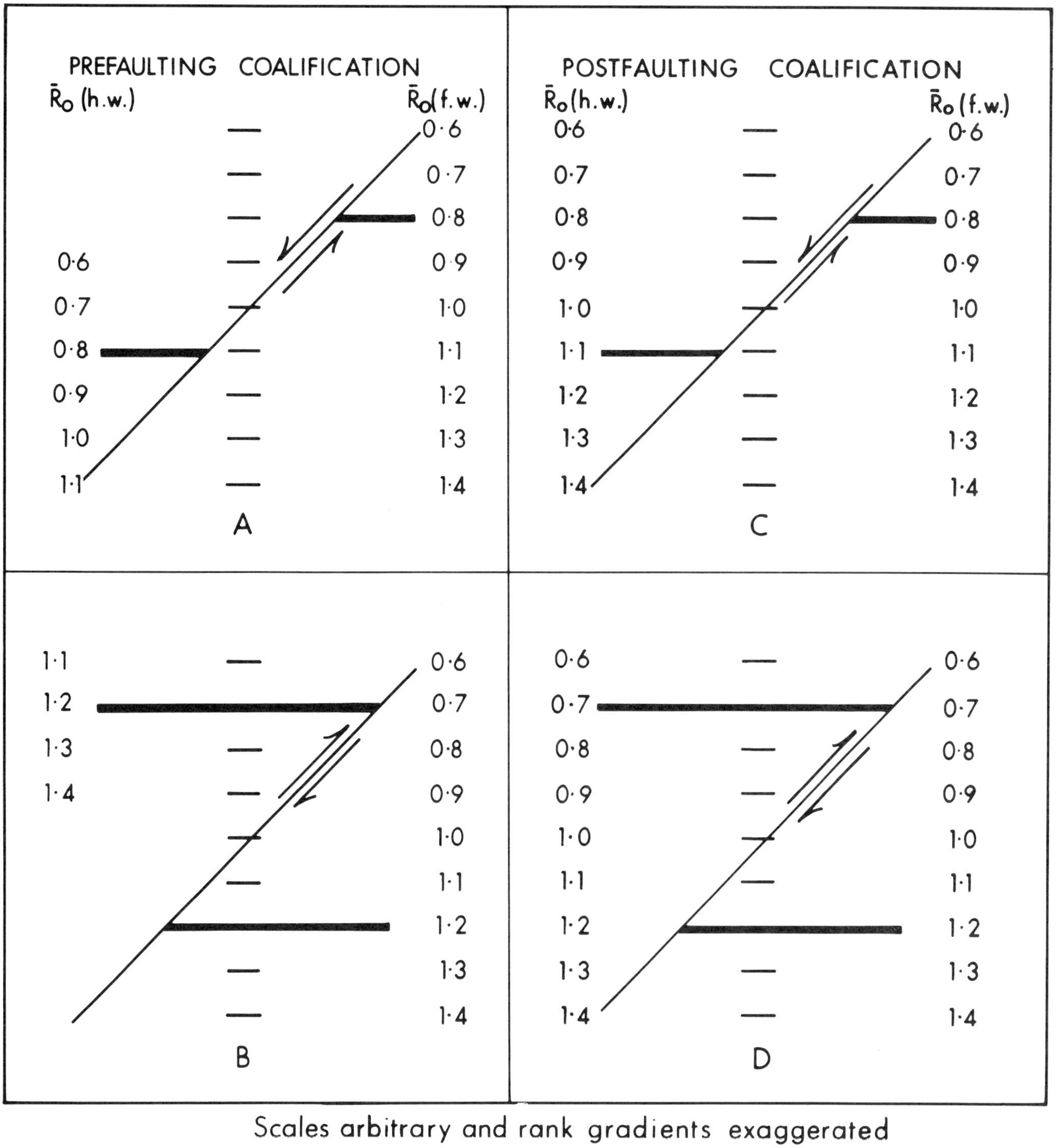

Figure 71. Idealized rank relationships in hangingwall (h.w.) and footwall (f.w.) of major faults: a - normal fault, coalification pre-faulting; b - thrust faulting, coalification pre-faulting; c - normal fault, coalification post-faulting; and, d - thrust fault, coalification post-faulting.

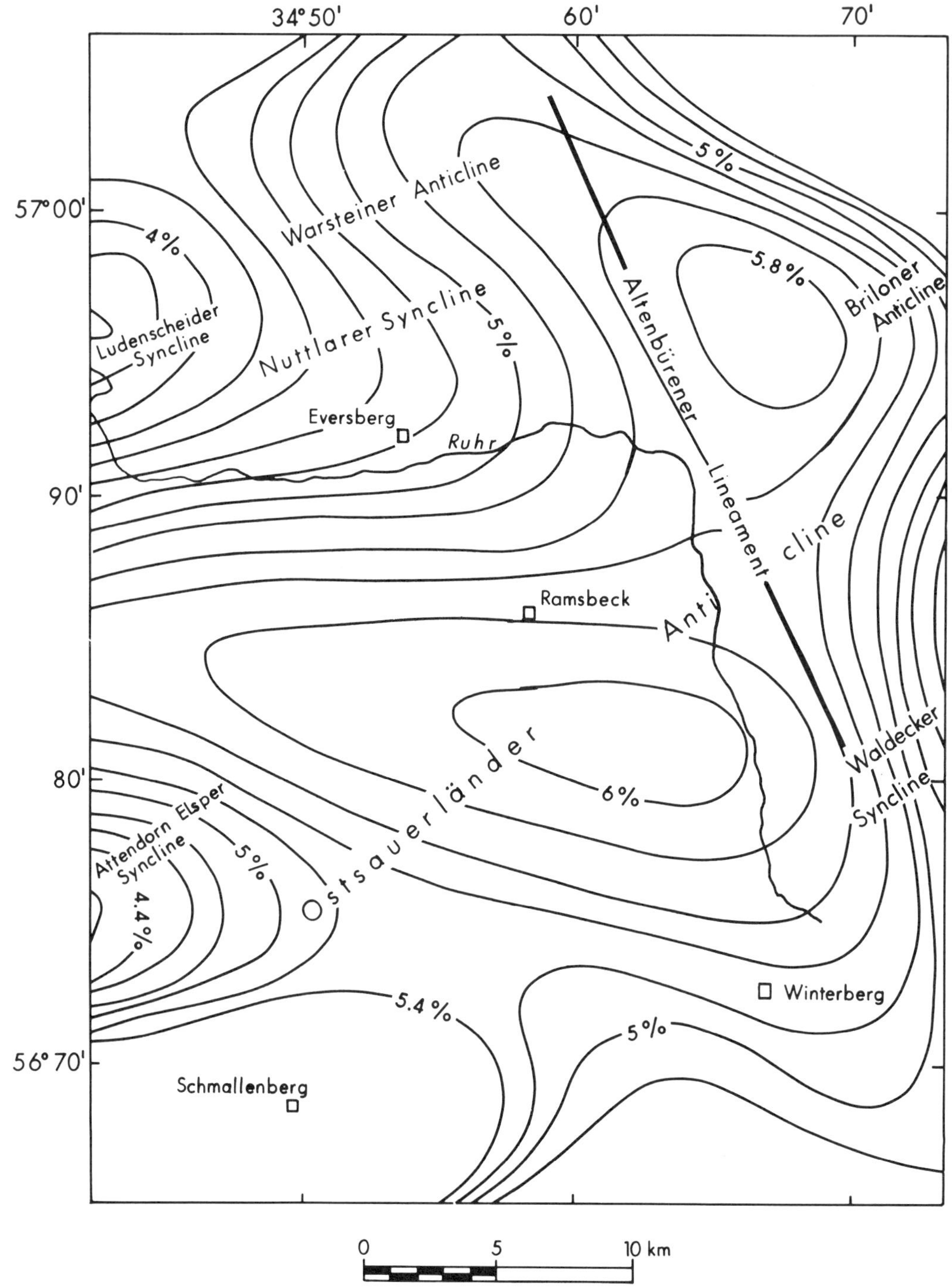

Figure 72. Isoreflectance lines in the Paleozoic in the region of the Ostsauerland anticline, Germany. A major break in the pattern of the isoreflectance lines coincides with the Altenburener lineament. West of the lineament coalification appears to have occurred pre-folding, in that the isoreflectance lines generally coincide with the fold structures. East of the lineament coalification appears to have occured syn- and post-folding, in that isoreflectance lines run transverse to the fold structures (from Kalkreuth, 1979).

9.2.2 *Structural Interpretation Based on Timing of Coalification*

Once the timing of coalification and deformation are known, it is possible to use rank distributions to interpret major structures. In folded terrane, if exposure is not adequate to allow mapping of fold axes and limbs, and if it is known that a significant portion of the coalification occurred pre-folding, then, as a general rule, relatively lower rank coals will be found in the cores of synclines and higher rank coals will be found in the cores of anticlines (Teichmüller, 1982). Relatively low rank values associated with the Attendorn-Elsper and Ludenscheider synclines in Germany (west part of Figure 72) demonstrate this principle (Kalkreuth, 1979). It should also be possible, in a similar situation, to distinguish stratigraphic 'tops,' because stratigraphically higher seams will have lower ranks. Coal ranks of an overturned panel of the Gething Formation in northeastern British Columbia, in an area of pre-folding coalification, demonstrate this principle (Figure 73; Kalkreuth and McMechan, 1984).

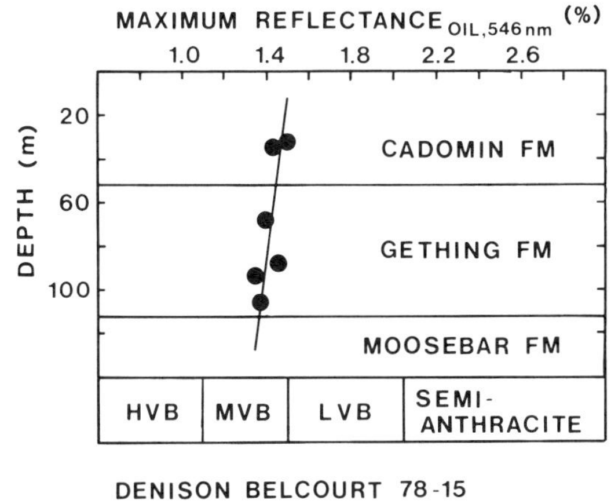

Figure 73. Decrease in reflectance with increasing depth in overturned strata in the Peace River Coalfield, British Columbia (Kalkreuth, unpublished).

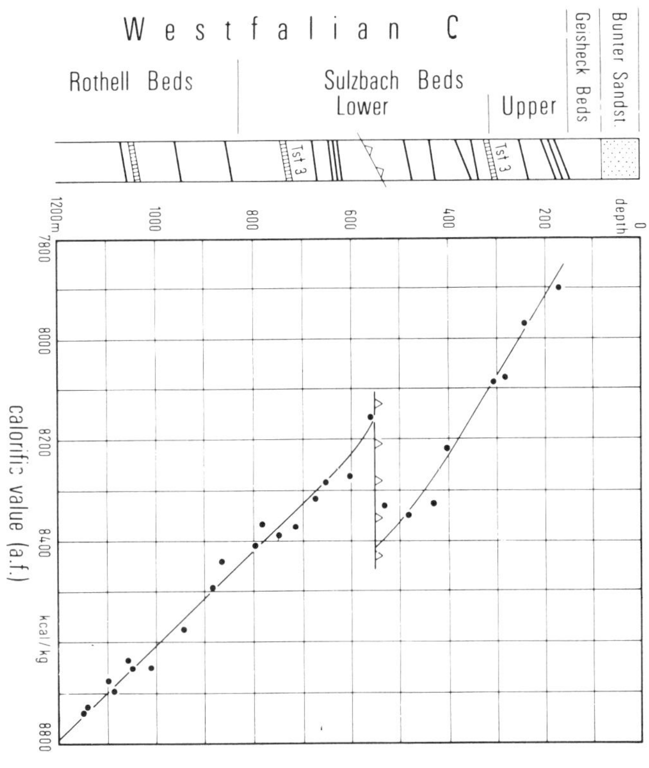

Figure 74. Thrust fault in an area of pre-faulting coalification detected in borehole by a repetition of rank values (calorific values in this case). This example from the Saar district, West Germany (from Teichmüller, 1982).

Direction and approximate amount of movement on a major post-coalification fault should be detectable by a gap or repetition of rank values. A thrust fault in the Saar district of West Germany was detected in borehole using calorific values of coals (Figure 74; Teichmüller, 1982). Stratigraphic evidence (presence of the same tonstein band above and below the fault) confirmed the interpretation. In the case of pre-coalification faulting, stratigraphic evidence will also be essential in interpretation of the structure, because it has to be demonstrated that the same seam has different rank values above and below the fault.

9.2.3 *Other Applications*

Maceral or lithotype analysis has potential of aiding solution of structural problems in cases where stratigraphic evidence is required. In areas where individual seams, portions of formations, or entire formations can be characterized by coal compositions, for example, the nature and amount of fault movement may be determined. An example on a macroscopic scale exists in parts of the Fernie Basin in southeastern British Columbia, where the humic coal-bearing Mist Mountain Formation of the Kootenay Group is thrust over the younger sapropelic coal-bearing Elk Formation (Pearson *et al.*, 1977). Field identification of the sapropelic coals (Section 9.1.3) allows accurate placement of the thrust fault trace.

Identifiable microtectonic features in coal, such as folds (Plate 14) and tension fractures, may provide a hint as to whether deformation in a certain zone was compressional or extensional (Teichmüller, 1982).

Presence of a shear zone is often evidenced by the presence of anomalously high bireflectance of vitrinite (Teichmüller, 1982). The biaxial nature of some vitrinites is thought to represent strain, and consequently the orientation of optic axes within an oriented sample is an indication of stress direction (Stone and Cook, 1979).

Tectonic forces are generally not believed to have an influence on coal rank (Hacquebard and Donaldson, 1974). A study by Bustin (1983) on organic material from thrust faults in the Rocky Mountains led to the conclusion that frictional heating during thrusting was generally not substantial or long-lived enough to enhance rank levels beyond those achieved during coalification. Only very local changes in rank were detected immediately adjacent to and within certain thrust fault zones.

Tectonic setting, on the other hand, has a potentially significant influence on coal rank. Strata beneath a major thrust, for example, may achieve an anomalously high rank because of the added thickness of overburden, and it has been suggested that the emplacement of relatively warm material from greater depth may enhance the rank of strata beneath a thrust fault (Bustin *et al.*, 1977; Bustin, 1983).

9.3 **Applied Coal Petrography — Environments of Coal Accumulation**

The petrographic composition of coal (as outlined in the chapter on origin of coal) is in the first instance a product of the depositional environment modified by the overprinting effect of post-depositional coalification. Detailed petrographic studies of the coal may thus provide information as to the coal facies (peat facies) at the time of deposition. Interpreting the coal facies from the petrographic composition requires knowledge of the coal maceral precursors and origin of the inorganic fraction. Studies such as those of Teichmüller (1962), Cohen and Spackman (1980), Casagrande *et al.* (1977), and Styan and Bustin (1983a, 1983b) of modern peats and lignites and their degradational products suggests, to some extent, that the coal macerals and microlithotypes can be associated with a coal facies (Figure 75). Such studies are most applicable to Tertiary coals where the flora is most similar to modern peats, but interpretations can be drawn from older coals as demonstrated by Teichmüller (1962), Hacquebard *et al.* (1967), Cameron (1971), and Marchioni (1980).

Environmental interpretation based on petrographic studies generally involves bench-by-bench petrographic analysis of polished coal blocks, such that the entire vertical extent of the seam is studied. It may also be desirable to examine a number of vertical sections to establish lateral coal facies changes.

Most depositional studies involve coal microlithotype and/or maceral and lithotype analysis (Teichmüller, 1982).

Plate 14. Macroscopic and microscopic (reflected light) views of folded coal from the Groundhog Coalfield, British Columbia.

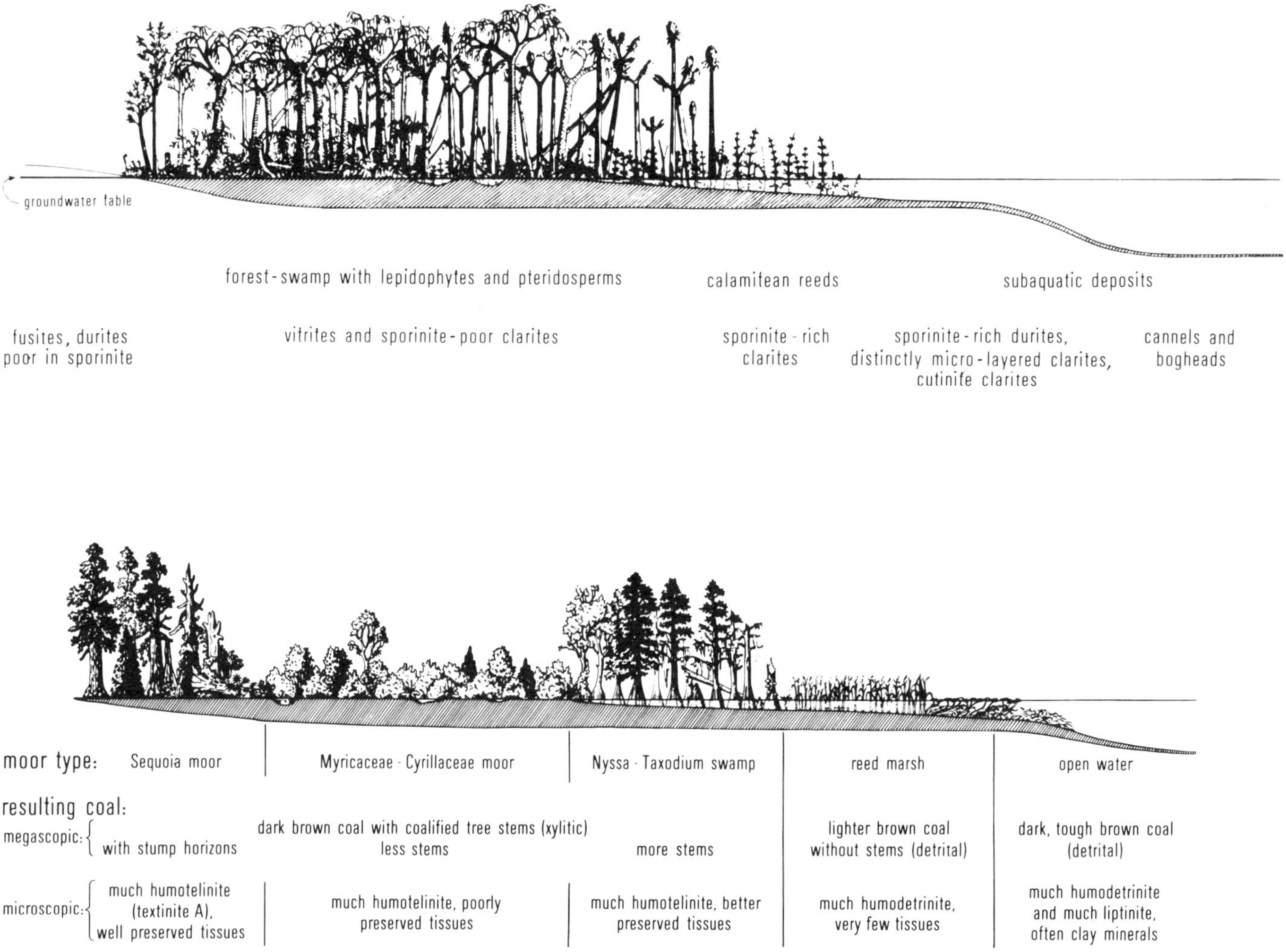

Figure 75. Coal and moor types and resulting coal microlithotypes (modified from Teichmüller, 1962 in Teichmüller, 1982).

9.3.1 *Microlithotype Environmental Interpretations*

Associations of macerals, the microlithotypes of coal, have been attributed environmental significance by a number of authors (Karmasin, 1952; Teichmüller and Thomson, 1958; Teichmüller, 1962; Hacquebard *et al.*, 1967, and others). The general interpretations of the microlithotypes are as follows:

> **Vitrite.** Vitrite is generally considered to be derived from woody tissues like stems, branches, and roots, and thus it indicates a forest swamp environment.
>
> **Clarite.** Liptinite-poor clarite originates from strongly decomposed forest litter (wood, bark, etc.) in a forest moor environment; liptinite-rich clarites originate from reeds and non-arborescent vegetation which decompose readily resulting in a concentration of the more resistant liptinite.
>
> **Durite.** Spore-rich durites are generally considered as subaquatic ooze deposits, whereas spore-poor durites likely represent deposition above the water table where extensive oxidation (and fires) occur (Teichmüller, 1982).
>
> **Trimacerite.** Vitrinite-rich trimacerites are generally considered forest swamp deposits; those rich in liptinite are generally considered subaquatic deposits; and, those rich in inertinite may represent accumulation in areas of low water table and thus increased oxidation.
>
> **Fusite-semifusite.** Thick lenses and bands of fusite and semifusite are the product of swamp fires and as such represent drying out of the peat surface and low water table levels.

In their study of the petrography of Canadian coals, Hacquebard *et al.* (1967) demonstrate many of the interpretative techniques applicable to sedimentology from coal petrographic studies. In the Harbour seam of the Sydney coal basin, Nova Scotia (Figure 76), they describe the occurrence: of forest moor facies characterized by fine to medium-banded clarite and vitrite with little inertinite, and high liptinite (well preserved); an open moor facies composed of durite with abundant inertinite, liptinite, mineral matter; and a mixed reed moor and open moor facies typified by high liptinite (with well-preserved spores), abundant inertinite and equal amounts of collinite and telinite. They consider the relative abundance of telinite and collinite to indicate the degree of degradation of the humic material — the greater the degradation, the greater the proportion of structureless vitrinite (collinite). In the No. 2 seam, Springhill Coalfield, Nova Scotia, Hacquebard *et al.* (1967) found thick sections of clarovitrite alternating with thin bands of durite. They interpret the seam as representing a pronounced forest moor facies with brief periods of desiccation during which fusinite and sclerotoids were formed (durite). In the Wellington seam of the Nanaimo Coalfield, British Columbia, Hacquebard *et al.* (1967) recognized seven petrographic intervals which they interpreted as telmatic forest moor facies and mixed open reed moor facies. The telmatic forest moor facies is composed of bright fine to medium-banded clarite and thick bands of vitrite. The low quantity of inertinite and telinite suggests that much of the seam was derived from woody material, whereas the high collinite content relative to telinite indicates much humic degradation and thus a relatively high water table. The mixed open moor and reed moor facies is a semi-bright coal consisting of clarodurite and durite with much massive fusinite, dispersed micrinite, and liptinite. The variable petrographic composition of the coal, together with low telinite and high collinite, was interpreted to indicate repeated fluctuations in the groundwater level and deposition in a mixed open and reed moor environment.

Hacquebard and Donaldson (1969) describe the use of coal facies diagrams. They recognize petrographic intervals which they consider as separate time-rock units, which are bounded at top and bottom by clastic partings or distinct dull layers and that have widespread continuity and a characteristic microscopic composition. In the petrographic intervals the total aggregate thickness of individual microlithotypes is expressed as a percentage and then plotted on a four-component facies diagram. The apices of the diagram represent microlithotypes or combinations of microlithotypes which are considered to represent depositional environments in the peat bog. In Hacquebard and Donaldson's (1969) study of the Lloyd Cove seam, Nova Scotia, they selected as end members (Figure 77): (A) spore clarite plus duroclarite with abundant spores; (B) fusito-clarite, comprising dense lenses of fusite in a matrix of claritic type coal; (C) vitro-clarite plus cuticle clarite which has abundant vitrinite and low sporinite; and (D) clarodurite plus durite plus carbargillite. Fusite, if not associated with clarite, is grouped within the group with which it occurs. The upper triangle on Figure 77 is the bright coal triangle and is used if the petrographic interval contains less

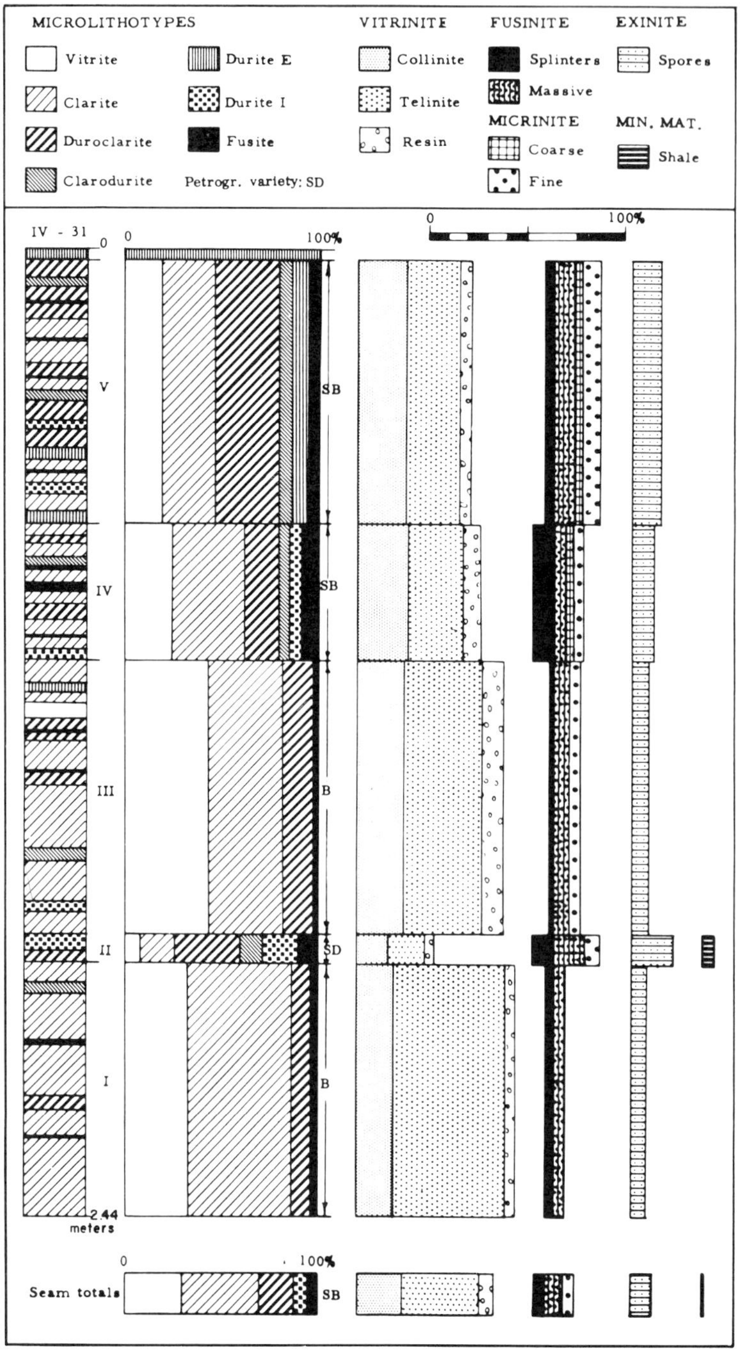

Figure 76. Variation in microlithotypes and macerals in a typical seam in the Sydney Coalfield (modified from Hacquebard *et al.*, 1967).

than 20 per cent of 'D'. If such is the case, the amount of D is added to A. If D comprises greater than 20 per cent of the interval, B is added to C and the dull coal triangle is used. Each apex of the triangle is attributed depositional significance according to the outline previously given. Within the Lloyd Cove seam (Figure 77) only one dull interval occurs which separates adjacent intervals of uninterrupted peat growth. During the first three intervals of the Lloyd Cove seam, the environment was telmatic reed moor and changed locally in interval three to forest moor. As a result of a rise in groundwater, the forest moor is followed by an open moor facies (interval 4) which is in turn followed by various forest moor facies (Figure 77; Hacquebard and Donaldson, 1969).

Smith's (1968) discussion of the Middle Coal Measures of the British Carboniferous demonstrates how coal petrology can be integrated with palynology in sedimentological studies. Smith (1968) defined that portion of a coal seam associated with a given sequence of miospores as a phase. The phases are named

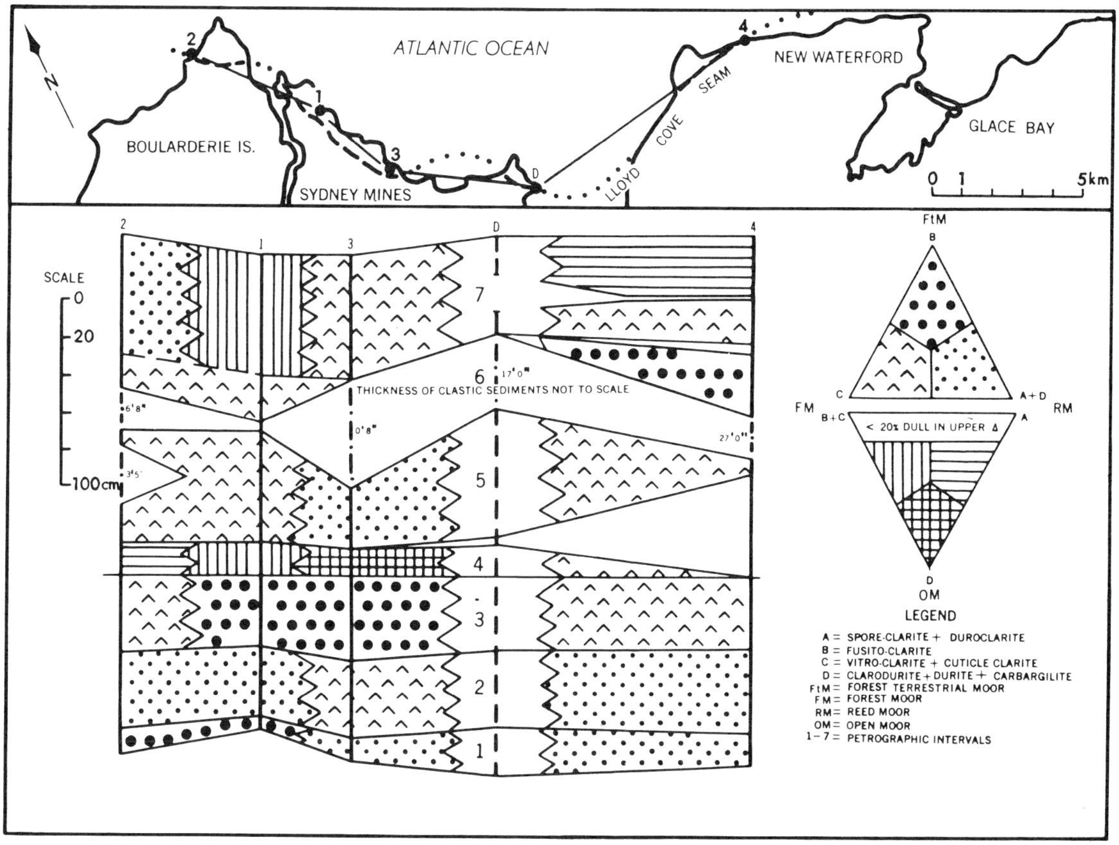

Figure 77. Coal facies in the Lloyd Cove seam, Nova Scotia (modified from Hacquebard and Donaldson, 1969).

Lycospore, Densospore, Transition, and Incursion. The Lycospore phase was found to be generally associated with coals rich in vitrite and clarite. The Densospore phase is almost always associated with durite, and the Transitional phase occurs in coals of varied composition. Durites in which spores, semi-fusinite, and mineral matter are common occur in the Incursion phase. From study of the vertical distribution, Smith (1968) found the sequence of phases generally followed the same pattern: Lycospore-Transition-Densospore- Transition-Lycospore. Species of the Incursion phase generally occur in association with the Lycospore and Transition phases and the Incursion phase was found not to be a consistent phase of the sequence. Smith (1968) interpreted the sequence to indicate that in a stable environment of peat deposition there is an accompanying succession of plant communities in response to changing groundwater table as the peat increases in thickness and raises the surface of the bog (Figure 78). The Densospore phase occupies the culmination position (low water table) prior to additional subsidence and reversing of plant and associated coal succession.

9.3.2 *Lithotype Environmental Interpretations*

Lithotypes, like microlithotypes, are a product of the depositional environment and thus may yield valuable sedimentological insights into coal deposition. The lithotypes present in a coal seam have been considered to be a product of the subsidence rate. The following interpretations of the lithotypes have been proposed by Tasch (1960).

fusain - low subsidence rates with shallow water cover and periodic emergence;
clarain and vitrain - flooding to a relatively shallow depth;
durain - deeper water conditions than clarain or vitrain;
carbonaceous shale - deeper water conditions than durain.

With increasing water depth the expected lithotypes are: fusain, vitrain, clarain, durain, carbonaceous shale, inorganic shale.

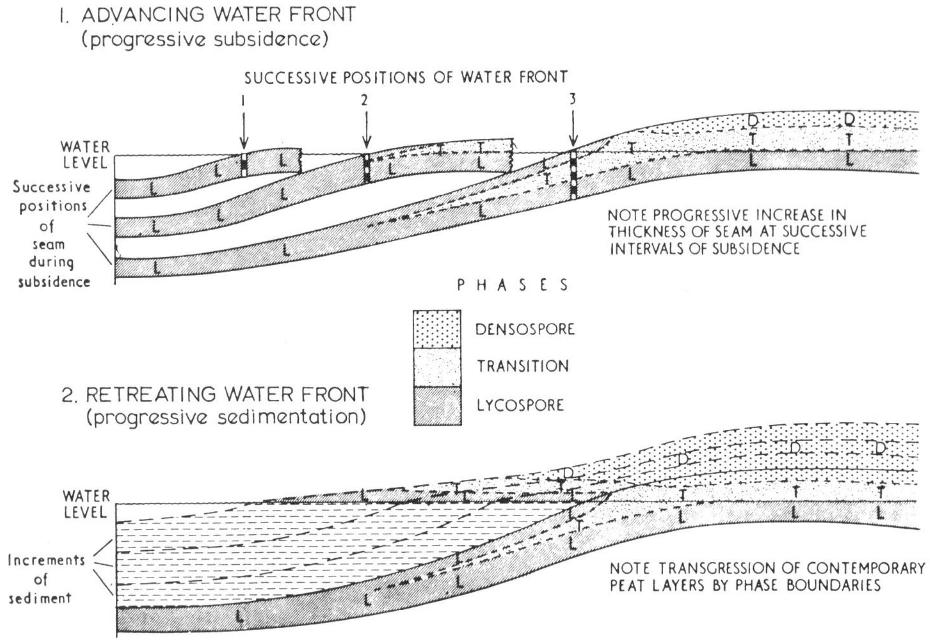

Figure 78. Development of miospore phases in relation to formation of a split in a coal seam. Modified from Smith (1968).

Tasch (1960) devised a 'seam formation curve' based on his interpretations of the lithotypes. An example of a seam formation curve is shown on Figure 79. Construction of the curve is as follows. If the lithotype at the base is overlain by a lithotype of a higher level in the diagram, a line is drawn from the lower boundary of the basal lithotype to the upper boundary of the higher lithotype. The horizontal component of the line corresponds to the thickness of the basal lithotype. If the next following lithotype is lower in the diagram, then an oblique line is drawn to the base of the following lithotype and the horizontal distance corresponds to the thickness of the second lithotype. If the succeeding lithotypes are not adjacent lithotypes on the diagram (for example, vitrain to durain), then a vertical line is drawn to the base or top of the lithotype of interest. If the next following lithotype is a higher lithotype, the vertical line is drawn to the top; if the next lithotype is lower, the vertical line is drawn to the bottom. The seam formation curve resulting from such a plot may be interpreted in terms of the degree of subsidence (or emergence) and may indicate a sequence of phases during formation of the coal. In addition, such curves may be valuable for comparing coal seams or the lateral variation of a single seam. The example shown on Figure 79 shows the following sequence from the floor to the roof of the seam: durain-clarain-vitrain-durain-vitrain-shale. Plotting of the thickness of the lithotypes is demonstrated by the relatively thick durain interval.

9.4 Application of Coal Petrology to Geothermal and Basinal Studies

The level of coalification (or organic maturation) is a result of the time-temperature history of the organic matter. Consequently it is possible to interpret the level of coalification in terms of the time-temperature history of the rock and to make inferences regarding the geothermal history of a basin. Such studies are not restricted to strata that contain coal. Finely dispersed organic material (phytoclasts) occur in almost all sedimentary rocks. The phytoclasts can be isolated from most rocks by techniques that involve dissolving the carbonate (HCL acid) and silicates (HF acid) or by specific gravity separation methods.

The level of coalification of the organic material can be determined in a number of ways of which vitrinite reflectance is the most widely used. Vitrinite reflectance has a number of advantages over other methods, including: (1) vitrinite reflectance is sensitive to even minor changes in level of coalification and vitrinite matures regularly; (2) it is a microscopic method so determinations are always made on the same

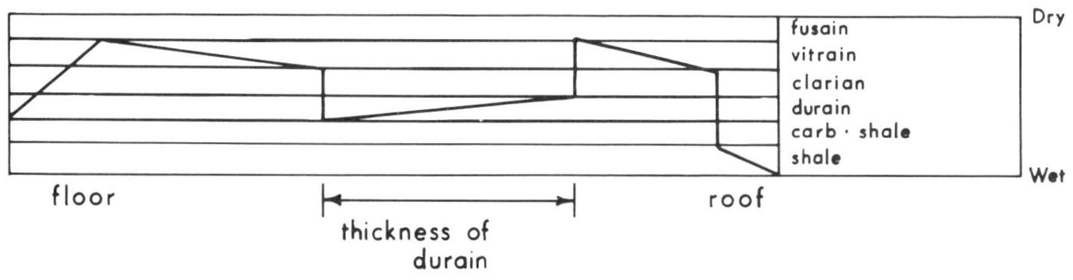

Figure 79. Seam formation curve after the methods of Tasch (1960). Refer to text for explanation.

component and thus composition of the total organics is not important; (3) vitrinite, with few exceptions, is the most common constituent; (4) it does not undergo retrograde coalification (with the exception of oxidation); (5) it does not readily react with fluids or other solids in the rocks; and (6) the method is applicable through the entire range of coalification.

Fluorescence microscopy of the liptinite macerals and colouration of the liptinite (Thermal Alteration Index, Staplin, 1969) are useful for coals of low rank, but as yet these methods are not as refined as vitrinite reflectance (refer to Section 7.3.2).

9.4.1 *Time-temperature Coalification Models*

A number of methods have been developed for correlating temperature, duration of heating, and coalification level. The most widely applied time-temperature-rank correlation is that of Karweil (1956) which allows prediction of the maximum temperature the strata have been exposed to based on level of coalification (assuming the heating duration is known). Karweil's diagram is based on first order reaction kinetics of the Arrhenius equation:

$$K = A \exp(-E/RT)$$

where E = activation energy (kcals/mole)
T = absolute temperature
A = frequency factor
K = reaction constant
R = universal gas constant

Karweil (1956) assumed an activation energy of 8.4 kcals/mole for the entire coalification range (Figure 80). More recent studies suggest this value is too low and that the activation energy may not be uniform for the entire range of coalification (Teichmüller, 1982). In their calculations, Hood and Castano (1974) consider the effective heating time, that is, the time during which the rocks were within 15 degrees Celsius of their maximum temperature. Other authors suggest the level of coalification is an integration of the entire time-temperature history of the strata. One such method is that of Lopatin (1971; see Waples, 1980, and Welte and Yukler, 1980). Lopatin's method is based on the premise that increasing the temperature 10 degrees Celsius doubles the maturation rate, whereas the dependency on time is linear. Lopatin's method considers each 10 degree Celsius increment and duration of each 10 degree Celsius interval to calculate interval maturations. The total maturation (TTI) of the sediment is then calculated as the sum of the interval maturations:

$$TTI = \sum_{nmin}^{nmax} (\Delta Tn)(r^n)$$

where nmax and nmin are the highest and lowest temperature intervals; and r = 2 if one assumes the maturation rate doubles for every 10 degree Celsius rise in temperature. The value of n is = 0 for the 100°-110°C interval. For each higher 10°C interval n = n + 1 and for each lower 10°C interval n = n − 1 (ie. for 90°-100°, n = − 1). The relationship between vitrinite reflectance and TTI is:

$$R_o max = -0.10588 \log TTI + 0.20647 (\log TTI)^2 + 0.5011$$

where TTI = temperature time index. Using this equation $R_o$max is not defined for TTI values less than 3.

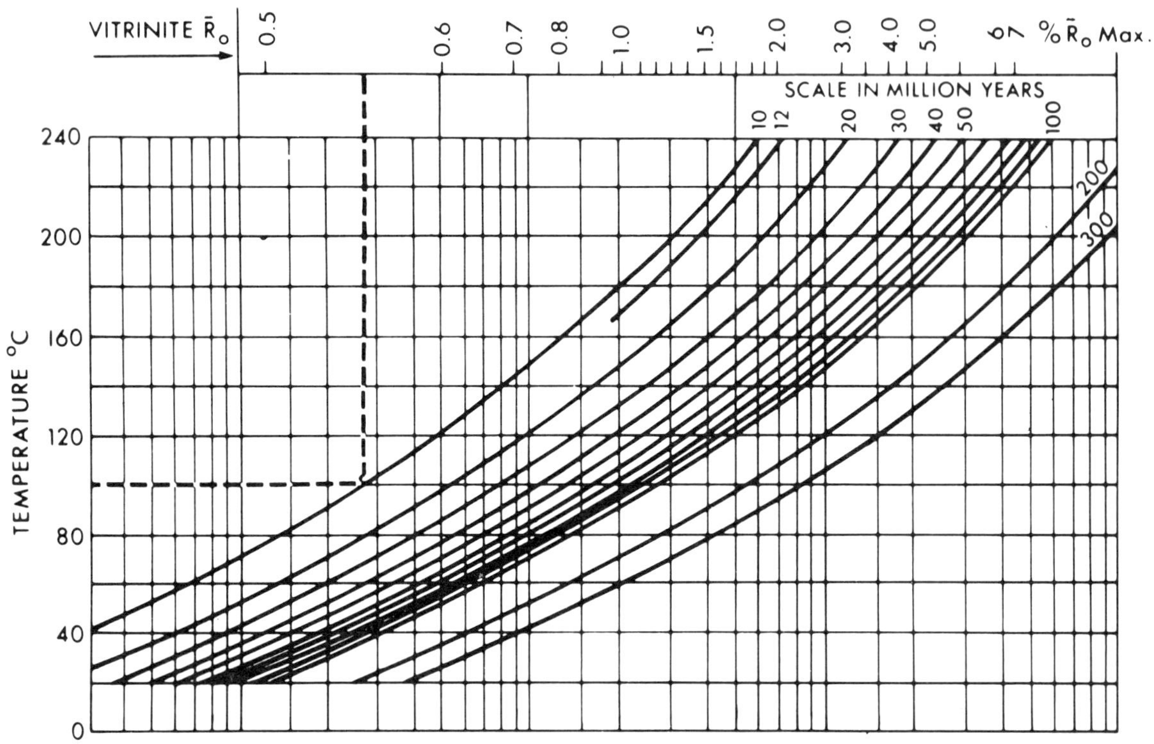

Figure 80. Relationship between coal rank (as determined by reflectance) and time of coalification and temperature (modification of Karweil's nanogram (1956) using Teichmüller's 1971 reflectance values). Dashed line indicates procedure by which a maximum temperature of 100°C was obtained at Fosheim Peninsula, Canadian Arctic, using the maximum reflectivity (0.56% R_o) and an estimated effective heating time of 10 Ma (modified from Bustin *et al.*, 1977).

For computer calculation the generalized continuous relationship is more suitable (McKenzie, 1981; Yorath and Hyndman, 1983):

$$TTI = \int_{to}^{tp} 2^{(T(t)-105)/10} dt$$

where T(t) = temperature in degrees Celsius as a function of time (t) from time of deposition (to) to the present (tp).

An example of Lopatin's method is outlined below (refer to Table 26 and Figure 81). Consider a stratigraphic horizon, A, which is 150 Ma old. From 150 Ma to 100 Ma it was buried to a depth of 3 000 metres; from 100 Ma to 80 Ma it was uplifted to 2 500 metres; and from 80 Ma to the present (time 0) it was buried to its present depth of 4 500 metres. Prior to uplift, we assume the geothermal gradient was 20 degrees Celsius/kilometre, while after (80 Ma) uplift the geothermal gradient is considered to have increased to 25 degrees Celsius/kilometre. The geological model depicting the burial history of horizon A is shown on Figure 81. In order to calculate the maturation (TTI), the interval maturations must be calculated and summed. Following Lopatin and others, we here assume that the maturation doubles for every 10 degree Celsius rise in temperature. For a given temperature interval, the temperature factor is tf = rn, where r = 2 and n = 0 for the 100 to 110 degree Celsius interval. Thus at lower intervals of 10 degrees Celsius, n = n - 1 and for higher intervals of 10 degrees Celsius, n = n + 1. The interval maturation is then calculated for each 10 degree Celsius interval by multiplying the appropriate r^n value by the time the stratigraphic horizon of interest (A) remains in that temperature interval. The appropriate r^n values, time durations, and interval maturations for horizon A are shown in Table 26. The total calculated maturation, TTI = 32' is equivalent to a vitrinite reflectance of about 0.66 per cent. In terms of source rock maturation (oil generation), horizon A would have entered the oil window (TTI = 15) at approximately 90 degrees Celsius or 16 Ma ago (150-134).

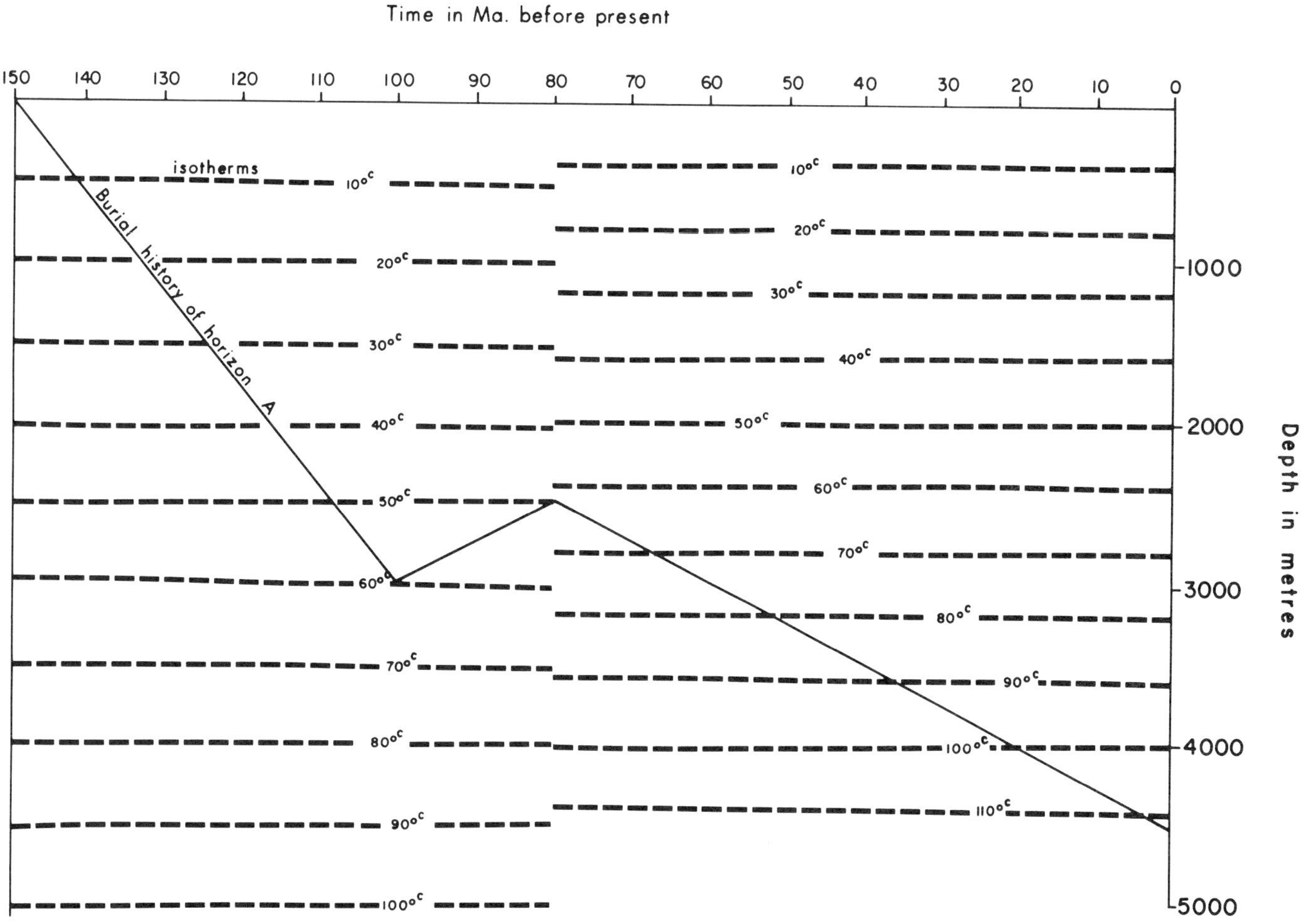

Figure 81. Application of Lopatin's (1971) method to evaluating the maturation (coalification) history of a basin. In the geological model the burial history of stratigraphic horizon A is shown with respect to time and depth of burial. The dashed lines and associated temperatures are a product of the geothermal gradient through time. The change in geothermal gradient at 80 Ma is considered instantaneous for ease of calculation. In the calculation an average surface temperature of 0°C was assumed.

An alternative method of modelling maturation has been recently outlined by Middleton (1982) based on previously established relationships (Dow, 1977). Middleton (1982) relates reflectance to the time-temperature history of the rocks:

$$R_o^a = R_o^a init + bt \exp(cT)$$

where Ro = vitrinite reflectance
Rinit = initial reflectance (0.20 or 0.15)
a = 5.5
b = 2.8×10^{-6}
c = 0.065
t = time (Ma)
T = temperature (°C)

This relationship results in an increase in maturation rate of 1.92 for an increase of 10 degrees Celsius as compared to 2 (doubling) of the Waples (1980) model of the Lopatin method.

TABLE 26. CALCULATION OF THE INTERVAL MATURATION AND TOTAL MATURATION OF HORIZON A. NOTE THE INTERVAL 50-60°C APPEARS TWICE BECAUSE OF THE UPLIFT AT 100 Ma.

Temp. Interval °C	r^n	ΔT(Ma)	Interval TTI	Time Cumulative	Total TTI
0-10	2^{-10}	8.33	.008	8.33	.008
10-20	2^{-9}	8.33	.016	16.66	.024
20-30	2^{-8}	8.33	.032	24.99	.056
30-40	2^{-7}	8.33	.065	33.33	.121
40-50	2^{-6}	8.33	.130	41.65	.251
50-60	2^{-5}	8.33	.260	49.98	.511
50-60	2^{-5}	20.00	.620	70.00	1.130
60-70	2^{-4}	16.00	1.000	86.00	2.131
70-80	2^{-3}	16.00	2.000	102.00	4.131
80-90	2^{-2}	16.00	4.000	118.00	8.131
90-100	2^{-1}	16.00	8.000	134.00	16.131
100-110	1	16.00	16.000	150.00	32.131

Integrating the above equation over the time-temperature history of the basin gives:

$$R_o^a = R_o^a init + b \int_{to}^{tp} \exp[cT(t)]dt$$

This equation can be solved numerically in a fashion similar to TTI as outlined above. From this equation, Middleton (1982) derives further equations for calculating maturation based on different subsidence and uplift history.

The methods of Lopatin, Waples (1980), and Middleton (1982) give slightly different results. The model which is most applicable should be considered the one which yields the most acceptable results for the particular study.

9.4.2 *Geothermal Gradient Determinations*

The paleogeothermal gradient in an area can be reconstructed by examination of the coalification gradient. The higher the coalification gradient, the higher the geothermal gradient. Castano and Sparks (1974) found in the Tejon area of California a modern geothermal gradient of 2.3 degrees Celsius/100 metres which correlated with a coalification gradient of 0.11 $\bar{R}_o$max per cent/kilometre. Bustin *et al.* (1977) found a similar coalification gradient in the eastern Canadian Arctic and assumed a similar geothermal gradient to calculate the thickness of sediment removed by erosion during subsequent uplift. In the Cerro Prieto geothermal field in Baja, California, Barker (1979) described geothermal gradients as high as 16 degrees Celsius/100 metres, and found the coalification gradient to be 0.27 per cent $\bar{R}_o$max/100 metres. Barker attributed the higher coalification gradient to high heat flow at depth related to magmatic activity along a rift zone. In the Groundhog Coalfield of central British Columbia, an anomalously high coalification gradient exists (per cent 0.12 $\bar{R}_o$max/100 metres) which indicates the presence of a very high paleogeothermal gradient, suggesting possible high heat flow along a major fault or magmatic intrusion (Bustin, 1984).

Facer *et al.* (1980), from studies of the variation in coalification gradients in the southern Sydney Basin, New South Wales, were able to demonstrate a relationship between the coalification gradient and thermal properties of the sedimentary pile, basement rocks, and depth of burial and depth of basement. On Figure 82 is shown an isopach map and rank gradients from a number of boreholes. In this basin the isoreflectance surfaces dip in a north-northeasterly direction at an angle slightly less than the dip of the strata. In the northern part of the Sydney Basin the coalification gradient is lower than that of the southern part which Facer *et al.* (1980) attribute to higher heat flow through Permian and Triassic rocks.

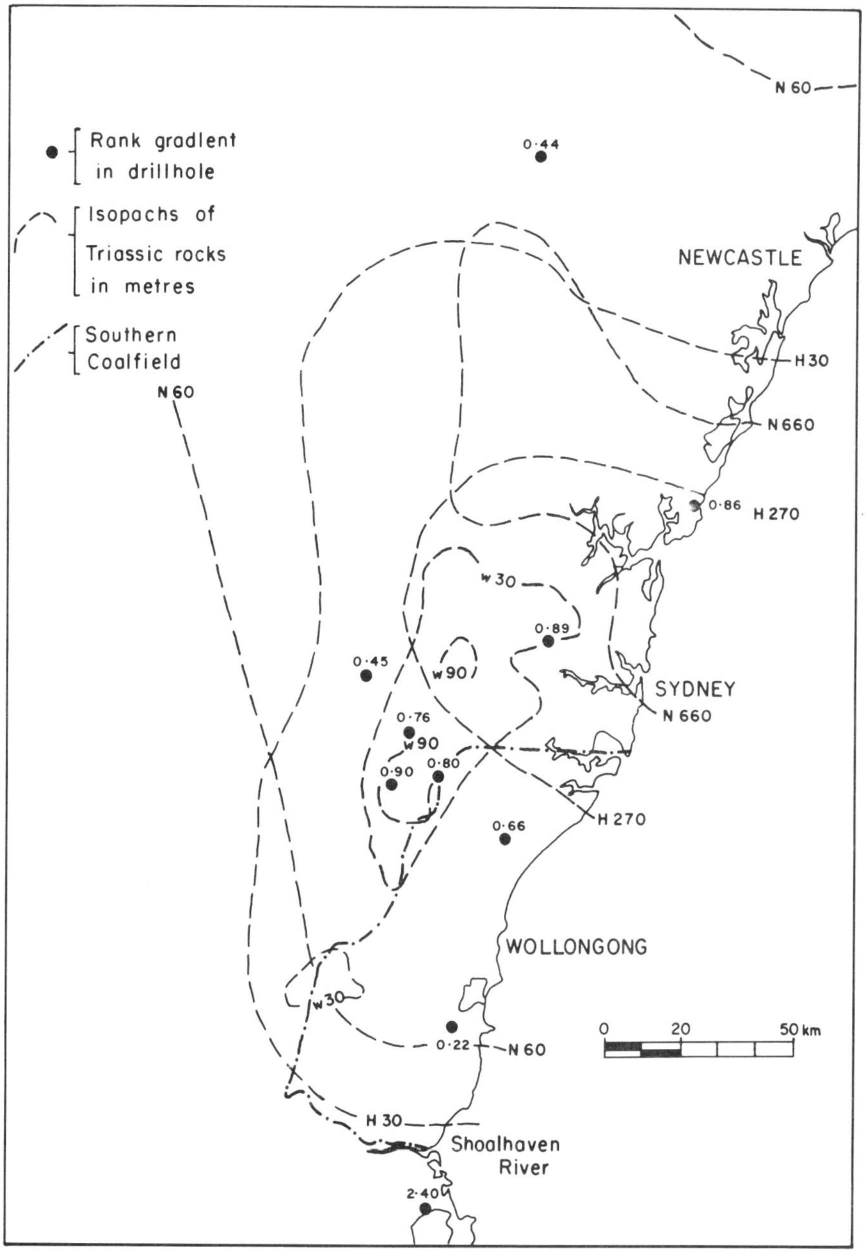

Figure 82. Rank gradient ($R_o$max/km) and isopach map showing original thickness (in metres) of Narabeen Group (N), Hawkesbury sandstone (H), and Wianomatt Group (W), all of which overlie the coal measures, southern Sydney Basin. Modified from Facer *et al.* (1980).

In some areas the geothermal gradient and thus the coalification gradient may not be linear with depth. Inasmuch as the geothermal gradient is a function of the heat flow and thermal conductivity of the rocks, variation in these two variables will influence the coalification gradient. Adjacent to salt diapirs, which have very high thermal conductivities, a high coalification gradient is to be expected. Intrusive bodies at depth may similarly result in high geothermal gradients and coalification gradients. Changes in geothermal gradients with time (Robert, 1971) and variation in subsidence-uplift history of the strata, will also influence the coalification gradient. The ability to interpret such data will depend largely on what additional geological information is available, particularly knowledge of the geological history.

From studies in the Upper Rhine Graben, Buntebarth (1978) devised an empirical method to determine the geothermal gradient when the rate of burial and coalification gradient is known. The relationship between coalification gradient and geothermal history is:

$$R_o^2 \propto \text{grad } T \int_{to}^{tp} z(t)dt$$

where R_o^2 = reflectance of vitrinite
grad T = temperature gradient
z(t) = subsidence history
to-tp = time of deposition to present

On Figure 83 are plotted the burial history of drill holes of known thermal gradient from the Rhine Graben. Considering the plane Rm = 1 per cent, the integral of the burial history and the time at Rm = 1 per cent is equal to I. The value of I, together with the known temperature gradients of the boreholes, yields the emperical relationship:

$$\text{grad } T = 98.7 - 14.6(\ln(I))$$

This relationship is valid only for constant geothermal conditions and the value of I must be determined for each borehole. Once the thermal gradient is determined from the above relationship and the thermal conductivity measured (or assumed), the heat flow may be obtained from:

$$Q = K \text{ grad } T$$

where Q = heat flow
K = thermal conductivity
grad T = thermal gradient

For tectonic models of constant geothermal gradient, Middleton (1982) showed that a plot of log (R_o) versus depth is linear and the slope (m) is:

$$M = (cG/a) \log_e (10) = G/194.83$$

where c = 0.065
a = 5.5
G = geothermal gradient.
(See discussion on Middleton's method in Section 9.4.1).

Thus, the paleogeothermal gradient can be determined if a near constant geothermal gradient during the basin history under investigation is assumed.

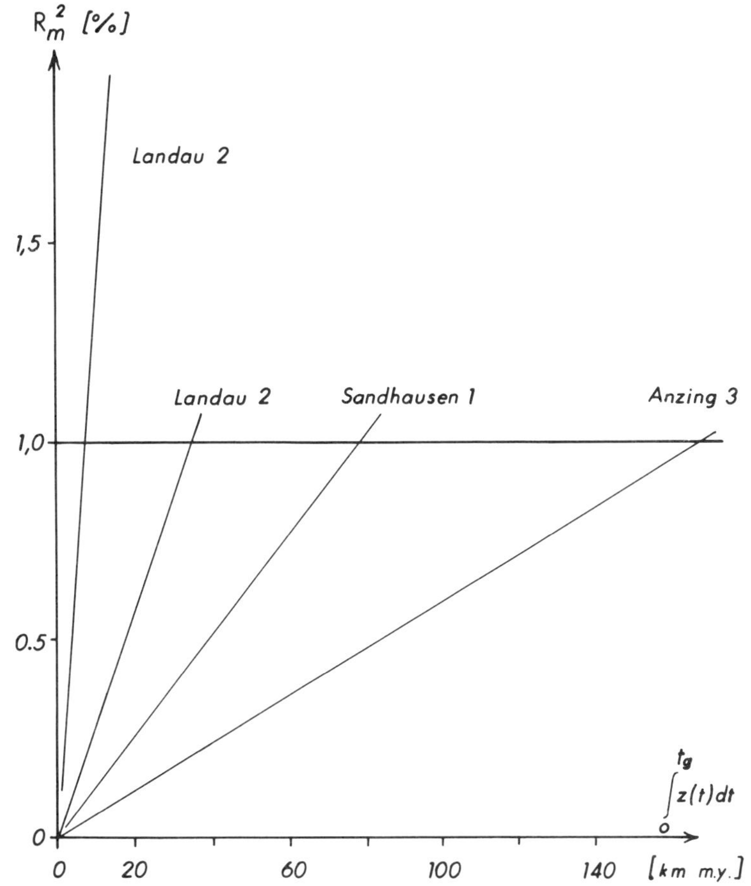

Figure 83. The integral of the burial history versus the square of the median vitrinite reflectance from borehole of known temperature in the Rhine Graben (modified from Buntebarth, 1978).

9.4.3 *Paleoisotherms and Isoreflectance Lines*

Calculation of paleoisotherms on a particular stratigraphic level, or conversely the stratigraphic level of a particular paleoisotherm, can provide much useful information regarding geothermal gradients, maximum depth of burial, diagenetic levels, and paleobasin geometry. If no magmatic or structural complexities exist, the paleotemperature or present level of coalification on a particular horizon are a product of the geothermal gradient and subsidence history. If a constant geothermal gradient is assumed, the levels of highest coalification indicate the greatest depth of burial. The difference between the highest and lowest obtained values is a measure of the difference in depth of burial — the amount of which can be extrapolated from the coalification or geothermal gradient if known. In their study in the eastern Canadian Arctic, Bustin *et al.* (1977) used such a method to determine the basin geometry during deposition of the Eureka Sound Formation. In the eastern Arctic, Tertiary strata of the Eureka Sound Formation presently occur mainly as erosional outliers. By plotting the levels of organic maturation of the base of the formation, it was possible to establish that the Remus Basin was an elongate deep basin with a much steeper western than eastern margin.

Hacquebard and Donaldson (1970) carried out a regional study of the variation in rank of upper Paleozoic rocks in the Atlantic provinces of Canada. They plotted isoreflectance maps of the entire area (Figure 84) and were able to demonstrate that: (1) a distinct pattern exists related to the tectonic development of the Appalachian region with the more highly metamorphosed sediments occuring in the south; (2) the lower values (V5, V6, V7) were found to correspond to the cratonic cover of Upper Pennsylvanian strata; and (3) there is little or no relationship between the degree of coalification and age of the strata, indicating the coalification was mainly post-deformational.

Calculation of paleoisotherms from isoreflectance data (or other measures of coalification) can be done using the method of Karweil (1956) or others. Determining the temperature from rank and heating duration is generally considered to be accurate within about 20 degrees Celsius (Bostick, 1979).

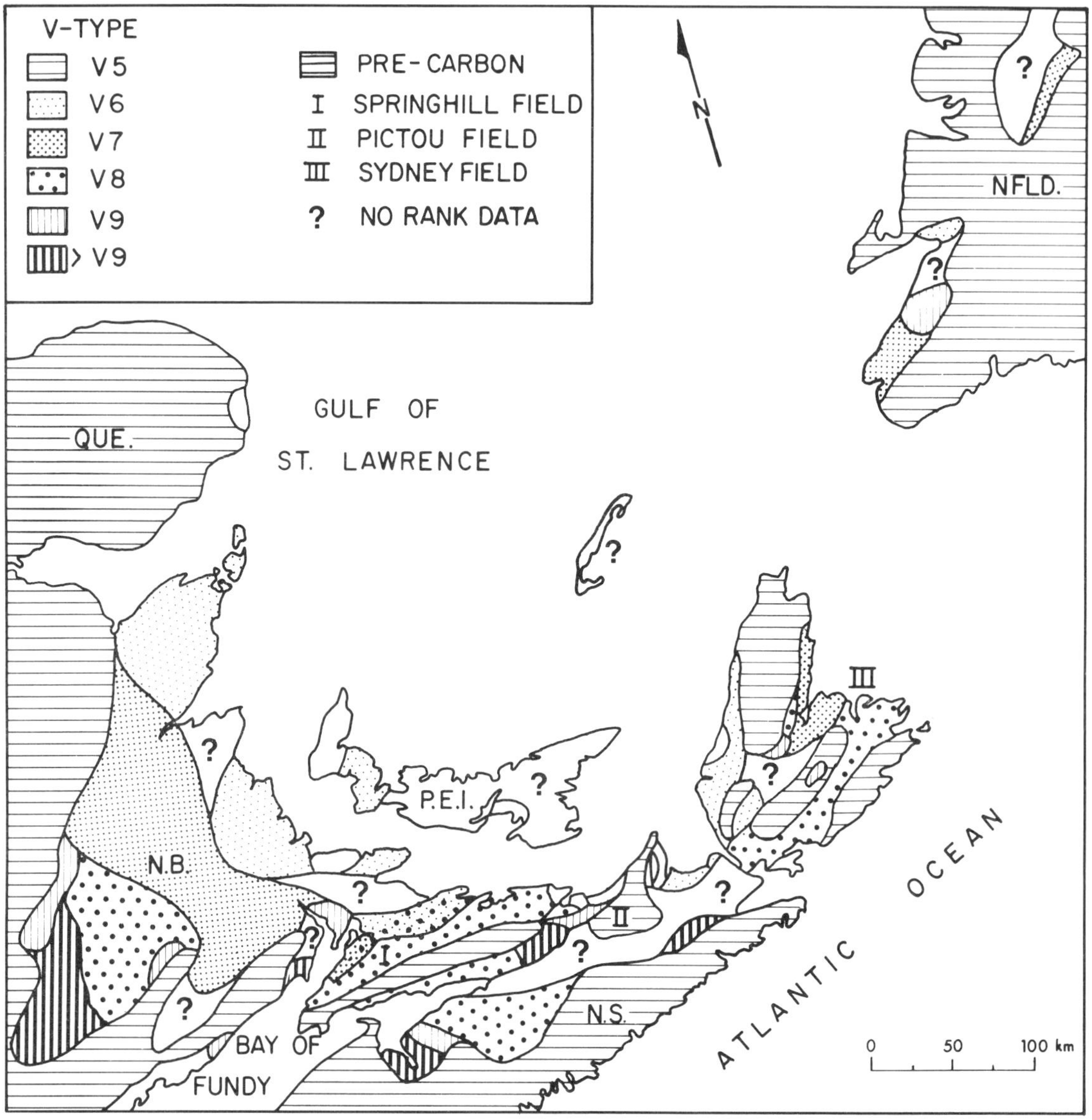

Figure 84. Isoreflectance map (in V types) showing regional variations in coalification of Upper Paleozoic rocks of the Atlantic Provinces (modified from Hacquebard and Donaldson, 1970).

9.4.4 *Magmatic Studies*

Adjacent to magmatic bodies the coalification level may be anomalously high. The lateral extent of the anomaly will depend on the thickness of the magmatic body and rate of cooling. Similarly, circulation of hot hydrothermal brines will result in anomalous coalification levels. As a rule of thumb, the thermal effects of an igneous body are generally negligible at a distance equivalent to several times the width of the body. Examination of the thickness of the thermal halo and local geothermal gradient around an igneous body may be used to establish the cooling history and temperature of intrusion using the established relationships between heating duration-temperature and coalification level.

Heating of low rank coals for short durations at high temperatures results in the formation of high rank coals. High rank coals formed in such a fashion may be recognized by careful petrographic examination. Creaney (1979) found that subbituminous coals near Kamloops, British Columbia, which were thermally metamorphosed by an igneous intrusion to a rank of semi-anthracite, still retained cell structure typical of subbituminous coals. Coals from a geothermal area in Washington State, which are otherwise of subbituminous rank, are locally thermally metamorphosed to anthracites. The anthracites also still retain visible cell structure typical of low rank coals. In the Groundhog Coalfield, where the coal seams are commonly crosscut by hydrothermal quartz veins, there is no evident thermal affect associated with the veins nor is there relic structure. Because the coals were not thermally influenced by the hydrothermal fluids, it is possible to establish the maximum temperature of the fluids by considering what minimum temperature would have resulted in thermal metamorphism of the coal. The highest temperature of the fluids must be less than the minimum temperature required for additional maturation of the coal.

In the Alston Block, northeast England, Creaney (1980) studied in detail the distribution of coal types and vitrinite reflectance adjacent to the Whin sill. Creaney (1980) found that the present day distribution of thermally metamorphosed vitrinites reflected the rank pattern prior to intrusion of the Whin sill. Creaney suggests that the presence of granular mosaic vitrinites indicate that the coals were coking coals prior to intrusion of the Whin sill, whereas vitrinites that lack the mosaic structure were above the coking coal rank prior to intrusion. As a result of intrusion and high heat flow from the Whin sill, the adjacent strata have been further metamorphosed and locally graphite has formed adjacent to the sill contact (Figure 85).

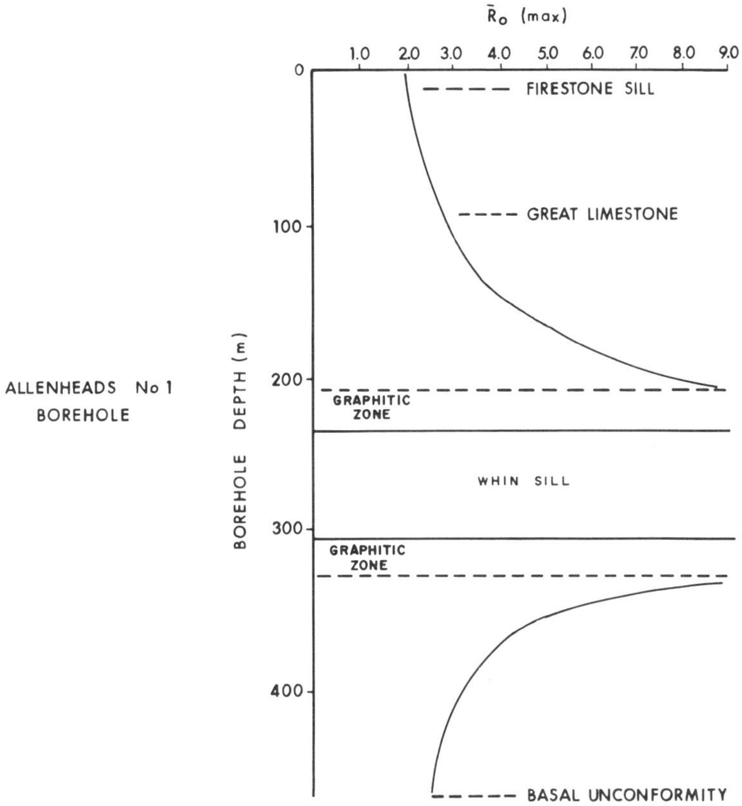

Figure 85. Vertical variation in vitrinite reflectance adjacent the Whin sill. The low reflectance values adjacent the sill are graphite (modified from Creaney, 1980).

10 TECHNOLOGICAL APPLICATIONS

10.1 **Coal Preparation**

10.1.1 *Principals of Coal Preparation*

The purpose of coal preparation is twofold; firstly, to upgrade (clean) the coal, in order to improve its quality by removing undesirable mineral matter, sulphur, and deleterious elements, and, secondly, to optimally size the coal for the intended purpose.

The natural size fractions of run-of-the-mine coal may vary considerably depending on mining methods, presence of cleat, extent of shearing, etc., as well as the composition of the coal. Everything else being equal, dull durain-rich coal will comprise a coarser natural size fraction of coal than clarain or vitrain. Thus the coarser size fraction will be somewhat impoverished in vitrinite relative to the finer fraction. In the coalfields of southeastern British Columbia where the coal seams are commonly tectonically sheared, the dull durain-rich coal is more competent and is thus relatively unsheared compared with adjacent clarain and vitrain which are very finely comminuted. The fine fraction of the coal would therefore be enriched in vitrinite relative to the coarser coal.

During coal preparation the coal is normally crushed to about minus 10 millimetres. The vitrain, which is brittle (if not mineralized), concentrates in the 10 to 0.5 millimetre fraction, whereas fusain, which is friable if it occurs in lenses or layers, is concentrated in the less than 0.5 millimetre fraction (Mackowsky and Hoffmann, 1960). If cleaning of the coal is required to reduce the sulphur or ash content, the 10 to 0.5 millimetre fraction is normally subjected to specific gravity separation techniques, while the less than 0.5 millimetre fraction is cleaned by froth flotation (for a full discussion of coal preparation techniques see Elliot, 1981). The specific gravity utilized for cleaning the coarse fraction is selected to provide the optimum recovery of coal with an acceptable ash content for the intended market. The ease with which coal can be cleaned is commonly expressed by the amount of 'near density material' — the lesser amount of material with a density close to the density required to clean the coal, the easier it will be to clean.

Each coal maceral varies in density with rank, and the coal macerals have different densities at any particular rank. The density of vitrinite (measured in helium) varies from 1.3 to 1.8 with the minimum density occurring in coal of 87 per cent carbon (Stach, 1982). The density of liptinite is somewhat less, being commonly in the range of 1.0 to 1.20, depending on the maceral. Inertinite varies in density from about 1.5 to 2.0, and semifusinite is in the range of 1.35 to 1.45 (Stach, 1982). Due to the variation in density, specific gravity separation should fractionate the coal macerals, with liptinite and vitrinite being concentrated in the light fractions and semifusinite and inertinite in the heavy fractions. If the coal is crushed to an extremely fine size, good separation of the coal macerals can be obtained.

During practical coal preparation, however, such fractionation of the macerals may not occur. From float-sink analysis and petrographic study of samples of two seams from the Fording River deposit (Bustin, 1982), for example, it was found that, on an ash-free basis, the vitrinite was concentrated in the less than 1.3 s.g. fraction, passed through a minimum in the 1.4 to 1.6 range, and progressively increased in abundance in the greater than 1.6 fraction (Figure 86). Such changes in the relative abundance of vitrinite are considered a product of both the relative density of the macerals and their mode of occurrence: in the less than 1.3 s.g. fraction, the vitrinite occurs as homogeneous fragments with only minor inclusions of fusinite and semi-fusinite; in the 1.3 through 1.6 s.g. fractions, larger fragments of fusinite, semifusinite, and vitrinite occur together; in the greater than 1.6 s.g. fractions, the vitrinite occurs intercalated or disseminated in argillaceous partings. The fusinite and semifusinite occur as inclusions in the vitrinite or as isolated fragments, but are only rarely intercalated or disseminated in the argillaceous partings. The concentration of vitrinite in the higher specific gravity fractions is thus considered to reflect its association with argillaceous partings, whereas the concentration of vitrinite in less than 1.3 s.g. fractions reflects its lower density than either fusinite or semifusinite. The results obtained from the Fording River deposit are not necessarily applicable to other seams, but they do suggest that the association of macerals and the intergrowth of the macerals and mineral matter may be as important in maceral fractionation as the maceral density.

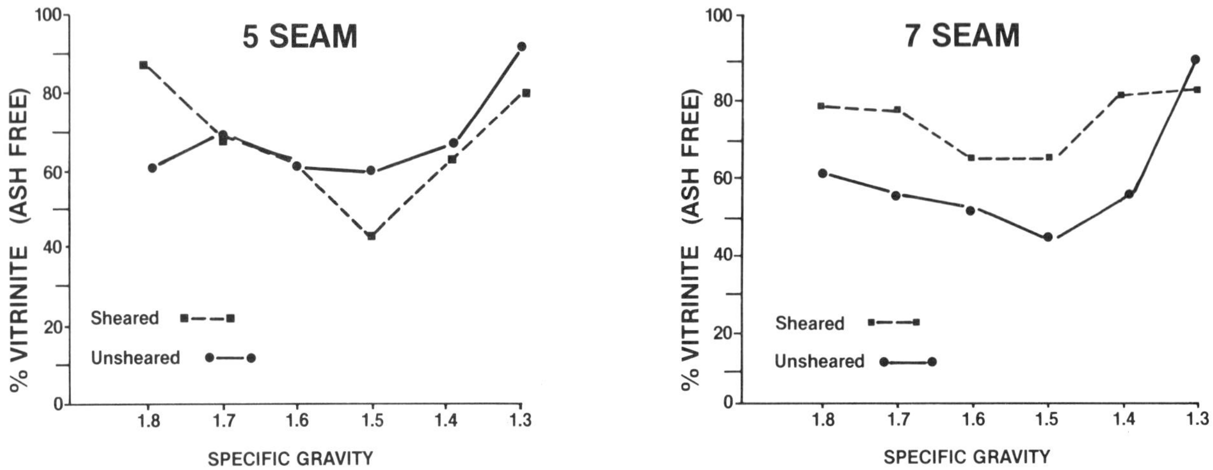

Figure 86. Vitrinite on an ash-free basis versus specific gravity for two Fording River coals (modified from Bustin, 1982).

Cleaning of coal by froth flotation or electrostatic methods depends on the chemical and physical properties of the surfaces of the coal particles rather than density. For a given coal rank, inertinite is considered to be less lycophobic than vitrinite, due to the higher oxygen and lower hydrogen content of the inertinite (Hornsby and Leja, in press). Electrostatic beneficiation methods, which are still experimental, involve charging the particles by trilboelectrification, conductive induction, or corona charging and subsequent separation based on charge-to-mass ratio (Inulet *et al.*, 1982). In electrostatic separation vitrinite charges positively, whereas inertinite, pyrite, and ash charge negatively. Experimental methods have been developed that may enable separation not only of the ash but also separation of the macerals (Inulet *et al.*, 1982).

10.1.2 *Application of Coal Petrology*

Coal petrographic studies used in conjunction with routine float-sink analysis may provide valuable information on coal preparation. Observation of the degree of 'intergrowth' of the organic and inorganic constituents in float-sink fractions may explain the observed quality of separation and indicate whether or not further crushing of the coal will increase the clean coal yield. Similarly, observing the type and distribution of sulphur may enable a better separation method to be designed. Of particular importance is the degree of intergrowth and the habit of the inorganic sulphur. If the pyrite or marcasite is coarse-grained and/or associated with the ash or present in fractures or cleat, then the sulphur content can be readily reduced. If the sulphides are finely disseminated throughout the organic fraction, they will be difficult, if not impossible, to remove. Also of importance is observation of the character of coal which sinks during froth flotation. Petrographic studies of the sinking coal may resolve whether the coal is oxidized or associated with mineral matter so as to enable corrective measures to be taken.

A method of optimizing the inert/reactive ratio in coking coals based on petrographic studies has been described by Burstlein (1957). This method, referred to as the Burstlein-Longwy method (Tompas, 1981), entails crushing the coal such that vitrinite plus liptinite is as coarse as possible and at the same time inertinite plus ash is as fine as possible. An example of the Burstlein-Longwy's method has recently been described by Tompas (1981) from a Hungarian coking coal. The precomminution particle size was initially determined, petrographic analysis performed, and the mass percentages of the components in different size grades determined. The selectivity of crushing is obtained both by separating undersized material and crushing the oversized material. The percentage of petrographic components getting in the various size grades during crushing can be calculated from:

$$mi = vi\ (ci/ai)$$

where i = the petrographic components (vitrinite, liptinite, etc.)
c = component assays in per cent
m = component recoveries
a = average petrographic assay of the coal in per cent

The efficiency of the separation is:

$$E = 100(m - v/100 - vo)$$

where v = yield in per cent
vo = ideal yield of petrographic components in per cent

The results of the calculations are shown in Tables 27 and 28 and Figures 87 and 88. In Table 27 the petrographic components are shown as mass percentages in the different size grades. In Table 28 the data in column 1 shows the reactive components (vitrinite and liptinite) and inerts plus ash. Columns 2 and 5 show mass percentages (v) of petrographic components with respect to total amount of coking coal from Figure 87. The data in columns 3 and 6 (Table 28) show petrographic component assays (c) in products finer and coarser than those of separation. Columns 4 and 7 show petrographic recoveries, while the efficiency of separation (E) is shown in column 8. On Figure 87 the recoveries of vitrinite plus liptinite and inertinite plus ash are shown as a function of screen size in product finer than the size x. Variation of the efficiency of separation, shown on Figure 88, indicates the maximum separation between inerts and reactives occurs at 3 millimetres with the oversized material being broken to less than 3 millimetres.

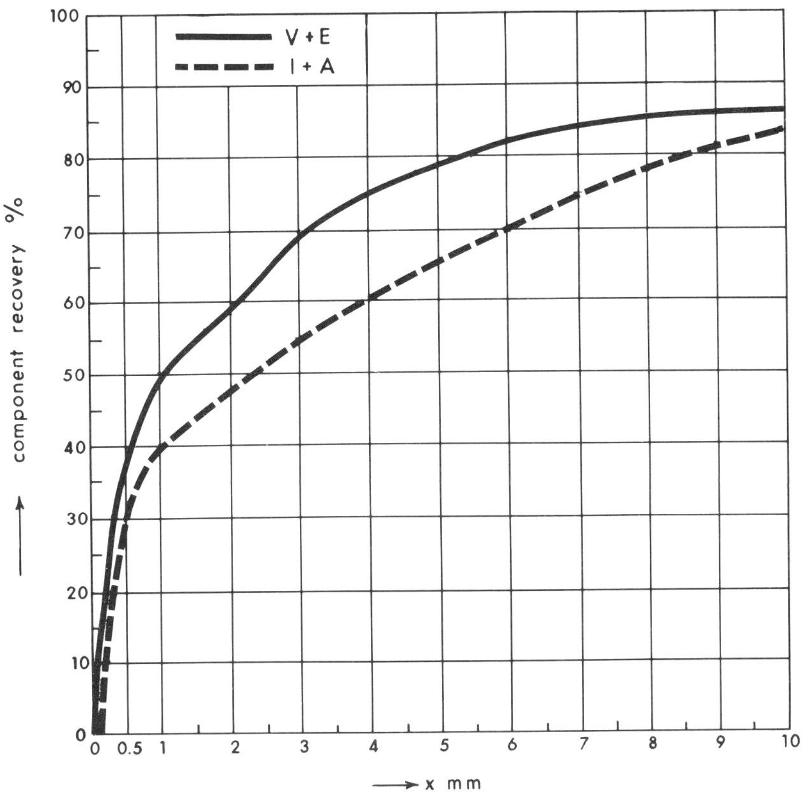

Figure 87. Recovery of petrographic components vitrinite plus liptinite (V + E) and inertinite plus ash (I + A) in the undersized product (modified from Tompos, 1982).

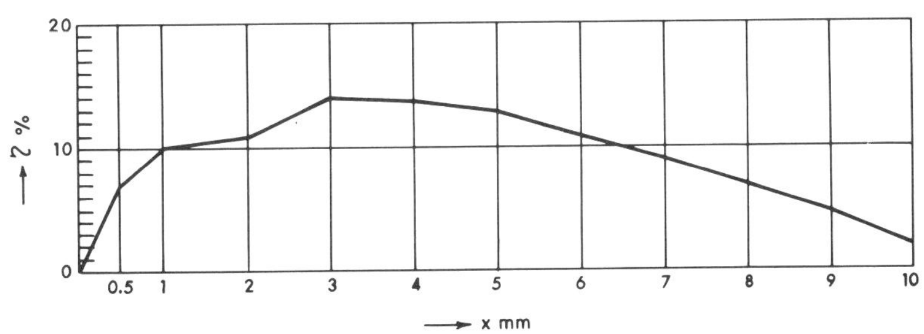

Figure 88. Variation in separation efficiency of vitrinite plus liptinite (V + E) and inertinite plus ash (I + A). See text for explanation. (Modified from Tompos, 1982).

In some coals it may be desirable to eliminate or concentrate certain trace or minor elements. For example, it is imperative that the phosphorous content of coking coal be kept to a minimum. In other coals it may be feasible to recover valuable trace or minor elements such as uranium or zinc (for example, Cobb *et al.*, 1980). By observing the affinity of the elements, it may be possible to design a recovery program in order to take advantage of the distribution of the elements. The affinity of elements for organic or inorganic fractions can be deduced by float-sink methods (Gluskoter *et al.*, 1977) or by petrographic studies. Examination of polished coal pellets using a scanning electron microscope equipped for element analysis or an electron microprobe enables visual observation of the distribution of the elements (including sulphur; see also section 4.36). Preliminary studies by Raymond (1977) have demonstrated that elements in coal may occur in different concentrations in different macerals. Such studies presently have only limited practical application; however, refinement of coal processing technology may enable enhanced elemental separation and recovery in the future.

TABLE 27. PETROGRAPHIC COMPONENTS AND MASS PERCENTAGES IN THE DIFFERENT SIZE GRADES (MODIFIED FROM TOMPOS, 1982)

Petrographic Components	Size Grade, x, mm											
	>10	6-10	4-6	3-4	2-3	1-2	0,5- - 1	0,2- -0,5	0,1- -0,2	0,06- -0,1	<006	0
1	2	3	4	5	6	7	8	9	10	11	12	13
Vitrinite	3,65	0,61	2,26	1,67	4,80	3,74	5,79	10,28	2,55	1,94	2,30	39,59
Liptinite	7,58	2,73	2,82	2,43	3,77	3,25	4,86	6,16	3,40	0,95	1,32	39,28
Inertinite Durite	2,26	2,50	1,32	0,61	0,54	0,61	0,64	2,02	0,84	0,28	0,17	11,79
Fusite	0,85	0,14	0,30	0,23	0,46	0,19	0,34	0,24	0,20	0,06	0,29	3,30
Inorg. Comp. Pyrite	0,09	0,08	0,17	0,15	0,35	0,61	0,88	0,44	0,29	0,14	0,28	3,48
Comb. Shale	0,21	0,11	0,12	0,12	0,28	0,23	0,27	0,36	0,13	0,08	0,14	2,05
Waste	0,05	--	0,03	0,01	0,02	0,02	0,09	0,17	0,08	0,03	0,01	0,51
	14,69	6,17	7,02	5,22	10,22	8,66	12,87	19,67	7,49	3,48	4,51	100,0

10.2 Combustion

Close to 90 per cent of the coal consumed in North America is used in combustion and most of that is used for the generation of electricity. In spite of the long-standing importance of combustion in the utilization of coal, petrography has played a rather insignificant role in identifying the qualities that characterize a good combustion fuel. One of the reasons for this is that the important properties of coal for combustion are not measurable directly by petrographic means or else they can be tested for more readily by other methods. Another related factor is that operating conditions and equipment at combustion facilities offer more options for change than do those of the other major coal-consuming industry (coking), and hence in combustion variations in coal quality can be compensated for more easily.

TABLE 28. CALCULATED PETROGRAPHIC COMPONENT RECOVERY AND EFFICIENCY OF SEPARATION BY SIZES (MODIFIED FROM TOMPOS, 1982). V+E IS THE REACTIVE COMPONENTS (VITRINITE AND LIPTINITE), I+A IS THE INERT COMPONENTS (INERTINITE AND ASH), V% IS THE MASS PERCENTAGES OF THE PETROGRAPHIC COMPONENTS WITH RESPECT TO TOTAL COKING COAL SHOWN IN TABLE 27, C% IS THE PETROGRAPHIC COMPONENT ASSAYS IN THE PRODUCT FINER THAN THE SIZE OF SEPARATION AND COARSER THAN THE SIZE OF SEPARATION, M% IS THE PETROGRAPHIC RECOVERY, AND E% IS THE EFFICIENCY OF SEPARATION.

Petr. Comp.	v %	c %	m %	v %	c %	m %	E %
1	2	3	4	5	6	7	8
	<0,06 mm			>0,06 mm			
V+E	3,6	80,3	4,6	75,2	78,8	95,4	
I+A	0,9	19,7	4,2	20,2	21,2	95,8	0,4
	4,5	100		95,5	100		
	<0,1 mm			>0,1 mm			
V+E	6,5	81,5	8,2	72,4	78,6	91,7	
I+A	1,5	18,5	7,0	19,6	21,4	93,0	1,2
	8,0	100		92,0	100		
	<0,2 mm			>0,2 mm			
V+E	12,5	80,5	15,8	66,4	78,6	84,2	
I+A	3,0	19,5	14,3	18,1	21,4	85,7	1,5
	15,5	100		84,5	100		
	<0,5 mm			>0,5 mm			
V+E	28,9	82,2	36,6	50,0	77,0	63,4	
I+A	6,2	17,8	29,6	14,9	23,0	70,4	7,0
	35,1	100		64,9	100		
	<1 mm			>1 mm			
V+E	39,5	82,4	50,1	39,3	75,6	49,8	
I+A	8,5	17,6	40,1	12,7	24,4	59,9	10,1
	48,0	100		52,0	100		
	<2 mm			>2 mm			
V+E	46,5	82,1	59,0	32,3	74,6	41,0	
I+A	10,1	17,9	48,0	11,0	25,4	52,0	11,1
	56,6	100		43,3	100		
	<3 mm			>3 mm			
V+E	55,1	82,4	69,9	23,7	71,7	30,1	
I+A	11,8	17,6	55,7	9,3	28,2	44,2	14,1
	66,9	100		33,1	100		
	<4 mm			>4 mm			
V+E	59,2	82,1	75,0	19,6	70,5	24,9	
I+A	12,9	17,9	61,0	8,2	29,5	38,9	14,0
	72,1	100		27,9	100		
	<6 mm			>6 mm			
V+E	64,3	81,2	81,5	14,6	69,9	18,5	
I+A	14,8	18,7	70,2	6,3	30,1	29,8	11,3
	79,1	100		20,9	100		
	<10 mm			>10 mm			
V+E	67,6	79,3	85,8	11,2	76,4	14,2	
I+A	17,8	20,7	83,6	3,5	23,5	16,4	2,1
	85,4	100		14,7	100		

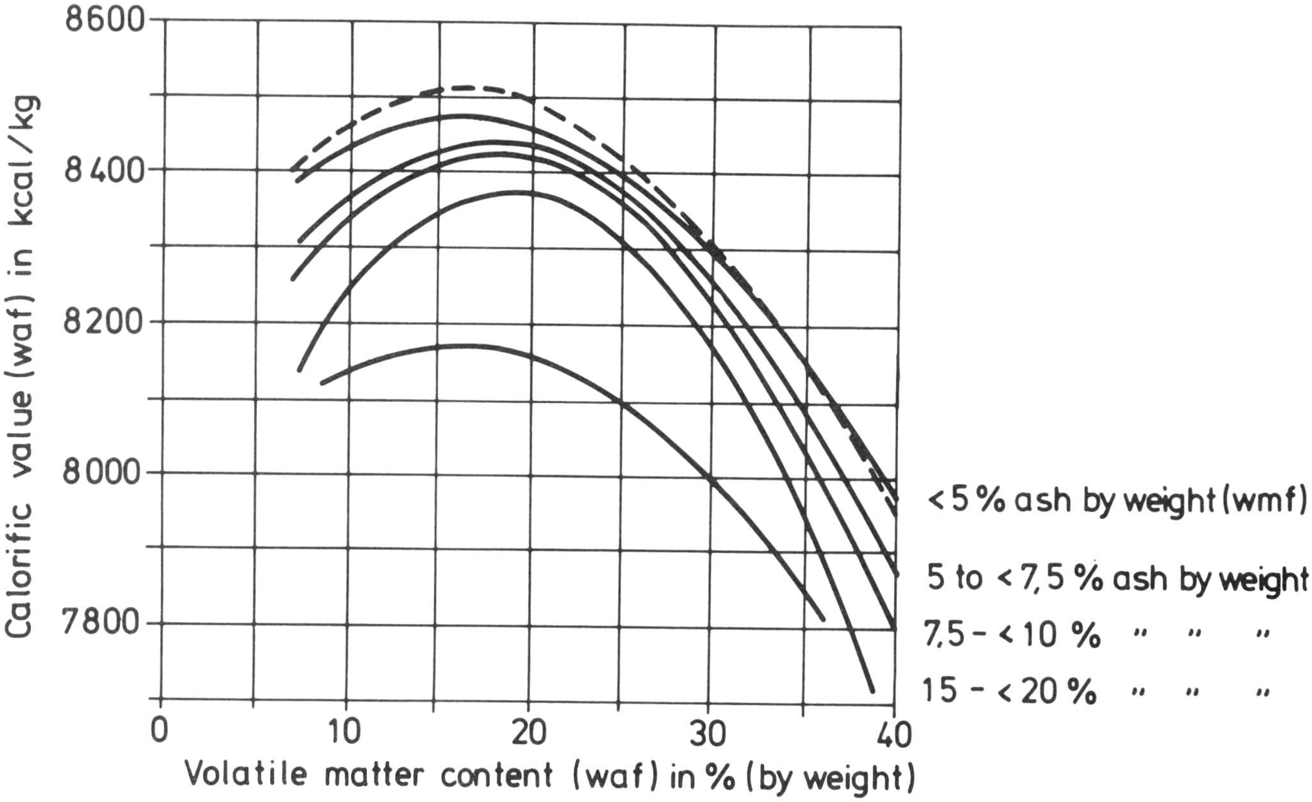

Figure 89. Relationship between volatile matter and calorific value for coals of different ash content (from Mackowsky, 1982).

Nevertheless there are meaningful relationships between combustion characteristics and petrographic composition. Neavel (1981), summarizing an earlier review by Spicer (1959), identified four coal characteristics related to rank and petrographic composition which are important in combustion. These are:

(1) Calorific value
(2) Grindability
(3) Combustibility, which includes such properties as swelling and ignition behaviour
(4) Ash properties

Calorific value can be related to maceral composition. Table 29 shows the differences in calorific value among the three main maceral groups for four German coals ranging in rank from high-volatile A bituminous to the upper part of the medium-volatile bituminous group. This table shows that the differences between maceral groups are greatest for seam R which is the lowest rank coal in the table. Examination of the data for the other coals shows that the variations decrease as rank increases. Marshall (1954) reported on calorific values for fusinite, one of the macerals of the inertinite group, and found a range of values from 7 500 to 8 800 calories per gram. There did not appear to be a rank relationship. According to Mackowsky (1982c), attempts to calculate the mean calorific value of a coal from data on the sum of its constituent macerals have not been very successful. A better established relationship is that between rank and calorific value as shown on Figure 89 for coals of different ash contents. Here rank is expressed in terms of volatile matter but it could equally well be expressed in reflectance values.

TABLE 29. CALORIFIC VALUES FOR MACERAL GROUPS (modified from Kroger *et al.*, 1957)

Seam	Volatile* Matter (daf)	Calorific Value cal/g (Btu/lb)		
		Vitrinite	Liptinite	Inertinite
R	36.13	7925 (14265)	8680 (15625)	7841 (14114)
Zollverein	31.97	8109 (14597)	8696 (15652)	8038 (14468)
Anna	28.36	8343 (15017)	8619 (15514)	8343 (15017)
Wilhelm	23.50	8368 (15062)	8360 (15048)	8216 (14788)

*Volatile matter determined on vitrinite only.

Coal used in pulverized fuel installations is ground to a particle size that is mainly below 65 microns (about 230 mesh). Therefore grindability of a coal is an important characteristic because of the additional energy required to grind hard tough coals. This property of a coal is expressed by the Hardgrove Grindability Index (HGI), the test for which is described in ASTM Standard D-409 (ASTM, 1981). The HGI varies considerably among coals depending on rank and maceral composition. Coals which are difficult to pulverize have low Hardgrove values, while high values of HGI indicate the opposite. Low-volatile bituminous coals are generally much easier to pulverize than high-volatile coals. Lithotypes such as durain, that are rich in liptinite and certain forms of inertinite, are often much more difficult to grind than brighter more vitrain-rich coals, or coals which may have a high content of fusain.

Combustibility includes such properties as the ignition temperature point of coal particles. This is the temperature at which a sample of finely ground coal begins to glow when heated in a stream of oxygen. It appears to be in part rank-dependent. According to Mackowsky (1982c) this temperature is about 200 degrees Celsius for coal with a volatile matter content of 40 per cent and about 250 degrees Celsius for a coal with 5 per cent volatile matter. There also appears to be a relationship with maceral content. Shibaoka (1969), in a series of interesting experiments using a microscope with an attached heating stage, showed that the order of ignition was liptinite-vitrinite-inertinite. Among individual macerals resinite was especially reactive, expanding explosively and then quickly burning with virtually no residue remaining. On the other extreme, fusinite, in the early stages of heating, changed very little except to become more fractured. Eventually, as heating progressed, the fusinite was consumed, but at a slower rate than the other macerals. Shibaoka's experiments showed some of the factors which influence the combustion behaviour of macerals, and showed how the maceral groups differed with respect to these factors. In his opinion the most important factors were:

(1) the amount of coke remaining after devolatilization of the macerals;

(2) the formation of vesicles in the carbonization stage; and

(3) the reactivity of the coke.

These studies suggest that the concepts of 'reactive' and 'inert' apply to the performance of macerals in combustion as well as in carbonization or liquefaction. Studies reported on by Nandi *et al*. (1977) confirm this view. They tested three coals in a pilot-scale pulverized fuel facility, and found that with two of the coals a high percentage of combustible material was carried over in the fly ash. Microscopic examination of the ash showed unreacted particles of fusinite, semifusinite, and oxidized vitrinite. The amount of unburned carbon in these fly ashes ranged from 16.8 to 59.8 (weight per cent) after initial runs. A second run after altering some of the operating conditions such as the fineness of the coal as charged and the supply of primary air showed a lower content of combustible material in the fly ash. These data are shown in Table 30. Such a carry-over of unburned fuel in a commercial plant would represent a big loss of a coal's calorific output. In normal power plant operation a carbon content of 5 per cent in the fly ash is the maximum that would be tolerated.

Nandi and his co-workers found that proximate and ultimate analyses of these coals (see Table 31) were not useful in predicting the reactivity behaviour. However, a petrographic examination of the coal before burning was more productive in this respect (see Table 32; note the high content of fusinite and semifusinite in the western Canadian coals). Nandi and his colleagues suggested that such problems could be alleviated by using higher combustion temperatures and longer residence times.

TABLE 30. COMBUSTIBLE CONTENT OF FLY-ASH SAMPLES (after Nandi *et al*., 1977).

Samples		Combustible in fly-ash (wt %) Initial	After optimization
Western Canadian bituminous	(Sample 1)	59.8	41.9
Pennsylvania bituminous	(Sample 2)	16.8	-
Western Canadian subbituminous	(Sample 3)	22.1	7.4

TABLE 31. ANALYSES OF COALS USED (after Nandi *et al.*, 1977)

	PROXIMATE ANALYSES		
Sample No.	1	2	3
	Western Canadian bituminous coal (wt %)	Pennsylvania bituminous coal (wt %)	Western Canadian subbituminous coal (wt %)
Ash	18.2	9.4	14.2
Volatile matter	24.8	36.3	33.8
Fixed carbon	57.0	54.3	52.0
	100.0	100.0	100.0
	ULTIMATE ANALYSES		
Carbon	70.4	75.5	67.0
Hydrogen	4.0	5.5	4.4
Sulphur	0.3	3.2	0.4
Nitrogen	1.1	1.4	1.2
Oxygen	6.0	5.0	13.0
Ash	18.2	9.4	14.0
	100.0	100.0	100.0

TABLE 32. PETROGRAPHIC ANALYSES (after Nandi *et al.*, 1977).

Sample No.	1	2	3
Coal	Western Canadian bituminous	Pennsylvania bituminous	Western Canadian subbituminous
Maceral			
Vitrinite (vol %)	40.6	73.2	52.2
Exinite (vol %)	0.0	11.6	7.8
Micrinite (vol %)	6.4	5.8	5.8
Fusinite (vol %)	35.2	3.4	17.4
Semifusinite (vol %)	17.8	6.0	16.8
	100.0	100.0	100.0
Reflectance ($\bar{R}_o$max) (Mean max.)	1.12	0.78	0.63

Petrography can also be used to study deposits and corrosion on boiler tubes and other heating surfaces (Mackowsky, 1982c). Such deposits reduce boiler efficiency and are mainly related to the combining of alkalis from clay minerals with sulphur oxides derived from pyrite and other sulphur forms in coal. Apparently the formation of such deposits may also bear some relationship to maceral content and rank of the coal and the interaction of such compositional characteristics with temperature conditions in the boiler. The presence in the coal of high contents of tar-producing macerals, liptinite especially, and the presence in the crushed coal as charged to the boiler of many small fusinite splinters may result in temperature gradient conditions within the boiler which promote the adhesion of partly fused mineral matter to heating surfaces.

The content and type of mineral matter is one of the important compositional characteristics of coal intended for combustion. For example it affects the heating value of the coal; variations in type of mineral matter critically affect ash fusion properties, ash disposal is always a problem, and, as mentioned previously, mineral matter affects the development of deposits and corrosion. In providing data relevant to such problems, petrography may only have supplementary importance; the main analytical tools are chemistry, X-rays, and of late, the SEM. Petrography may play an indirect role in the control of mineral matter by the study of raw coal and the products of preparation plant procedures. For example, Gray *et al.* (1963) used petrography to study the liberation of pyrite during coal washing experiments.

10.3 **Carbonization**

10.3.1 *Coke Quality*

Coke is an essential commodity in steel making and its quality has long been of prime concern to the industry. Coal is the raw material from which coke is made and its quality characteristics, to a large extent, control the quality of the resulting coke. The interplay of numerous factors which contribute to coking capacity is diagrammatically illustrated on Figure 90. Some adverse coal properties can be minimized by varying the conditions of the coking process itself, for example, bulk density, coking time, etc. Of the coke consumed by the steel industry, over 90 per cent is used in the blast furnace; a small amount is used in the foundry processing of metals. In the blast furnace, coke is both a reductant of iron ore and a source of process heat.

Over the years certain quality characteristics of good coke have come to be defined. These include both chemical and physical properties.

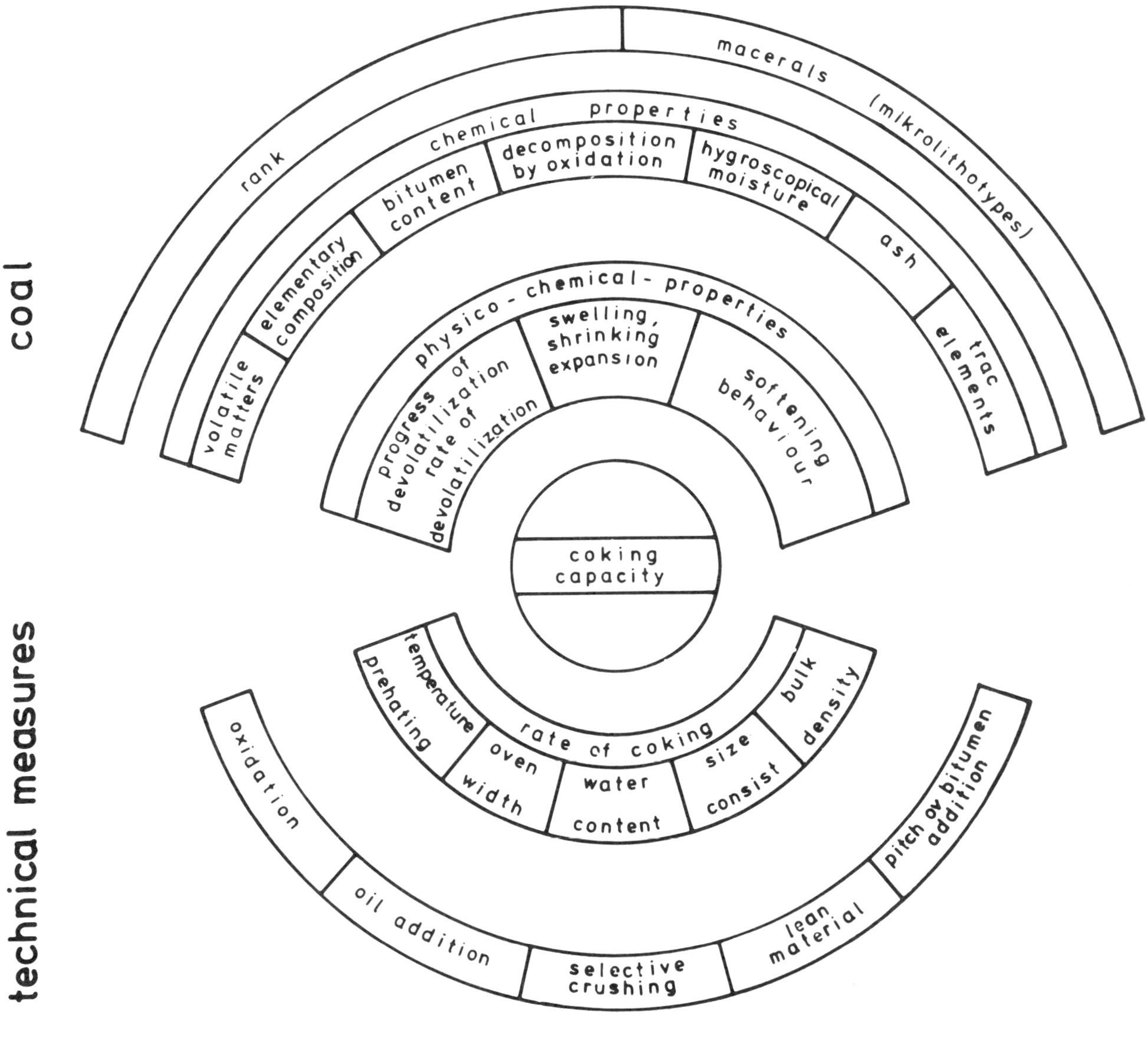

Figure 90. Diagram illustrating interaction of coal properties and coking conditions and influence upon coking capacity (from Mackowsky, 1982c).

10.3.1.1 **Chemical Properties**

Chemically the following compositional limits are recommended for good metallurgical coke (Berkowitz, 1979; Zimmerman, 1979).

(a) Volatile matter < 1.5 per cent
(b) Ash <10.0 per cent
(c) Sulphur < 1.0 per cent
(d) Calorific value >14 500 Btu/lb (33 725 kJ/kg)
(e) Phosphorus < 0.12 per cent

Of these chemical properties ash and sulphur are of particular importance because as they increase in coke productivity in the blast furnace decreases. In blast furnaces coke rate is a variable that is constantly monitored by the operators. Coke rate is defined as the weight of coke required to produce a ton of metal. It is a variable that is readily translatable into cost of production and it is vulnerable to increases in ash and sulphur. Industry experience indicates that a 1 per cent increase of ash in the coke reduces metal production by 2 or 3 per cent. Sulphur has a serious effect on the quality of iron and much of the sulphur in a coke will eventually report in the iron unless removed by an appropriate amount of limestone flux in the blast furnace. Increases of sulphur in the coke require the use of additional limestone and increased slag volumes, and again this is something which can be readily translated into cost of production. It is estimated that an increase of 1 per cent in the sulphur content of coke will mean an increase of 32 kilograms (70 pounds) in the coke rate, an unwelcome addition to the cost of iron production.

Phosphorus is another element with adverse effect on iron quality. Unlike sulphur, its final placement in the iron product is not so easily controlled by adjustment of slag volume. Therefore, care must be taken that the coals used in coke making have a low initial phosphorus content. Zimmerman (1979) quotes 0.05 to 0.06 per cent as a safe limit. Gray *et al.* (1978) quote an even lower safe limit of 0.03 per cent. The latter authors list limits for a number of other elements in the components of a blast furnace burden (see Table 33).

TABLE 33. PROPOSED LIMITS FOR VARIOUS CHEMICAL COMPONENTS OF BLAST-FURNACE BURDENS INCLUDING ORE, STONE, AND COKE (from Gray *et al.*, 1978)

	Component	Maximum Ratio of Component to Fe	Maximum Amount of Component Expressed as lb per ton Fe
Group I	Cu	0.0001	0.2
	Ni + Co	0.0004	0.8
	Mo	0.0003	0.6
	Sn	0.00015	0.3
	Cr	0.0004	0.8
	V	0.0001	0.2
Group II	Zn	0.0004	0.8
	Pb	0.0008	1.6
	TiO_2	0.01	20.0
	$Na_2O + K_2O$	0.002	4.0
	As	0.0001	0.2
	Sb	0.0001	0.2
	P	0.002	4.0
	S	0.001	2.0
	Cl	0.0001	0.2
	F	0.0002	0.4
Group III	SiO_2	0.1	200.0
	Al_2O_3	0.04	80.0
	Mn	(Each application must be considered separately)	

Group I - Limits for the components are based on the chemical specifications for residual elements for many grades of steel.

Group II - Limits for the components are based on the maximum amounts that can be tolerated without encountering severe operating difficulties or serious environmental pollution problems.

Group III - Limits for the components are based on amounts in excess of which higher than normal operating cost would be encountered because of additional fuel requirements or refining times.

10.3.1.2 **Physical Properties**

Most of the physical properties measured on coke relate in some way to coke strength. The strength of coke should be sufficient to support a heavy burden of ore and flux in the blast furnace while sustaining a high degree of permeability in the bed. The latter is necessary for the passage of gases during the iron-smelting process. A number of empirical tests have been devised to measure strength. Two of these used in North America are the Shatter test and the Tumbler. The former measures the resistance of coke to impact, while the latter measures resistance to abrasion. Specifications for these tests are described under ASTM Standards D 3038 (Shatter) and D 3402 (Tumbler), (ASTM, 1981). Strength is the most important physical property of coke. It is also a property which can be evaluated from the petrography of the coal of which the coke is made. This relationship will be dealt with in more detail later. Other physical properties of coke which are often measured are density, porosity, and size distribution. These can also be related to strength in varying degrees.

Because indices derived from strength tests are referred to later on, it is appropriate to describe these tests briefly at this point. The prescribed procedure for the Shatter test requires the dropping of 50 pounds (23 kilograms) of plus 2 inches (50 millimetres) coke on to a steel plate from a height of 6 feet (1.83 metres). The coke is then screened through a series of sieves, namely 2 inches (50 millimetres), 1 ½ inches (37.5 millimetres), 1 inch (25 millimetres), and ½ inch (12.5 millimetres). The results, however, are usually only reported as plus and minus 2 inches (5.0 cm) coke.

The Tumbler test is used more frequently than the Shatter test. The ASTM Tumbler is one of several similar tests in use throughout the world which are based on the size degradation of a specified quantity of coke of a specified size range tumbled in a revolving steel drum for a prescribed number of revolutions. In the case of the ASTM test the coke is screened after tumbling on 1 inch (2.54 cm) and ¼ inch (0.63 cm) sieves. The weight percentage of the original sample remaining on the 1-inch (2.54 cm) screen is designated the stability factor, while the cumulative percentage retained on the ¼-inch (0.63 cm) screen is termed the hardness factor.

The Micum test is similar to the ASTM Tumbler, differing only in details such as size of original coke sample, drum dimensions, and number of revolutions. The Micum test is widely used throughout the world and two indices are derived from it, namely Micum 40 (M40) and Micum 10 (M10). The M40 is the percentage of coke remaining on a 40-millimetre sieve after testing, while the M10 is the percentage of original sample which passes a 10-millimetre sieve. There are several minor variations in the Micum test, for example, the German standard for this test varies somewhat from that prescribed by ISO. A third type of drum test is the Japanese JIS test and it is also quite similar to the ASTM Tumbler, though differing in some details of prescribed procedure. A commonly reported index from this test is:

$$DI \frac{30}{15}$$

meaning the percentage of coke remaining on a 15-millimetre sieve after 30 drum revolutions. A second index,

$$DI \frac{150}{15}$$

is sometimes reported; this means the percentage of coke remaining on a 15-millimetre sieve after 150 revolutions.

The IRSID and Sundgren procedures are two additional drum tests and data from them are sometimes reported in the literature.

The relationships among indices of these various drum tests have been examined by a number of authors. Some of these relationships are shown on Figure 91. Waters *et al.* (1971) published the following equation for conversion of Micum data to ASTM:

$$\text{ASTM 1-inch tumbler index} = 0.913\,(M_{40}) - 13.88.$$

Matsuoka (1975) indicated a relationship of the Japanese DI $\frac{30}{15}$ with ASTM stability factor as:

$$DI \frac{30}{15} = 56\,(\text{ASTM stability})^{0.14}$$

(although this would seem to be at variance with the data plotted on Figure 91B.)

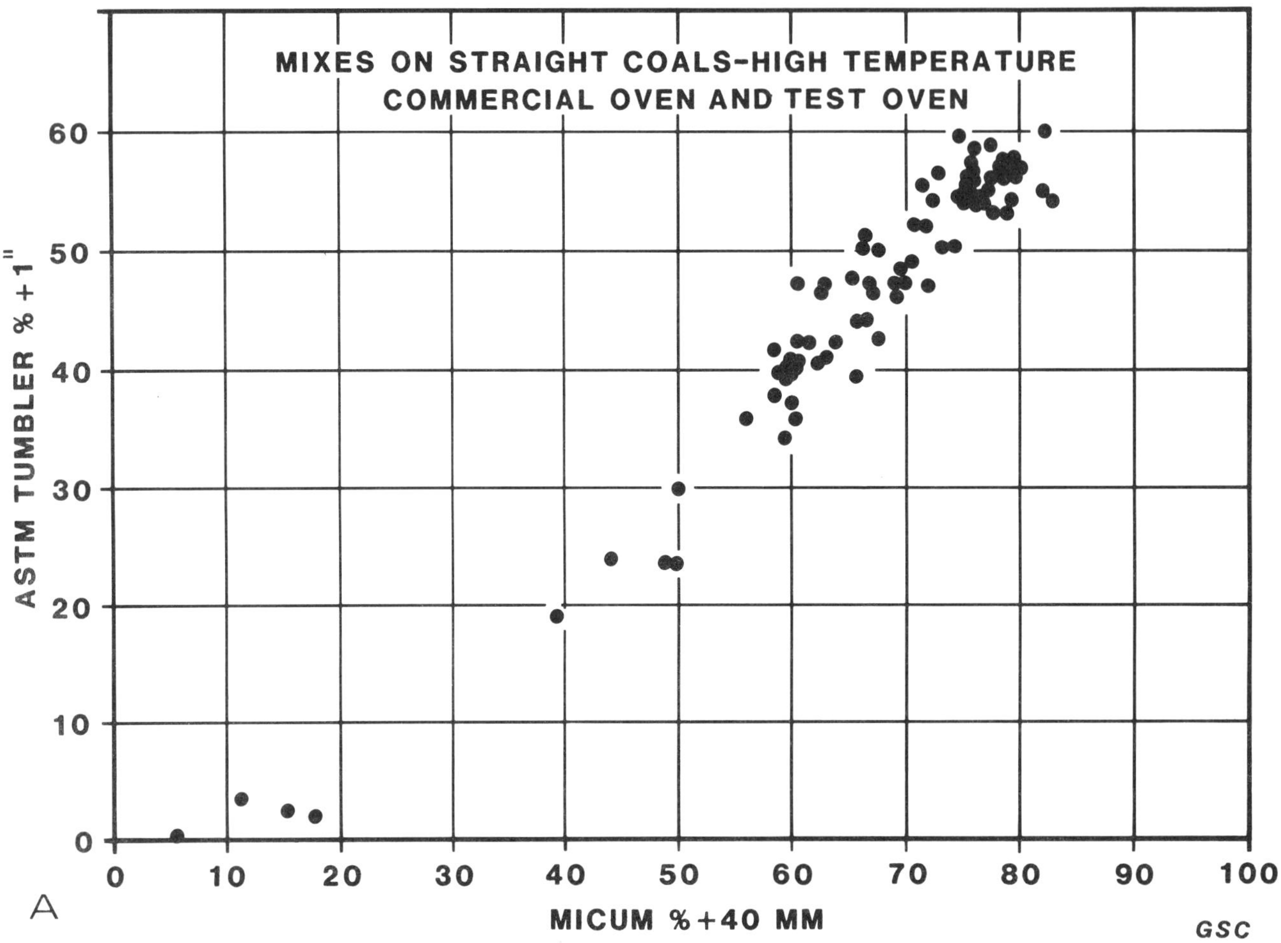

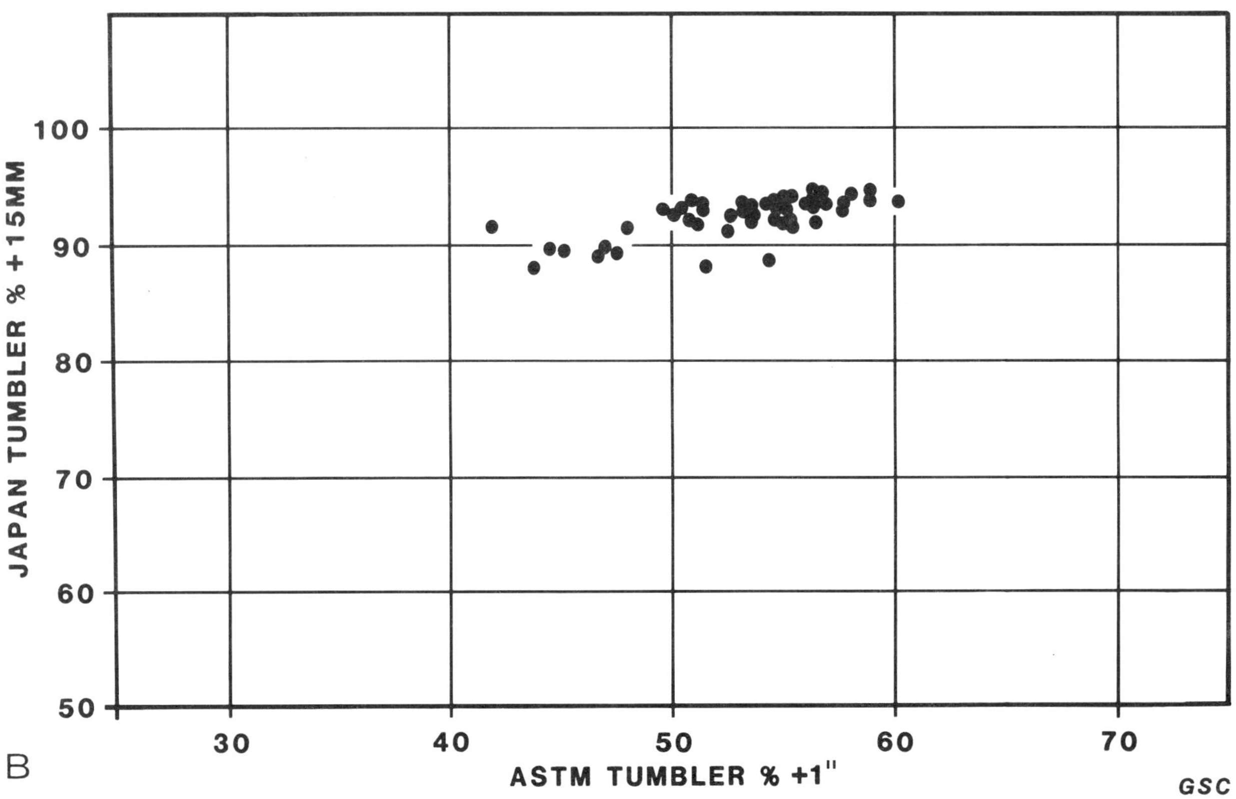

Figure 91. Relation between Micum 40 and JIS 15 indices with ASTM stability factor (from Zimmerman, 1979).

Another often mentioned coke property, partly physical and partly chemical, is reactivity. Reactivity is defined as the ability of a coke to react with substances such as carbon dioxide and steam. In laboratory tests for this property, the amount of carbon monoxide produced under prescribed conditions along with weight loss in the subject coke is taken as an index of reactivity. There is some question as to the relevance of such tests when compared with the actual behaviour of coke in the blast furnace. Nevertheless, its evaluation is considered important by coke technologists. Because of an inverse relationship with coke strength, high coke reactivity is not desirable in blast furnace operation (Benedict and Thompson, 1973).

10.3.2 *Qualities of Coking Coal*

Coke properties are a reflection of the properties of the parent coal or coals. One of the most important is *rank*. During the carbonization process coal softens, fuses, and resolidifies to form the porous carbon-rich material we call coke. The capability of a coal to behave in this manner is essential, and only the bituminous part of the coal-rank series has such capability. However, not all bituminous coals coke equally well. The medium-volatile coals within the bituminous group appear to possess the most optimum natural association of properties to make good coke on a single coal basis. Low-volatile bituminous coals can also produce strong cokes but they, and many of the medium-volatile coals, generate excessive pressures during coking, and if carbonized by themselves, can damage coke oven walls. Because of this and because medium and low-volatile coals are relatively scarce, the usual practice is to blend high-volatile bituminous coals with varying amounts of medium and/or low-volatile coals so as to reduce the development of high pressures but at the same time to incorporate some of the inherent propensity of the higher rank coals to produce strong coke.

Mention has been made previously of the softening, fusibility, and agglomerating behaviour of a coking coal. These phenomena fall under the general heading of *rheological properties* and are sometimes referred to as the swelling or dilatation, fluidity or plasticity attributes of a coal. These vary from coal to coal, depending partly on rank, and on the maceral composition (see Section 8.3). A certain amount of plasticity and swelling is essential in a coking coal because the whole process of coking requires that some of the constituents melt and form a cement which bonds together materials or particles which fuse poorly or not at all. A number of empirical tests or devices measure rheological properties including the Free Swelling Index (FSI), the Gray-King test, the Gieseler Plastometer, the Audibert-Arnu and Ruhr dilatometers, and the Roga test. Some of these tests are described in section 8.3 (see also Zimmerman, 1979). In North America the FSI and Gieseler test are commonly used to measure swelling and fluidity respectively.

Mention has been made already of the acceptable limits of chemical parameters in coke. Suggested limits of *physical and chemical* parameters for coking coals are presented in Table 34. Petrographic analyses can be used indirectly to control such chemical characteristics as ash and sulphur, by identifying the mode of occurrence of pyrite and other minerals, thus assisting in the solution of problems in the coal preparation stage. Another characteristic of coking coal evaluation which properly belongs under chemistry is coal oxidation and especially its detection. Except for a few cases where excessive fluidity may be depressed by controlled oxidation, this phenomenon has a deleterious effect on coking properties such as swelling, fluidity, and coke strength. Various methods for detecting it have been proposed and used including staining, flotation behaviour in light oils (Gray *et al*., 1976), solubility in NaOH solutions (Gray *et al*., 1978), reflectance (Benedict and Berry, 1964), microhardness (Nandi *et al*., 1978), and a combination of reflectance variation, staining, and FSI (see Chapter 5).

Grindability, or ease of pulverization, is another physical property of coal to be taken into consideration. Because size-consist in a coking coal charge is quite important and can have an influence on coke strength (Smith, 1973; Mackowsky, 1982c), close attention to pulverization characteristics in the preparation of coking coals is desirable, especially if blends of different rank are being produced. Reference has already been made to the work of Burstlein in this connection (see Section 10.1.2).

Determination of the *petrographic composition* of coal has come to be an increasingly important element in the evaluation of a coal's coking potential. Not only can it be used to predict certain coke properties such as strength, it can also be correlated with coal properties such as fluidity, FSI, and volatile matter content. Petrographic methods can also be used to study the coke itself as will be discussed later.

TABLE 34. RATING OF COKING COALS FOR BLENDING (from Gray *et al.*, 1978).

	COAL CLASSIFICATION								
	High-Volatile A			Medium-Volatile*			Low-Volatile		
	Rating			Rating			Rating		
Property	Good	Medium	Poor	Good	Medium	Poor	Good	Medium	Poor
1 Volatile Matter, %	31.0-33.0	33.0-36.0	+36	21.0-24.0	24.0-27.0	27.0-31.0	18.0-21.0	15.0-18.0	<15.0
2 Vitrinite Reflectance, %**	0.92-1.09	0.85-0.95	0.68-0.85	1.40-1.50	1.20-1.40	1.10-1.20	1.51-1.70	1.70-1.85	>1.85
3 Fluidity, ddpm***	+20,000	5000-20,000	<5000	500-8000	300-20,000	<300->20,000	100-300	30-1000	<30->1000
4 Free-Swelling Index	9	6-8	<6	9	7-8	<7	9	7-8	<7
5 Hardgrove Grindability Index	48-75		32-70	80-135		60-90	90-120		85-105
6 Composition Balance Index**	0.40-0.80	0.80-1.40	>1.4	1.0-1.50	1.50-2.00	>2.0	2.00-3.50	3.50-5.00	>5.00
7 Rank Index	3.4-4.3	3.0-3.4	2.2-3.0	6.0-6.5	4.3-5.5	<4.3	>6.8	6.0-7.5	<7.5

*Those properties such as volatile matter content, reflectance in oil, and rank index have little bearing in the ranking of medium-volatile coals because the rank required for a medium-volatile coal is dependent upon the rank and amount of the other coals used in the blend.
**Determined petrographically.
***Dial divisions per minute.

Differences in the behaviour of the petrographic constituents of coal in coking tests and other experiments related to utilization have long been recognized (Stopes and Wheeler, 1923; De Boosere, 1926; Kuhlwein and Hoffmann, 1952). As information from such studies accumulated, and as more precise petrographic methods evolved, it became clear that the constituents of coal could be classified as 'reactive' or 'inert' in a given utilization process, and further that the degree of 'reactiveness' or 'inertness' could be quantified, at least to the point where such information could be used to predict the behaviour of a coal. The greatest advances in this area have been made in the field of carbonization with the result that most steel companies now routinely carry out petrographic analyses to monitor blend quality and to evaluate new coals.

In the use of coal petrography for the prediction of coking properties, the macerals are generally grouped into reactives and inerts in the following manner.

Reactives - Vitrinite, liptinite, variable amounts of semi-inertinite (mainly semifusinite).
Inerts - Variable amounts of semi-inertinite, fusinite, macrinite, micrinite, inertodetrinite, sclerotinite, mineral matter.

The term 'generally grouped' is used advisedly, because not all laboratories follow this grouping in detail. Some use a one-third reactive, two-thirds inert breakdown for the semifusinite (Ammosov *et al.*, 1957; Schapiro *et al.*, 1961), others use a similar breakdown for semifusinite plus other inertinite macerals of low reflectance (Benedict *et al.*, 1968). Some include very little semifusinite in their reactives (Brown *et al.*, 1964), while others may include relatively large and variable amounts (Steyn and Smith, 1977). At least one laboratory identifies a semi-inert variety of vitrinite and includes part of it in the inerts (pseudovitrinite of Benedict *et al.*, 1968).

Based on such concepts, a variety of mathematical techniques have been developed to predict coke properties, especially strength. Each seems to work provided they are applied to coals which are more or less similar in petrographic composition to the coals for which the model was developed, and provided that coke potential is being evaluated on coals which might be tested under the same experimental conditions used for the model. It is possible that the composition and hence technological behaviour of materials grouped under maceral headings (for example, semifusinite) may differ from region to region or age to age, so that for one coal such materials may be largely inert while in another they may be reactive. Further it is likely that the mathematical manipulations leading to the development of correlation curves between petrography and coke properties help to compensate for conceptual variations among laboratories as to what is 'reactive' and 'inert.'

Despite the apparent reservations expressed previously, the experience of the past 25 years shows that coke quality prediction, especially strength, by coal petrography has been quite successful. All of the various techniques have two premises in common. The first is that for any given rank of coal there is an optimum mix of reactives to inerts which will give the best coke, and the second follows from the first in recognizing that the percentages representing this optimum mix or ratio vary with rank. The optimum mix concept has been aptly compared to the making of concrete (Schapiro *et al.*, 1961). The reactives in coal supply 'cement' while the inerts fill the role of sand or gravel. Just as too much cement or too much aggregate produces a weak concrete, so also will too much reactives or too much inerts produce a weak coke. In line with such concepts Schapiro and his associates speak of a Composition Balance Index (CBI) which is the ratio of the optimum inerts that a coal of given rank should have to the amount of inerts which it actually contains. On determination of this index it is possible to say that a given coal is oversaturated (CBI >1.0) or undersaturated (CBI <1.0) with inerts. A second index which they determine is called a Strength Index (SI); in later publications they also refer to it as a Rank Index. This index is the sum of strength factors for each vitrinite reflectance class (V type, see section 7.1.6) in the coal related to the amount of inerts present. With the calculation of these two indices coke strength can then be read off a chart or table. All of the various coke prediction methods in general follow the same approach, that is, they condense rank data and other petrographic compositional data into two or more indices which can then be employed to predict coke strength from a graph or via a computer program. Some introduce additional information such as coking conditions or size of inertinite masses into the calculations to further refine their predictions. Table 35, taken from Mackowsky (1977), summarizes different approaches which have been followed in terms of analysis, treatment of coking conditions, and particular index of coke strength determined.

TABLE 35. SUMMARY OF COKE PREDICTION METHODS (FROM MACKOWSKY, 1977)

	I	II	III	IV	V	VI
AUTHORS	AMMOSOV, EREMIN *et al.* (1957)	SCHAPIRO *et al.* (1961), BCR, US STEEL	BROWN, COOK, TAYLOR (1964)	APPLEBY, BCRA, NCB, ITALSIDER	SMITH (1973)	MACKOWSKY & SIMONIS (1969)
A TYPES OF ANALYSIS						
1 REFLECTANCE	▨	▨	▨	▨	▨	▨
2a MACERAL ANALYSIS	▨	▨		▨	▨	
2b MACERAL GROUP ANALYSIS						▨
3 MICROLITHOTYPE ANALYSIS			▨			
4a OPTIMUM RATIO: REACTIVE/INERTS	▨	▨		▨		
4b STRENGTH INDEX*	▨	▨		▨		
4c COMPOSITION BALANCE INDEX*	▨	▨		▨		
5 M40/V-STEP AND INERTS					▨	
6a VOLATILE-MATTER* CONTENT (FB_{RM})						▨
6b COKING POWER*						▨
6c G_{RM}*						▨
B COKING CONDITIONS						
7 CONSTANT	▨	▨	▨	▨	▨	
8 VARIABLE						▨
C COKE STRENGTH						
9 1 INCH ASTM STABILITY CALCULATION FROM 4		▨		▨		
10 1 INCH ASTM STABILITY CALCULATION FROM 1 AND 3			▨			
11 JAPANESE DRUM INDEX FROM 1,2,4 RESP. 1 AND 3			▨			
12 SUNDGREN DRUM INDEX FROM 1,2a, AND 4	▨					▨
13 M40 (ISO) FROM 1,2b,6 AND 8		▨ ?		▨		
14 M40 FROM 9						
15 M40 (AFTER SMITH) FROM 1,2 AND 5					▨	
16 ASTM STABILITY FROM 13						▨

ARC/83 GSC

10.3.3 *Petrographic Methods for Calculation of Coking Properties*

10.3.3.1 **Strength**

As mentioned earlier, strength is one of the most important properties of coke and much effort has gone into defining the correlation of coal petrographic composition to strength in the resulting coke. Several of the methods for doing this will now be described.

10.3.3.1.1 Method of Ammosov *et al.* and Schapiro *et al.*

In 1957 Ammosov *et al.* in the Soviet Union described a method to predict coke strength from the petrography of the parent coals. Schapiro *et al.* (1961) modified this method and applied it to coals being tested or used by the United States Steel Corporation. It is used mainly for the prediction of the stability factor. Some aspects of this approach have been described already in the discussion of Composition Balance Index and Strength Index (Rank Index). In the method of Schapiro *et al.*, reactives include all the vitrinite and liptinite and one-third of the semifusinite; inerts include the remaining two-thirds semifusinite, the remaining inertinite macerals and mineral matter. The one-third - two-thirds breakdown of semifusinite is a fixed ratio in the Schapiro method. There is reason to believe that with many coals this ratio may be in error. Indeed later publications (Gray *et al.*, 1978) of the U.S. Steel group indicate that they also may not hold rigidly to it for they suggest that in low-volatile coals semifusinite is four-fifths inert. Because the U.S. Steel method was developed primarily on Appalachian coals, which normally have low semifusinite contents, variations in the way in which it is proportioned probably make little difference in predicted stability values.

The U.S. Steel method was based on the correlation of actual test-coke data with the petrographic composition of the coked coals. Some of these tests were carried out in 500 pound (1100 kg) ovens and the authors pointed out that the test conditions were kept constant, that is: bulk density 55 pounds per cubic foot; moisture 2 per cent or less; ash 12 per cent or less, and charges crushed to 80 per cent minus one-eighth inch. The application of this method for the prediction of stability on coals coked under conditions different from those cited may be subject to discrepancies between calculated and actual stabilities.

The method of Schapiro *et al.* requires the calculation of the two indices, CBI and SI. The first step in this determination is to divide the reactives into types based on reflectance measurements. These are the 'V' types mentioned earlier. First the vitrinite is apportioned to 'V' types based on the actual reflectance values obtained on this maceral. Reactive semifusinite is proportioned according to its reflectance spectrum and divided into 'V' types as was done with the vitrinite. Liptinite is prorated according to the amounts of the other reactives but restricted to the 'V' type classes V8 to V15. Once the amount of reactives in each V type has been determined, the two indices may be calculated as follows:

$$\text{CBI} = \frac{\text{Total Inerts (volume per cent)}}{\frac{R_1}{M_1} + \frac{R_2}{M_2} \text{-----} \frac{R_n}{M_n}}$$

where R_1, R_2, R_n = volume per cent of reactives in V types 1, 2 ----- n.

M_1, M_2 ----- M_n = optimum ratio of reactives to inerts in V types 1, 2 ----- n.

The values of M_1, M_2, etc. are obtained from the chart shown on Figure 92. In this figure note that there is a wide variation in optimum ratio among V types, for example, for V9 it is 2.6 while for V17 it is 12.3. These ratios are based on the earlier coking tests carried out at U.S. Steel.

The calculation of the SI goes according to the formula:

$$\text{SI} = \frac{(K_1 \times P_1) + (K_2 \times P_2) + \text{------} + (Kn \times Pn)}{\text{Total Reactives (volume per cent)}}$$

where K_1, K_2 ------ K_n = strength indices of reactive V types l, 2 ------ n
P_1, P_2 ------ Pn = volume per cent reactives in V types 1, 2 ------ n (same values as R_1, R_2, etc. in CBI equation)

The strength indices (K_1, K_2, etc.) are shown graphically on Figure 93 for each V type in relation to the inerts content of a particular coal. For example, in a coal with 20 per cent inert content the K value for V10 would be 3.8. These values are inserted in the equation for the calculation of SI. With the CBI and SI determined, the predicted stability can then be read off the plot shown on Figure 94.

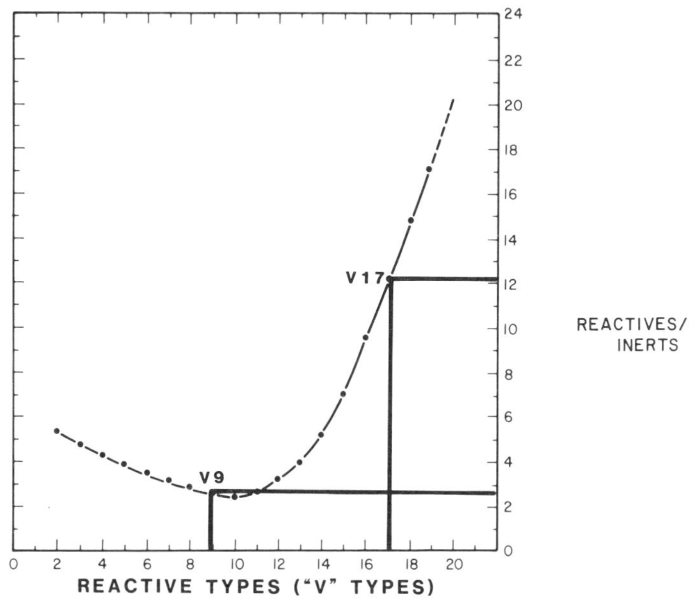

Figure 92. Graph of optimum ratios; reactive to inert components (modified from Zimmerman, 1979, after Schapiro *et al.*, 1961).

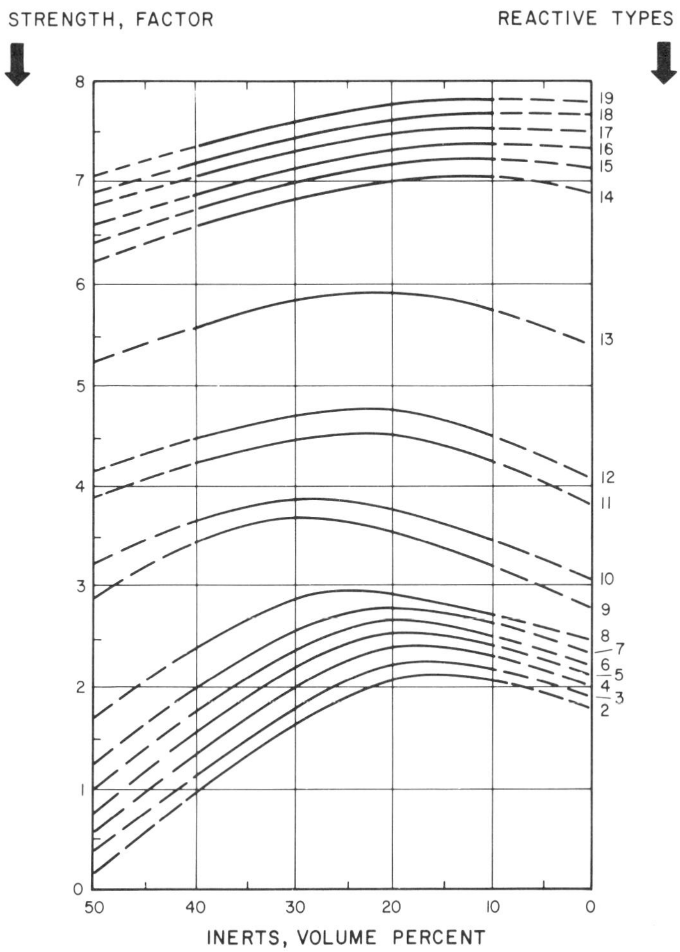

Figure 93. Strength index graph; coke strengths for different V types with variable quantities of inerts (from Zimmerman, 1979, after Schapiro *et al.*, 1961).

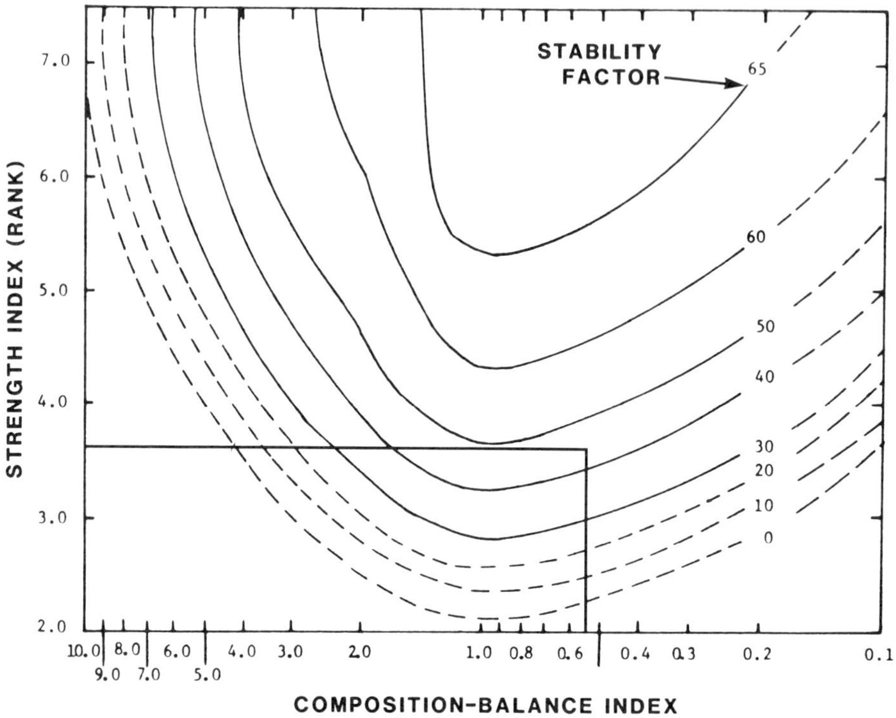

Figure 94. Relationship of coke stability factor to Strength Index and Composition Balance Index (modified from Zimmerman, 1979, after Schapiro *et al*, 1961).

The application of this prediction method is best shown by the following sample calculation on a Canadian high-volatile A bituminous coal.

Step 1. Determination of maceral and reflectance data.

A. Maceral content (volume per cent); macerals already sorted into reactives and inerts.

REACTIVES			INERTS		
Vitrinite	Liptinite	⅓ Semi-fusinite	⅔ Semi-fusinite	Other Inertinite	Mineral Matter*
72.7	8.6	0.7	1.4	12.8	3.8
Total Reactives = 82.0			Total Inerts = 18.0		

*Mineral matter calculated by Parr's formula (see Section 7.2).

B. Reflectance data on vitrinite and semifusinite (per cent by V types, based on 100 measurements on vitrinite and 27 on semifusinite).

Macerals	V Types						
	8	9	10	11	12	13	14
Vitrinite	2	42	56	0	0	0	0
Semifusinite	0	0	0	8	39	23	30

Step 2. Prorating of reactives by V types.

Macerals	V Types							Totals
	8	9	10	11	12	13	14	
Vitrinite	1.5	30.5	40.7	0	0	0	0	72.7
Semifusinite	0	0	0	0.1	0.3	0.1	0.2	0.7
Liptinite	0.2	3.6	4.8	0	0	0	0	8.6
TOTALS	1.7	34.1	45.5	0.1	0.3	0.1	0.2	82.0

Step 3. A. Calculation of Composition Balance Index (CBI)

$$\text{CBI} = \frac{\text{Total Inerts (from Step 1)}}{\frac{1.7}{2.8^*} + \frac{34.1}{2.6} + \frac{45.5}{2.4} + \frac{0.1}{2.7} + \frac{0.3}{3.2} + \frac{0.1}{4.0} + \frac{0.2}{5.2}} = \frac{18.0}{32.89} = 0.55$$

*These values (M1, M2, etc.) from Figure 92.

B. Calculation of Strength Index (SI)

$$\text{SI} = \frac{(1.7 \times 2.90^*) + (34.1 \times 3.50^*) + (45.5 \times 3.72) + (0.1 \times 4.49) + (0.34 \times 4.73) + (0.1 \times 5.91) + (0.2 \times 7.01)}{\text{Total Reactives (from Step 1)}} = \frac{297.40}{82.0} = 3.63$$

*These values (K1, K2, etc.) from Figure 93.

Step 4. Determination of stability factor.

With a CBI and SI calculated (0.55 and 3.63 respectively), a predicted stability factor of 44 can be read from Figure 94. An actual stability of 43 was determined on coke made from this coal. The predicted stability factors, as well as the M coefficients used in the CBI equation and the K coefficients used in SI equation, can also be read from appropriate tables (Moses, 1976).

Schapiro and Gray (1964) achieved a linear relationship between the calculated or predicted stability and that determined on actual coke with a correlation coefficient of 0.98 (see Figure 95).

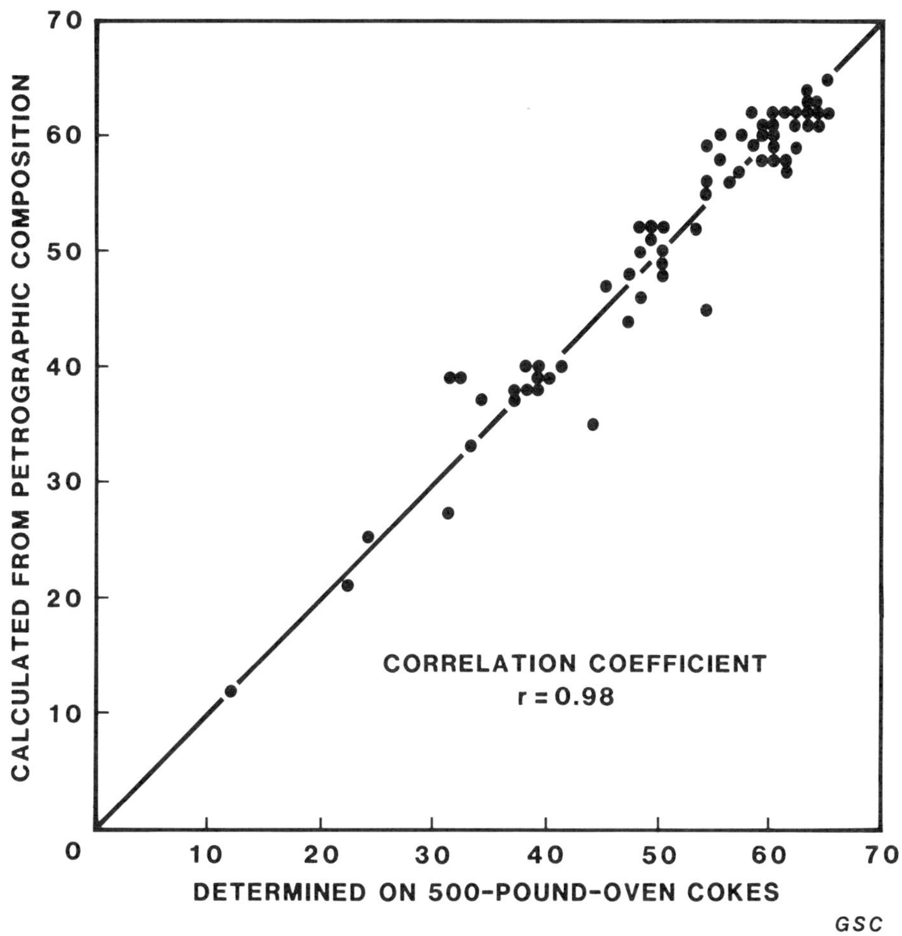

Figure 95. Correlation of predicted and determined coke stability factors (from Schapiro and Gray, 1964).

A number of coke properties relative to strength and use can be summarized on the stability diagram as is shown on Figure 96. Isorank lines have been superimposed on this figure to separate strength domains for different ranks of coal. Note that low-volatile coals can be expected to yield the strongest cokes except when the Composition Balance Index is very high. In comparison, high-volatile coals must have a rather severely restricted CBI centering around 1.0 before they will produce cokes with expected stabilities of 50 or better. Note that the area of optimum metallurgical coke properties is a small part of the total diagram with rather limited CBI and SI ranges. This is one reason why blending for metallurgical coke production is very important because there are few coals which singly will fall into the optimum window.

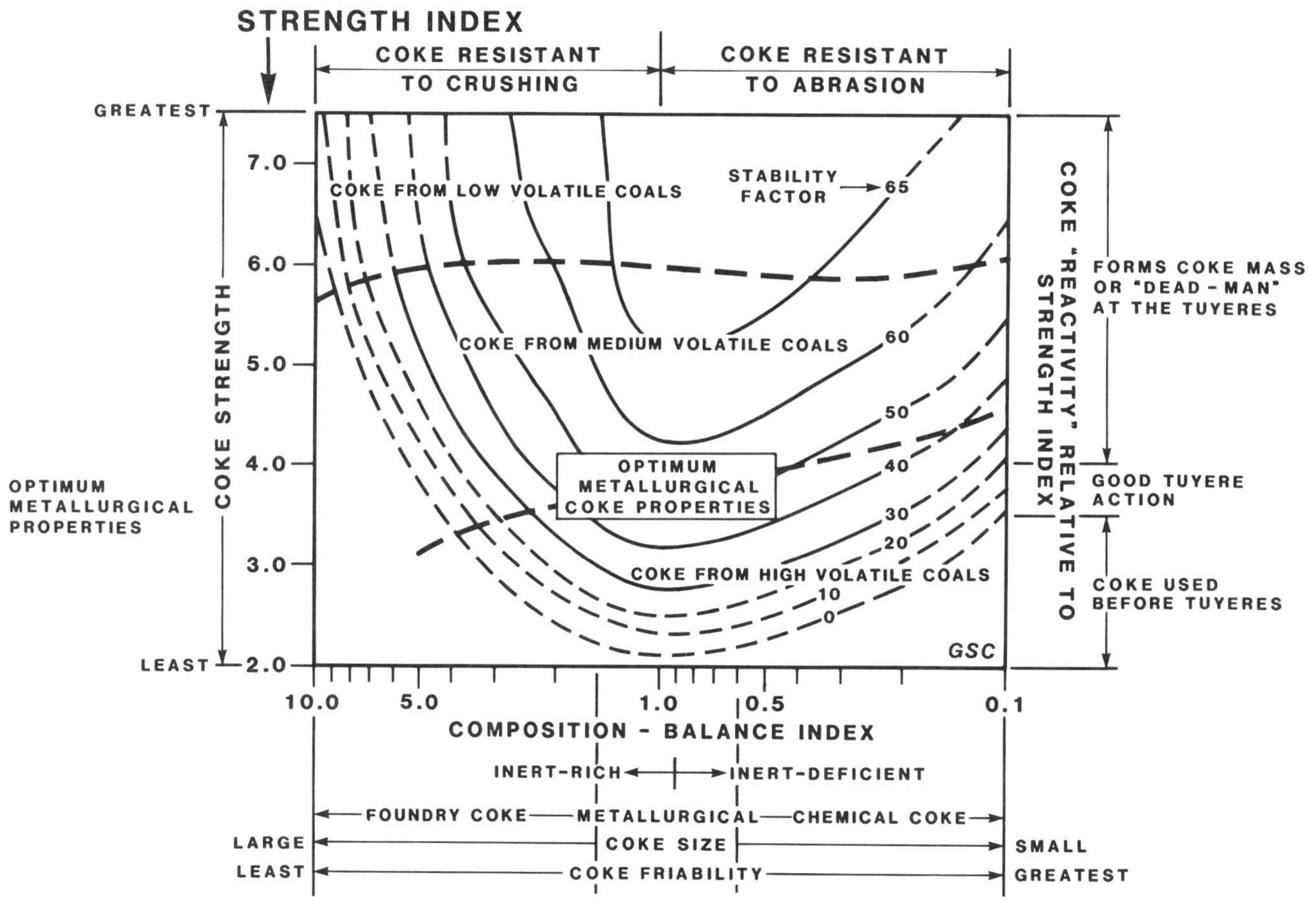

Figure 96. Relationship of coke size, coal rank, and utilization characteristics to stability diagram (modified from Zimmerman, 1979).

10.3.3.1.2 Method of Benedict *et al.*

Another method of predicting coke strength is used at the Bethlehem Steel Corporation and was described by Benedict *et al.* (1968). The method is similar to the approach of Schapiro and his colleagues in that it recognizes 'reactive' and 'inert' materials in coal, in that it utilizes a one-third - two-thirds breakdown of semifusinite and other semi-inerts, and in that maximum reflectance is the technique used for rank determination. An important departure from Schapiro and Gray is that part of the vitrinite maceral group is considered with the semi-inerts. The Bethlehem group called this material pseudovitrinite and they feel that some of it is inert because they have observed essentially unreacted particles carrying through into their coke (Plate 15).

The Bethlehem approach to grouping macerals by reactives and inerts is slightly different from that used by U.S. Steel and is presented in Table 36.

TABLE 36. PROCEDURE OF BETHLEHEM STEEL FOR GROUPING REACTIVES AND INERTS.

Reactives	'Normal' vitrinite Liptinite Pseudovitrinite (variable amounts) ⅓ semi-inerts (semifusinite and semi-macrinite)
Inerts	⅔ semi-inerts Pseudovitrinite (variable amounts) Fusinite Macrinite Micrinite Mineral matter*

*Determined from chemical analyses:
MM = 0.6 (wt % ash = wt % sulphur)

The inertness of pseudovitrinite is determined by reflectance in the following manner. Mean maximum reflectances are measured separately on pseudovitrinite and 'normal' vitrinite and, based on the difference between them, the degree of inertness of pseudovitrinite is read from a chart (Figure 97). Based on this degree of inertness, the percentage of pseudovitrinite to be added to the inerts is calculated while the remainder is consigned to reactives. The total amount of inert material, adjusted to include some pseudovitrinite, is designated by the Bethlehem group as *effective inert content* and this, combined with the reflectance of 'normal' vitrinite, is used to predict the ASTM stability factor as shown by the curves on Figure 98. It should be pointed out that the curves on Figures 97 and 98 were developed after a number of experiments in test coke ovens had been carried out on a variety of coals.

The use of the Bethlehem method is best illustrated by an example for a hypothetical coal.

Step 1. Determination of maceral and reflectance data.

	Maceral Content	**Reflectance**	
'Normal' vitrinite	60%		
Pseudovitrinite	15%	$\bar{R}_o$max of normal vit.	= 0.90
Liptinite	10%	$R_o$max of Psv.	= 1.02
Semi-inerts	6%	Difference in $\bar{R}_o$max	= 0.12
Other inertinite	4%		
Mineral matter	5%		

Step 2. Determination of reactive-inert breakdown for pseudovitrinite.

Using the difference in $\bar{R}_o$max of 0.12 and the reflectance (0.90) of the normal vitrinite, the appropriate curve on Figure 97 can be used to determine that 57 per cent of the pseudovitrinite should be inert. This means that 8.6 per cent (15 x 57) of the pseudovitrinite should be included in the inerts. Thus the effective inert content is:

4.0 (⅔ semi-inert) + 8.6 (Psv) + 4.0 (other inertinite) + 5.0 (MM) = 21.6 per cent

Step 3. Determination of stability factor.

Using the value of 21.6 per cent and the 'normal' vitrinite reflectance of 0.90, a predicted stability of 36 can be read on the appropriate curve on Figure 98a. Values which do not fall on the curves shown can be determined by interpolation.

Plate 15. Photomicrographs of coke showing unreacted fusinite, massive micrinite (macrinite) and pseudovitrinite (from Benedict and Thompson, 1980). Reflected light, oil immersion.

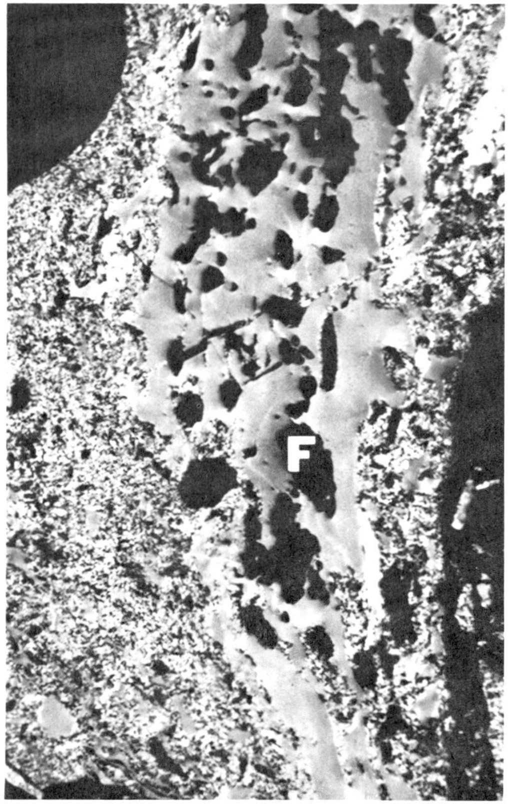

F = Fusinite

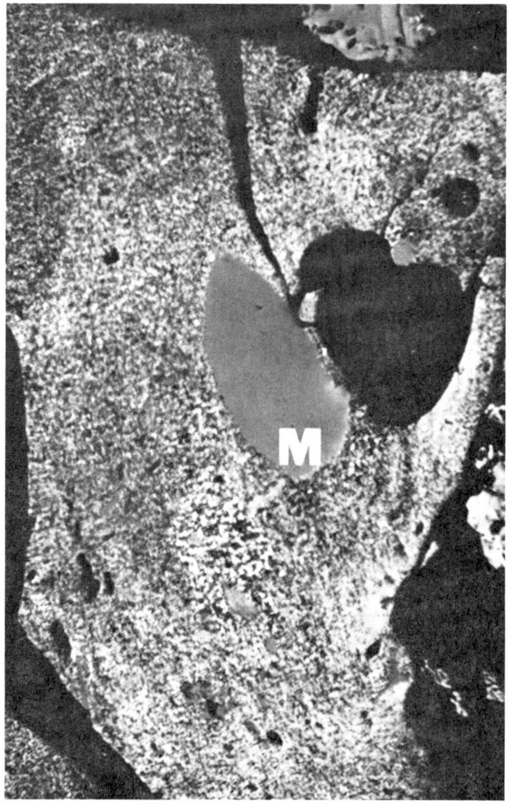

M = Micrinite

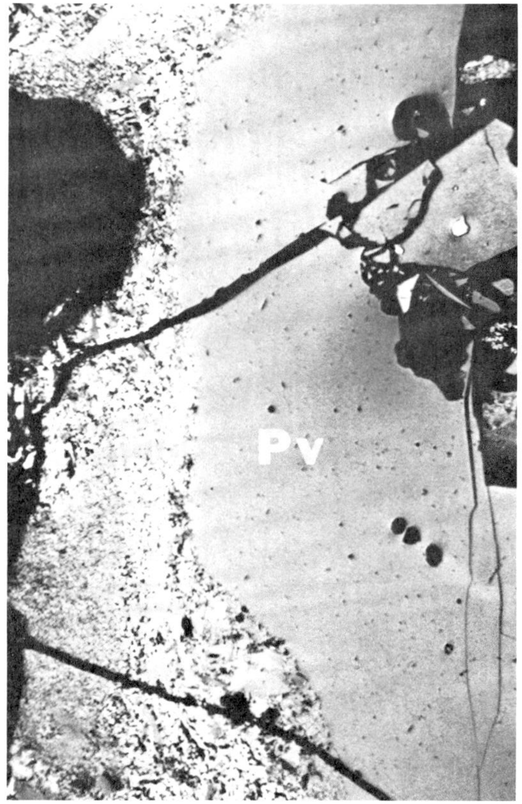

Pv = Pseudovitrinite

0 200μ

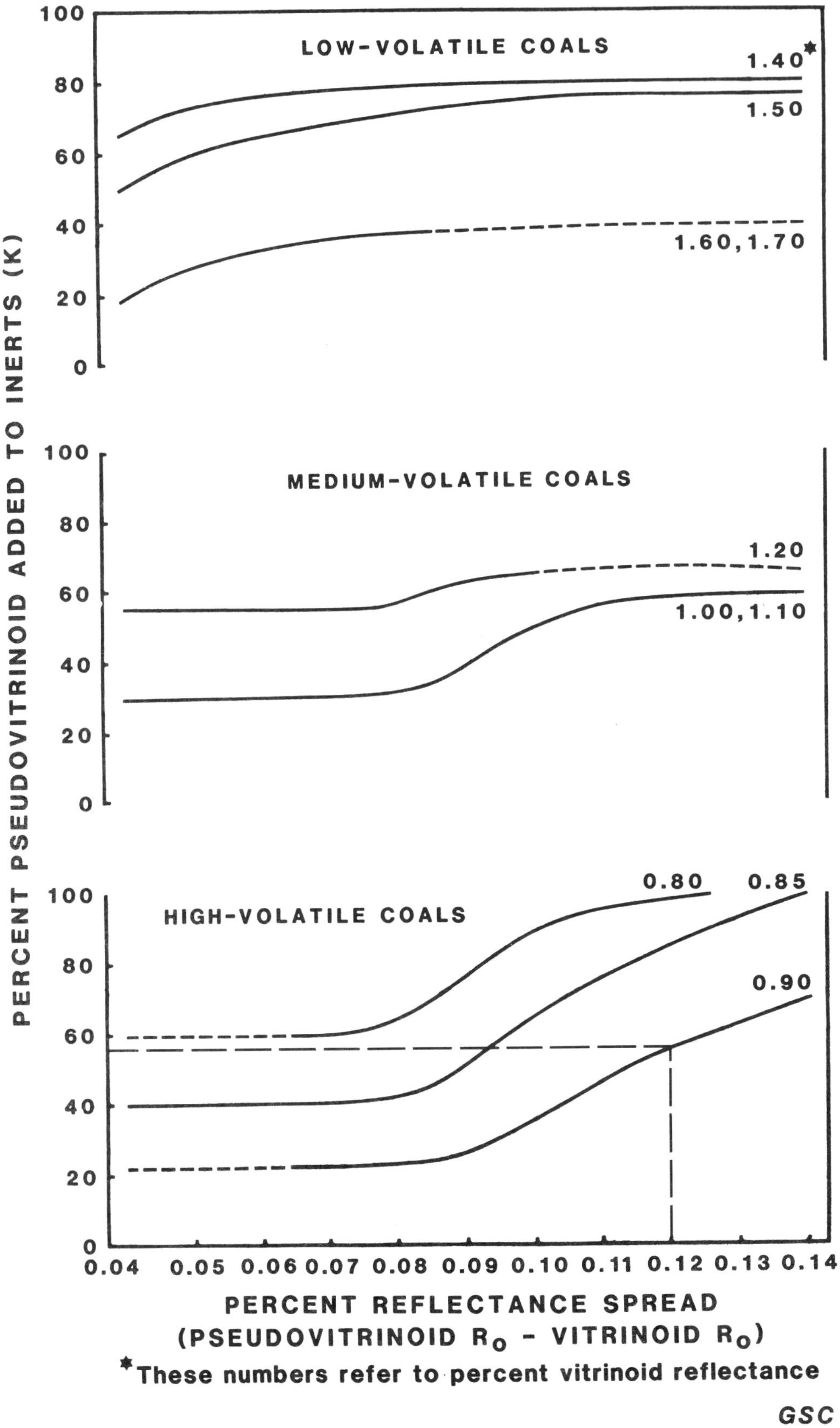

Figure 97. Curves for determination of inert character of pseudovitrinite (from Benedict, Thompson, and Wenger, 1968).

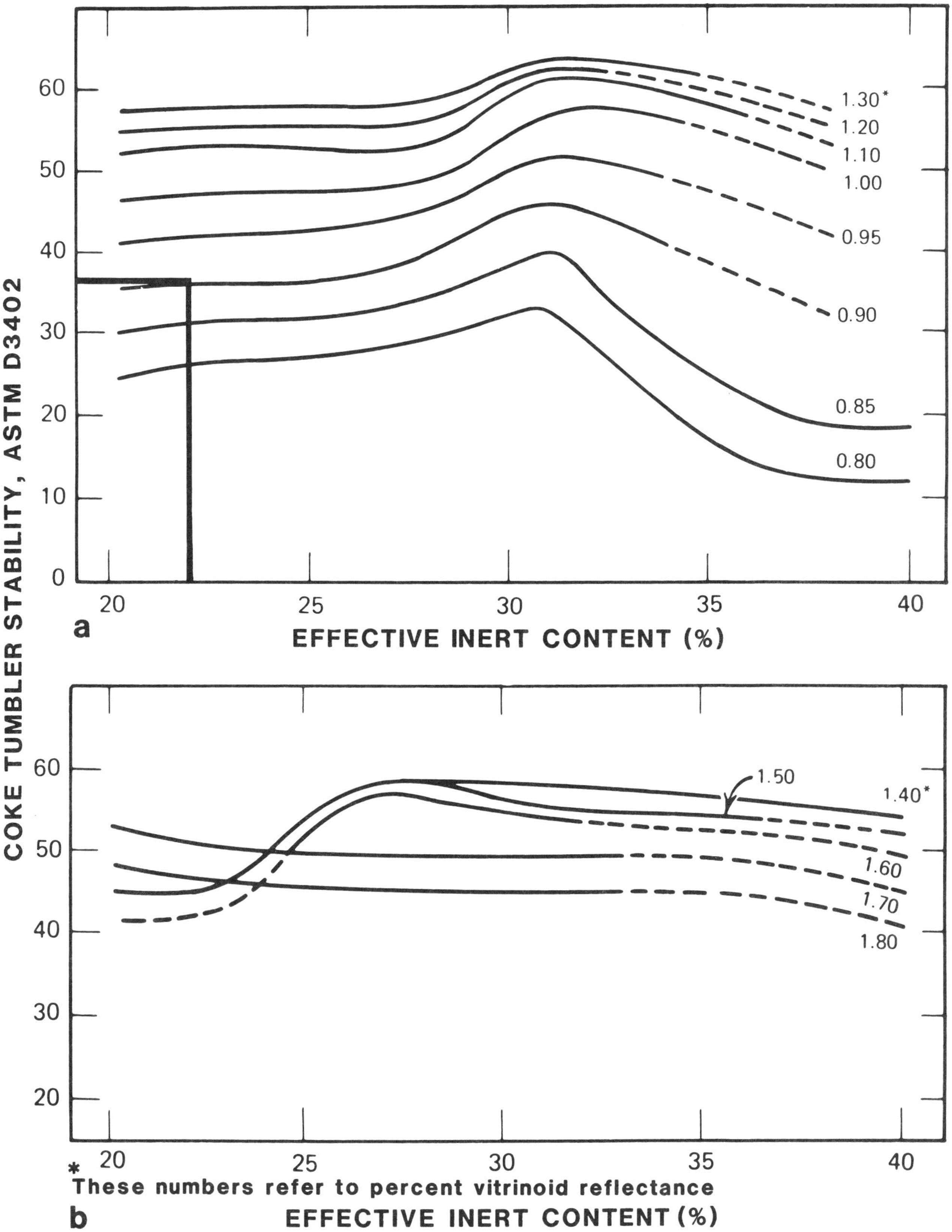

Figure 98. Relationship of effective inert content, reflectance, and ASTM Stability Factor, (a) high-volatile A and medium-volatile bituminous coal, (b) low-volatile bituminous (modified from Benedict, Thompson, and Wenger, 1968).

As with the U.S. Steel method, that of Bethlehem can also be adapted to prediction of coke strength from coal blends. According to Benedict *et al*. (1968) the stability of blends is affected to a major degree by the composition and percentage of the low-volatile coal in the blend. Nevertheless, the prediction of stability in such cases can be refined by taking into account the effective inert content of the high-volatile blend component as well.

A series of curves for various blend combinations has been developed at Bethlehem for this purpose according to the format shown on Figure 99. Again an example may serve to illustrate the procedure. Assume a blend composed of 15 per cent low volatile (assumed R_0max 1.50) and 85 per cent high volatile coal (assumed R_0max 0.80), each with an effective inert content of 30 per cent. Figure 98a shows that the stability factor of the high-volatile component by itself is about 32, while it is about 57 for the low-volatile component by itself (Figure 98b). Using these two values of 32 and 57 as end points, a curve is plotted on the format of Figure 99, with the 32 located above the 0 per cent point on the abscissa and the 57 above the 100 per cent point at the appropriate level. From the latter position a horizontal line is drawn to intersect a vertical projection from the 23 per cent point on the abscissa. The 23 per cent point on the abscissa is taken as a critical value for low-volatile content in the blend because experience at Bethlehem suggests that for values above 23, strength characteristics are determined almost wholly by the low-volatile component and for values below 23 the high-volatile entity exerts an increasingly important influence. From there the line curves downward to the left to intersect the position of the 32 value. Using this curve and with 15 per cent low volatile in the mix, the blend strength can be predicted at about 51.

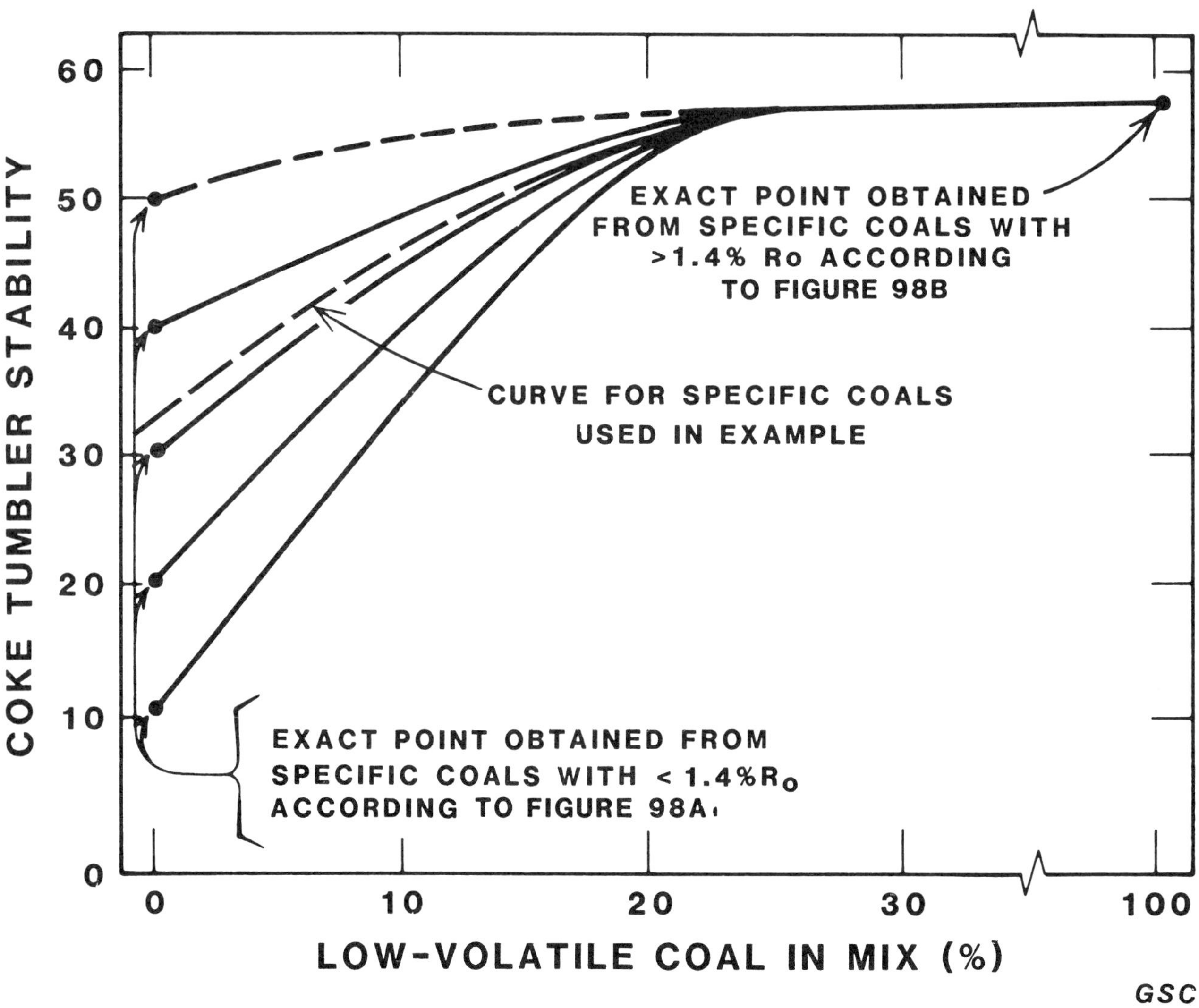

Figure 99. Curves for predicting ASTM stabilities on blends (modified from Thompson and Benedict, 1978).

10.3.3.1.3 Method of Mackowsky and Simonis

A third method for the prediction of coke quality has been developed in Germany by Mackowsky and Simonis (1969). It is somewhat more involved than the two just described, but according to the authors it can be used to predict not only coke strength (Micum 40 and Micum 10) but also yield of coke, gas, liquid by-products, and tar. In addition to coal compositional parameters (rank and maceral content), the relevant formulae also takes into account coking conditions, that is, size consist of the charge, bulk density, coking rate (in centimetres/hour), mean oven width, and coking time.

The reflectance and maceral data are not used directly but rather are employed to calculate the volatile matter and a parameter called the coking power or G value. The latter is usually obtained by the Ruhr dilatometer method, but experiments on the behaviour of the different maceral groups in the dilatometer showed that is was possible to calculate the G value for a given coal from its petrographic composition according to the equation:

$$G_{RM} = (\Sigma_7^{20} V_1C_1) - \Delta G_{RM}$$

where V_1 = percentage of vitrinite in individual V types
C_1 = G values for individual V types and
ΔG_{RM} = is an adjustment made to take account of the liptinite and inertinite content of the coal.

The numbers 7 and 20 associated with the equation represent V7 and V20, that is, the limits for which this method of calculating G values seem to be reliable.

The reflectance data expressed in terms of V types along with maceral data can also be used to calculate the volatile matter of the whole coal according to the equation:

$$VM_{RM} = VM(V) + VM(L) + VM(I) + \Delta VM$$

where VM(V), VM(L), and VM(I) represent the volatile matter of the vitrinite, liptinite, and inertinite respectively determined according to individual V types. ΔVM is a correction factor introduced to compensate for a relatively wide scatter in liptinite volatile matter within the same rank of coal. From the four parameters of size consist, volatile matter, G value, and coking conditions, Mackowsky and Simonis have developed the following formula for prediction of coke strength expressed in terms of the Micum 40 (M_{40}) index:

$$M_{40} = aK + b + M_sd$$

where K is a factor embracing the coking conditions
M_s represents the size consist
b is a rank (volatile matter) factor
a and d are coefficients based on volatile matter and G value.

Each of these factors and coefficients are developed by separate formulae which take account of the detailed compositional variations of a given coal and variation in coking conditions and size consist. These calculations require a computer and appropriate programs. Mackowsky and Simonis suggest that reliable results for strength indices M_{10} and M_{40} can only be obtained if the four main parameters are within the following limits:

<u>For prediction of M_{10}</u>

VM_{RM}	between 11 and 36 per cent
G_{RM}	between 0.95 and 1.10

<u>For prediction of M_{40}</u>

VM_{RM}	between 18 and 36 per cent
G_{RM}	between 0.95 and 1.10
K	between 18 and 25 per cent
Ms	between 5 and 10_o

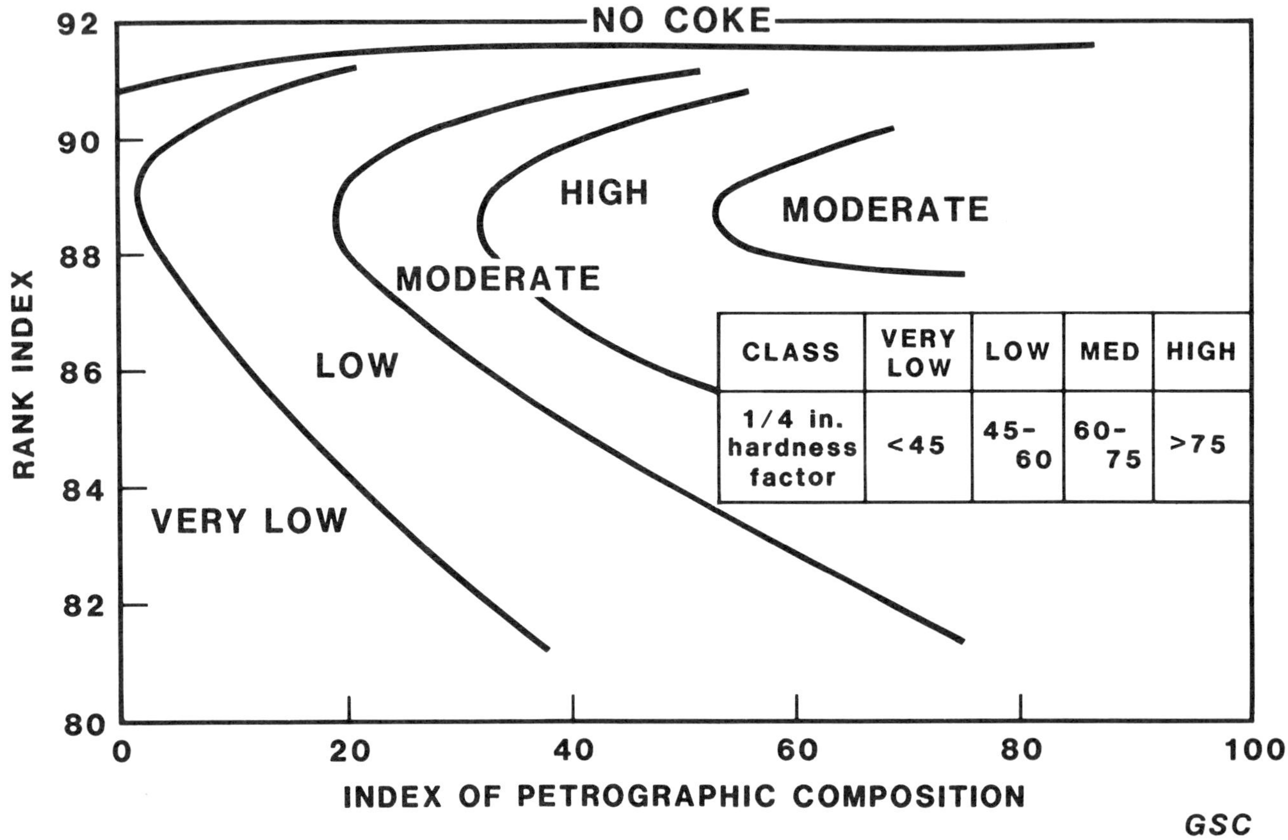

CLASS	VERY LOW	LOW	MED	HIGH
1/4 in. hardness factor	<45	45-60	60-75	>75

Figure 100. Relationship of petrographic composition and rank to ¼ inch hardness factor of coke (from Brown *et al.*, 1964b).

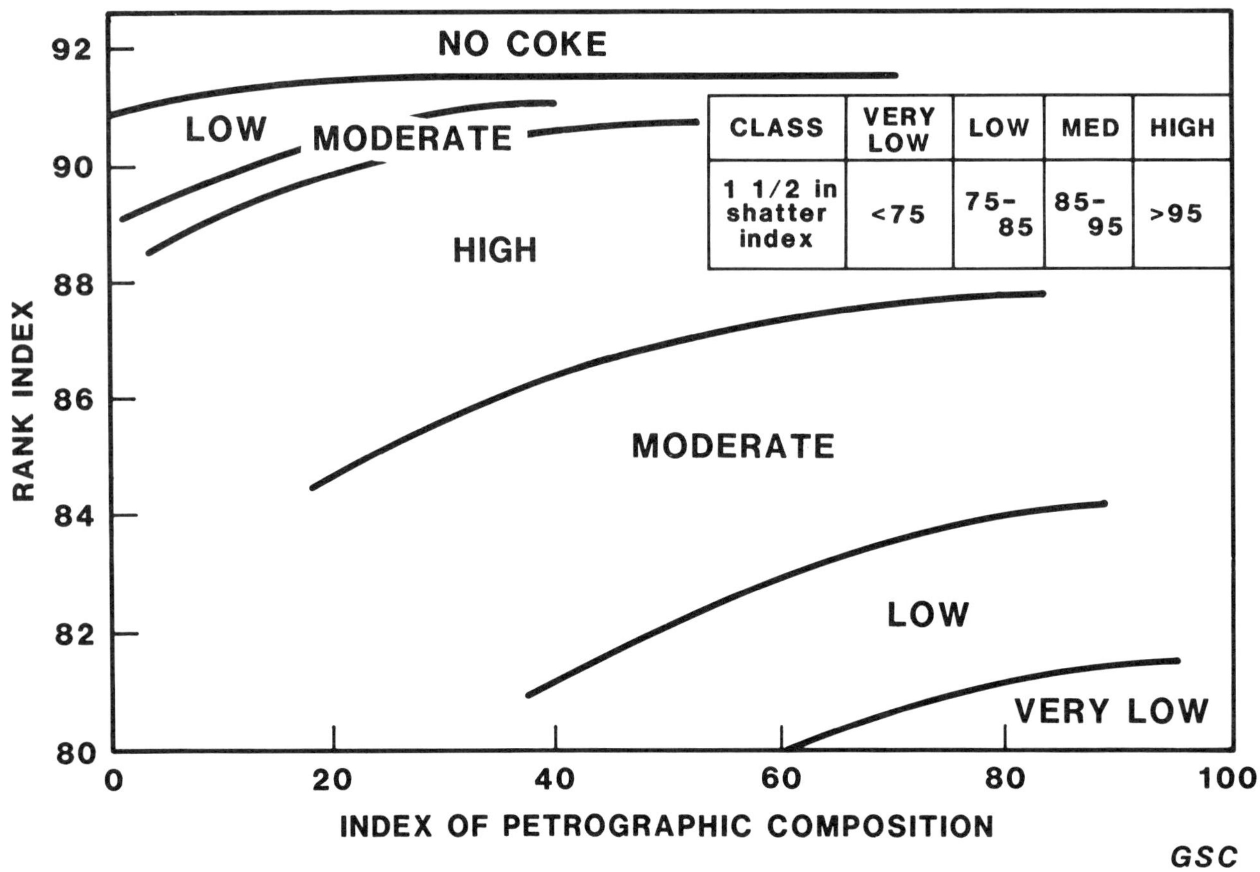

CLASS	VERY LOW	LOW	MED	HIGH
1 1/2 in shatter index	<75	75-85	85-95	>95

Figure 101. Relationship of petrographic composition and rank to 1½ inch shatter index of coke (from Brown *et al.*, 1964b).

10.3.3.1.4 Method of Brown *et al.*

The coke strength prediction method of Brown *et al.* (1964b) differs rather radically from those already described. Virtually all semifusinite is considered inert in this method. The predominantly inert character of semifusinite was elaborated upon by Taylor *et al.* (1967). As a result of small scale coking tests, they concluded that nearly all semifusinite remained infusible during heating to carbonization temperatures. Consequently Taylor and his colleagues suggested that as a general rule semifusinite with a reflectance higher than 0.1 to 0.2 compared to the mean reflectance of vitrinite in the same coal is inert. Brown *et al.* also considered two other factors which have not really been applied in formulating the methods already described. One of these factors is related to the size of the inertinite masses in the coal. If these are relatively large they tend to form centres of weakness in the resulting coke and therefore have a deleterious effect on strength. Smaller masses are more effectively bonded by the cementing action of the reactive constituents. The second and related factor concerns the way reactive and inert materials are admixed, that is, their association. For these reasons, the coke quality studies undertaken on Australian coals by Brown *et al.* defined coal composition in terms of microlithotypes with the reactives being expressed by the content of the microlithotypes vitrite and clarite. The pertinent relationships of this study are shown by Figures 100 and 101 in which the ¼ hardness factor and the 1½ inch shatter index have been correlated with rank index and petrographic index. Rank index is quoted in terms of ultimate carbon (daf) though the authors suggest that it might equally well be expressed by vitrinite reflectance. Petrographic index means content of the microlithotypes vitrite plus clarite. This is a rapid and simple method of correlating petrographic composition and coke quality and in many cases it provides sufficient information for an initial assessment of a coal's coking potential.

10.3.3.1.5 Method of Steyn and Smith

A coke quality prediction method developed in South Africa by Steyn and Smith (1977), considers variable proportions of semifusinite as reactive. Many South African coals contain large amounts of semifusinite yet behave as quite respectable coking coals, so that it is almost certain that some of the semifusinite does react.

A later amplification by these South African researchers (Smith *et al.*, in press) suggests three methods to determine reactive semifusinite. The first method directly utilizes physical characteristics as determined microscopically to separate vitrinite, liptinite, reactive inertinite, and nonreactive inertinite. Vitrinite and liptinite are relatively easy to identify and to assign to reactives. The difficult part is the distinguishing of reactive from nonreactive inertinite solely on the basis of microscopic criteria. It is pointed out that this method requires experienced petrographers. The second method is apparently based on a certain relationship between the content of vitrinite in a coal and the percentage of semifusinite present which can be regarded as reactive. The relationship is illustrated by the diagram on Figure 102. In this approach the total amount of semifusinite is determined in a maceral count and then a percentage of it, which is to be added to reactives, is determined from the curve according to the amount of vitrinite present. The third method utilizes the volatile matter (dry basis) of a given coal and the mean maximum reflectance of the vitrinite to determine the total reactives in the coal according to the relationship shown on Figure 103. In this case detailed maceral analysis is not required, although at least vitrinite content must be determined so as to establish which of the two paths (high or low vitrinite) is to be followed. Apparently in practice the results of the three methods often correspond closely and the authors suggest that if they are within 3 per cent of each other they can be averaged. The liptinite content of South African coals is normally low and is usually treated as vitrinite in coke quality calculations. Once the total reactives have been determined, coke strength is predicted by the use of two indices; a Composition Balance Index similar to that of Schapiro *et al.* (1961) and an RoR index which is an overall reflectance index, a weighted average which takes into account the reflectance and amounts of vitrinite, liptinite, and reactive semifusinite. These two indices are then used to predict two coke strength parameters, M_{40y} and M_{10y}, according to the relationships shown on Figure 104. The index M_{40y} is not the same as M_{40} mentioned earlier. The M_{40y} is on average about 7 to 10 points lower than an M_{40} determined on the same coke, while the M_{10y} is also lower than the corresponding M_{10} value.

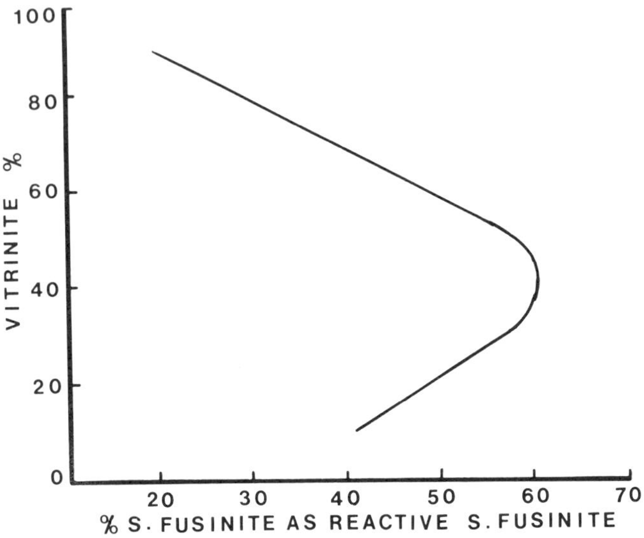

Figure 102.
Graph showing relationship of reactive semifusinite to vitrinite content (from Smith *et al.*, in press).

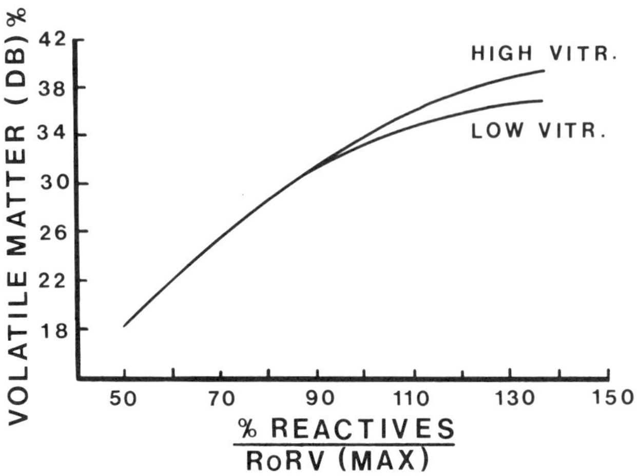

Figure 103.
Relationship of reactives/vitrinite reflectance to volatile matter content (from Smith *et al.*, in press).

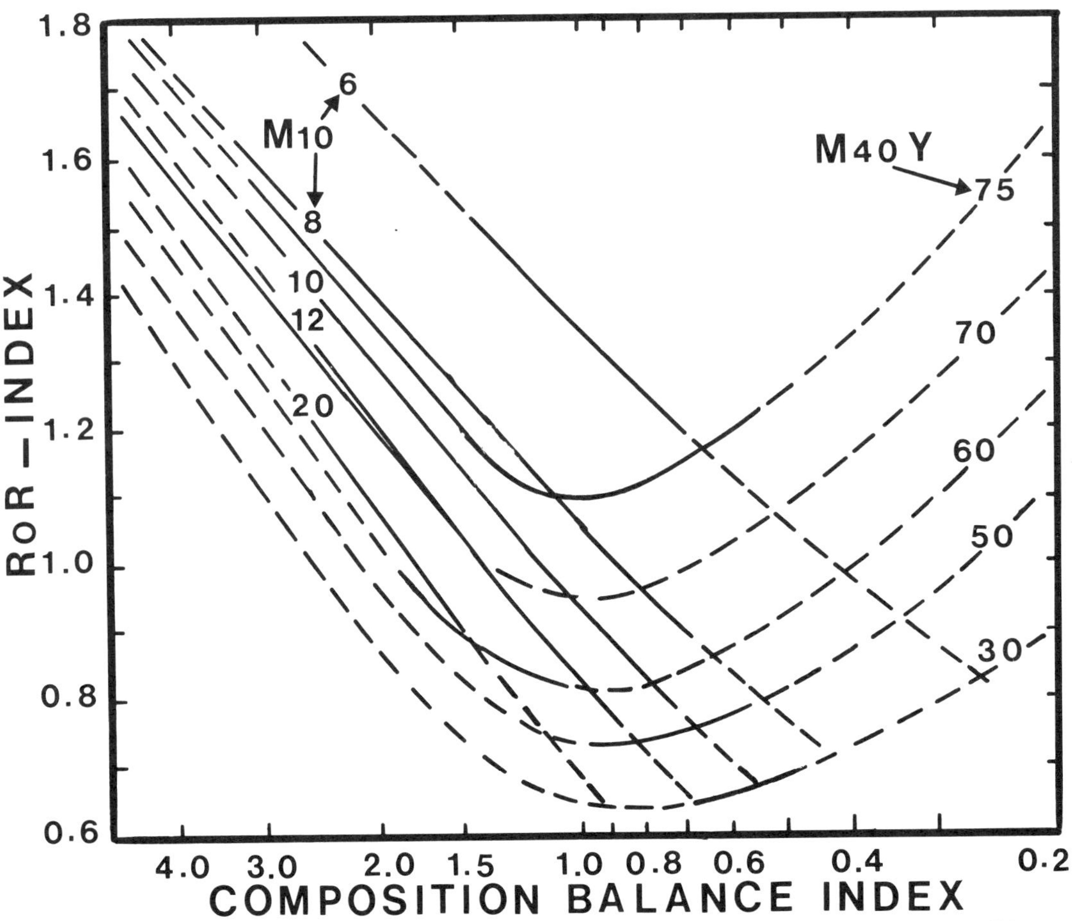

Figure 104. Graph for determining M40y and M10y indices from reflectance and Composition Balance Index (from Steyn and Smith, 1977).

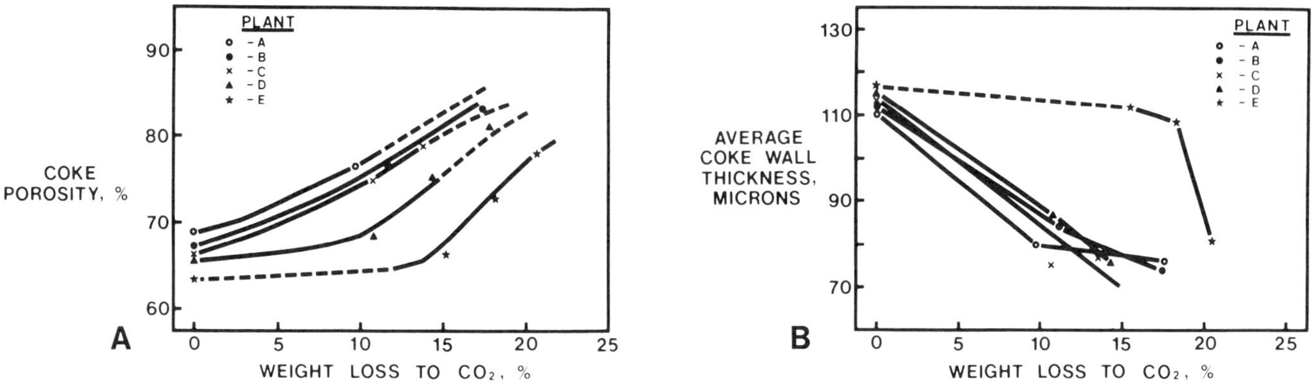

Figure 105. Relationship of coke porosity (A) and coke wall thickness (B) to CO_2 weight loss (reactivity) (from Benedict and Thompson, 1980).

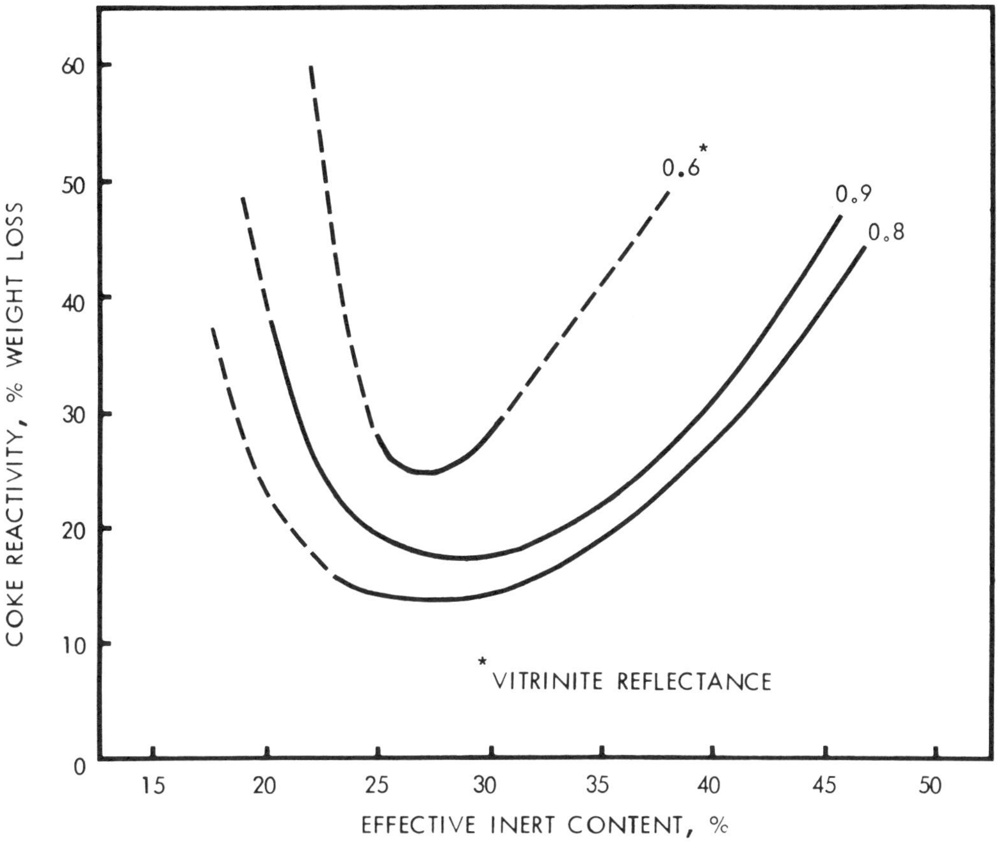

Figure 106. Relationship of effective inert content (Bethlehem Steel method) to coke reactivity (from Thompson and Benedict, 1978).

10.3.3.2 Relationship of Petrography to Other Coking Parameterse

Mention has been made of a number of other properties of coal and coke which reflect directly or indirectly on the quality of coke. One of these properties is reactivity. Schapiro and Gray (1963) investigated reactivity and related it to microscopically observed coke textures, particularly pore volume and reflectivity. High pore volume was correlated with high reactivity. The Bethlehem Steel group also have been quite active in studying the relationship between coke reactivity and coal petrography (Benedict and Thompson, 1973, 1980). They showed that reactivity is related to porosity and to average pore-wall thickness (Figure 105). They also showed a relationship between reactivity and effective inert content for high-volatile coals (Figure

106). Figure 106 suggests that, just as for strength, there may be an optimum effective inert content for reactivity as well. Benedict and Thompson state that an important use of reactivity measurement for them has been in the monitoring of coke uniformity. Coke uniformity is influenced strongly by the maximum temperature developed in the coke oven and maximum temperature in turn is closely correlated with reactivity.

Expansion/contraction (volume change) and oven pressure are two other important coking properties of coal which have been related to compositional characteristics. When coal is coked it is necessary that it contract so that it can be pushed from the oven. The somewhat related property of pressure on the oven wall developed during coking must not be excessive, otherwise oven walls may be damaged.

Coals of high-volatile bituminous rank and the low end of the medium-volatile range contract sufficiently to permit pushing, while higher rank coking coals exhibit net expansion and present difficulties in pushing. In addition to the rank relationship, volume change is affected by the inert content of the coal as shown by Figure 107 from Gray *et al.* (1978).

As indicated earlier, oven wall pressure is also strongly related to rank with the higher rank coking coals, particularly those in the low-volatile range, exhibiting dangerous pressures when used alone. There is also a relationship with inert content as shown on Figure 108, which shows that pressure decreases with increasing inerts.

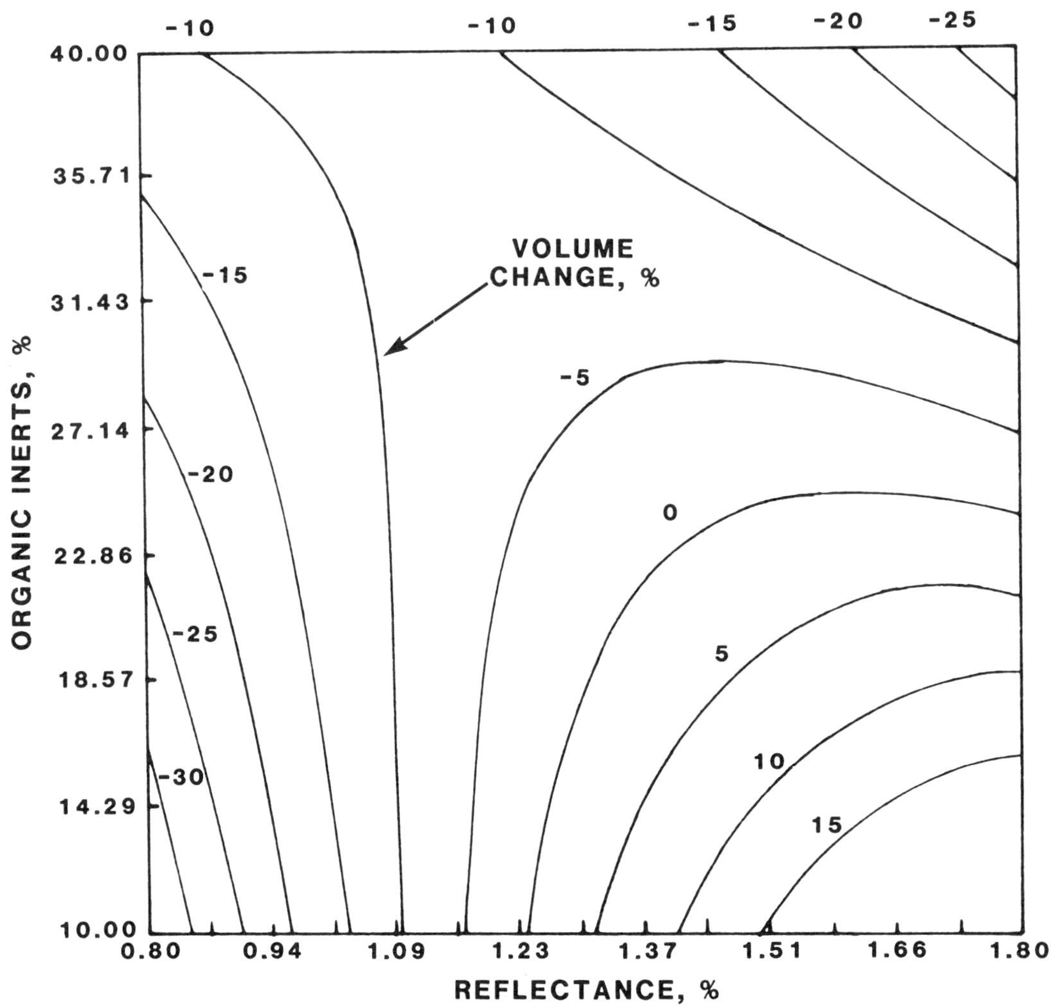

Figure 107. Relationship of rank (by reflectance), and content of organic inerts to volume change in coke oven charges (from Gray *et al.*, 1978).

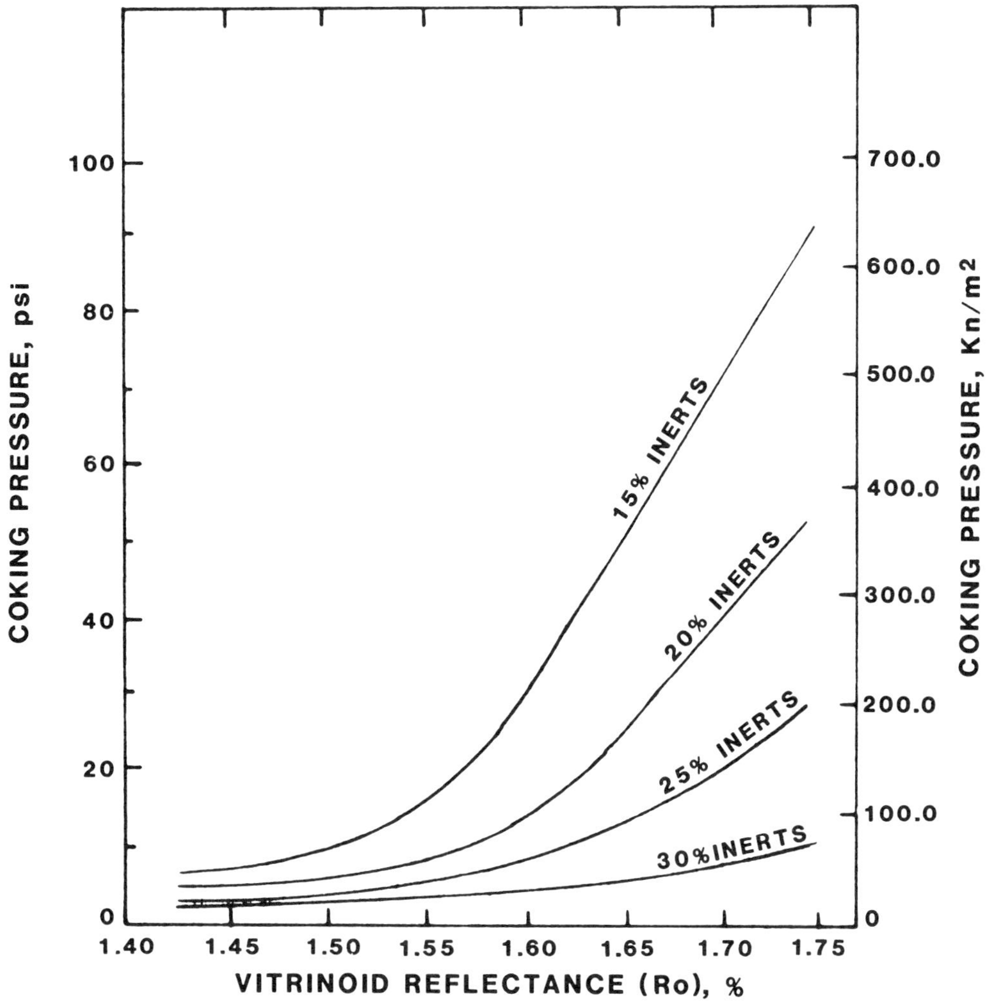

Figure 108. Relationship of rank (by reflectance), and inert content to pressures developed in coke ovens (from Gray *et al.*, after Thompson *et al.*, 1965).

Fluidity and FSI are two properties of coal which are often related to coking ability. These properties are very much influenced by oxidation, but in fresh coals the primary influences are rank and maceral composition as shown on Figure 109 (Paulencu and Readyhough, 1974) and Figures 110 and 111 (Pearson, 1980, see section 8.3). On Figure 109 note the position of western Canadian coals. They appear to follow a different track with lower fluidities when compared with Appalachian coals of the same rank. An examination of Figure 111 may show part of the reason for this. Nearly all of the Canadian coals show inert contents of 30 per cent or higher and as a result they plot in the areas of 100 dd/m or less. Figure 110 shows that higher FSI values are associated with lower inert contents and also generally tend to be associated with intermediate ranks. Table 37, adapted from Cameron and Babu (1968), shows the effect of petrographic composition on fluidity and FSI for samples of a western Canadian coking coal. Pairs of samples were tested for these two properties with one sample of the pair being the whole coal and the second a hand-picked vitrain fraction. The latter samples are composed, therefore, of essentially 100 per cent reactives. The table shows marked differences between the reactive-rich vitrain fractions and their whole-coal counterparts for these two rheological properties. Because of the sampling history, there is strong suspicion that sample pairs 1-10 were mildly oxidized, yet the vitrain fractions retain strong FSI capability.

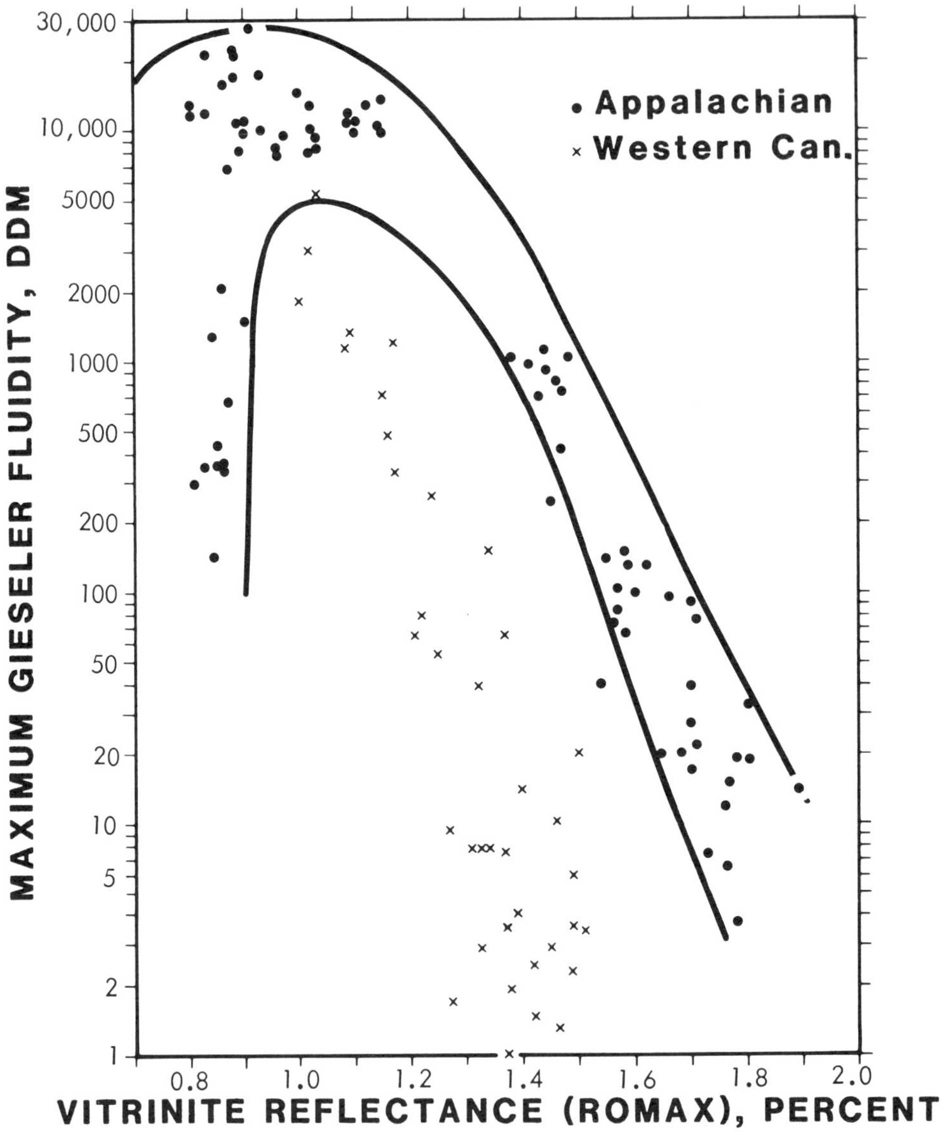

Figure 109. Relationship of vitrinite reflectance to Gieseler fluidity (from Paulencu and Readyhough, 1974).

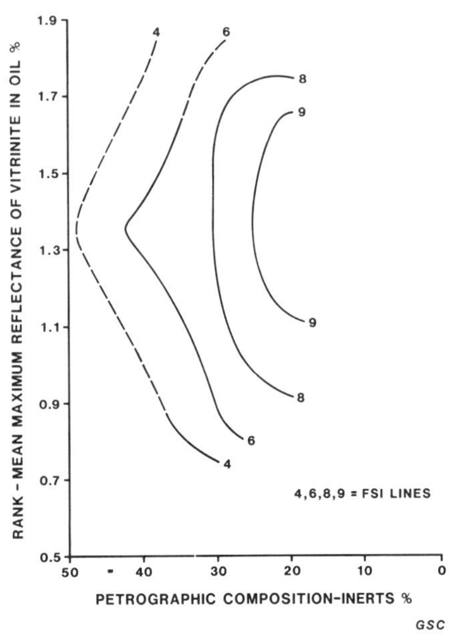

Figure 110.
Relationship of rank (by reflectance) and inert content to Free Swelling Index (FSI) (modified from Pearson, 1980).

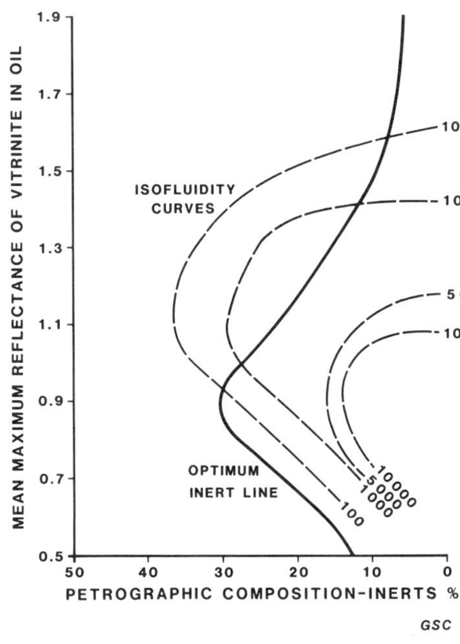

Figure 111.
Relationship of rank (by reflectance) and inert content to Gieseler fluidities (after Pearson, 1980).

TABLE 37. FLUIDITY (GIESELER) AND FSI DATA ON WHOLE COAL AND VITRAIN SAMPLES (from Cameron and Babu, 1968).

Sample	Start. temp. °C>1 DD/M[1]	Temp. of max. fluidity °C	Final fluid temp. °C<1 DD/M[1]	Temp. of solid. °C; no movement	Melting range in °C	Max. fluidity DD/M	Total movement in dial divs.	FSI[2]
1 Vitrain	466	487	504	510	38	8.3	59	9+
Whole Coal	-	480	-	497	-	0.4	3	1
2 Vitrain	477	486	503	512	56	8.5	84	9+
Whole Coal	480	480	486	501	6	1.2	11	4½
3 Vitrain	468	471	475	500	7	1.1	12	8
Whole Coal	-	469	-	498	-	0.6	6	3
4 Vitrain	466	492	505	514	39	11.8	73	9+
Whole Coal	-	484	-	502	-	0.9	8	4
5 Vitrain	462	480	495	502	33	3.2	32	9+
Whole Coal	-	480	-	501	-	0.5	5	2½
6 Vitrain	474	497	512	522	38	52	41	9+
Whole	477	477	483	502	6	1.2	10	5
7 Vitrain	455	480	504	510	49	10.0	80	9+
Whole Coal	-	477	-	501	-	0.9	8	3½
8 Vitrain	458	481	501	508	43	13.0	79	9+
Whole Coal	492	492	495	516	3	1.2	10	4½
9 Vitrain	470	493	508	514	38	16.5	87	9+
Whole Coal	463	481	488	503	7	1.3	14	3
0 Vitrain	465	486	501	510	36	3.8	32	9
Whole Coal	-	494	-	510	-	0.5	5	1½
1 Vitrain	456	483	502	510	46	20	121	8
Whole Coal	-	473	-	486	-	0.5	4	1
2 Vitrain	444	475	500	504	56	28	225	9+
Whole Coal	456	468	473	492	17	1.6	15	4
3 Vitrain	436	473	501	505	65	85	642	9+
Whole Coal	459	465	482	491	23	1.4	15	4
4 Vitrain	437	471	498	504	62	88	644	9+
Whole Coal	447	474	492	498	45	11	77	7.5
5 Vitrain	438	475	500	504	62	81	622	9+
Whole Coal	447	474	491	496	46	7.2	63	7+

[1] DD/M = Dial Divisions/Minute.
[2] FSI = Free Swelling Index.

Finally, a word should be said about coke microscopy. A detailed discussion on this subject was published as far back as 1952 (Abramski and Mackowsky, 1952). Much of the earlier work involved determination of pore texture in coke and its relationship to properties such as strength and reactivity. The evaluation of pore structure was made much more rapid with the advent of automated microscopy (Denton *et al.*, 1967). Additional studies on coke have been directed toward determination of its optical properties and particularly in the differentiation of isotropic and anisotropic domains and in the differentiation of the latter into various mosaic categories depending on size (Mackowsky, 1982c; Patrick *et al.*, 1973; Murchison, 1978; Goodarzi and Murchison, 1978; Marsh and Smith, 1978). Because development of mosaic texture varies with rank (see Plate 16), such studies are especially helpful in investigations made on blends. Bayer and Denton (1966) showed a relationship between the reflectance of cokes and that of the coals from which they were made. Such information was then used to judge blending efficiency, compatibility of coals, and presence of oxidized coal. In the studies of those researchers whose main interest is coke strength, it is also necessary to do some coke microscopy so as to judge the behaviour of the various macerals in the carbonization process.

Plate 16. Photomicrographs showing influence of coal rank on structure of coke walls. Coke made from low rank bituminous coal (A) is essentially isotropic; cokes made from higher rank bituminous coals (B and C) show increased anisotropic properties, while coke D from low-volatile bituminous coal shows pronounced anisotropy with large individual graphite-like anisotropic units. This graphite-like structure is believed responsible for the high strength of cokes made from low-volatile coals (from Thompson and Benedict, 1974). (Reflected light, oil immersion).

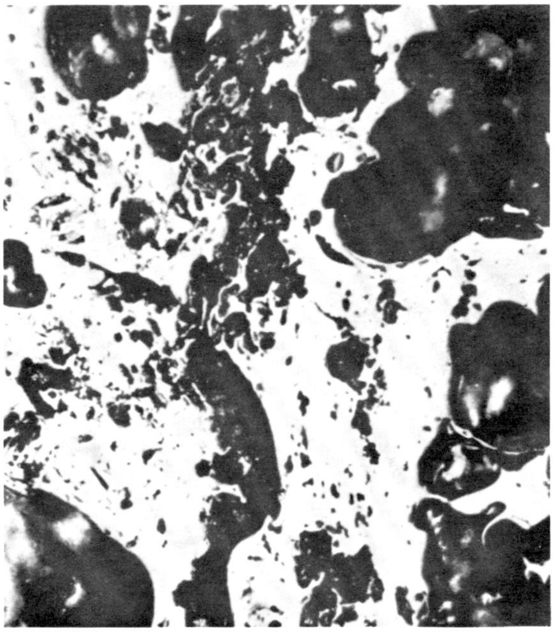

A. COKE FROM REACTIVE VITRINOID OF 0.6% REFLECTANCE

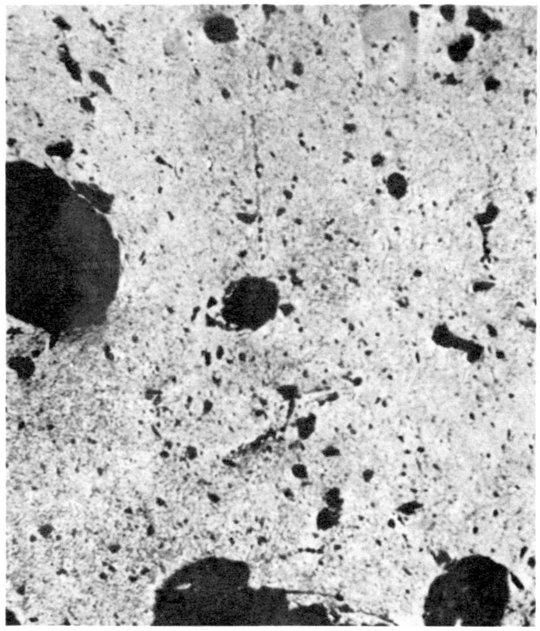

B. COKE FROM REACTIVE VITRINOID OF 0.8% REFLECTANCE

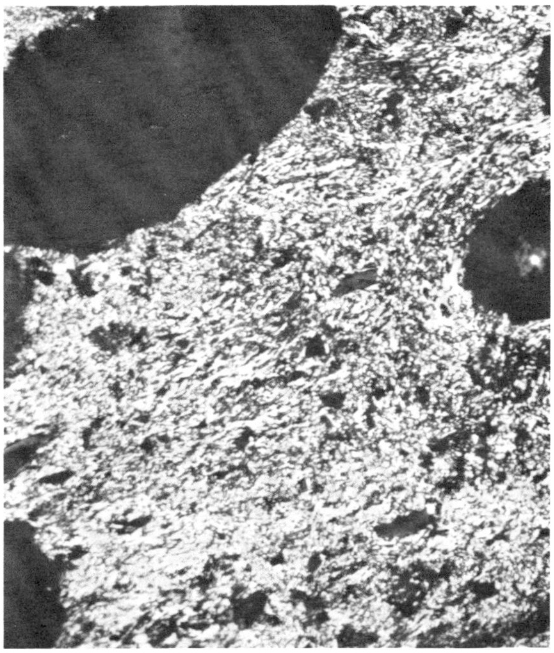

C. COKE FROM REACTIVE VITRINOID OF 1.1% REFLECTANCE

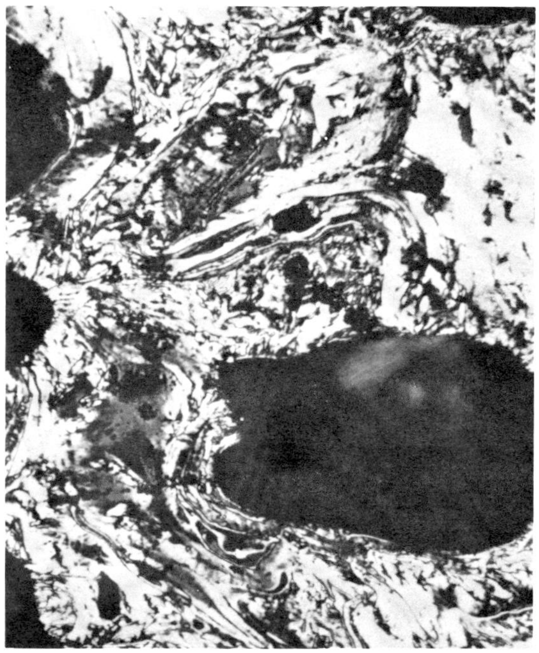

D. COKE FROM REACTIVE VITRINOID OF 1.4% REFLECTANCE

0 200μ

10.4 **Liquefaction**

10.4.1 *Principles and Methods*

In simplest terms the conversion of coal into liquids or gas calls for the adding of hydrogen to the coal. The hydrogen to carbon ratio by weight in coal is below 0.1, while in crude oil and gasoline it is close to 0.2, and in natural gas it is above 0.3 (see Figure 112).

The complex molecules of coal, consisting mainly of aromatic ring-structures and functional groups (Figure 113), have to be broken down by conversion processes to form lighter hydrogen-rich products (Figure 114).

Liquid fuel is derived either from a degradation process using hydroliquefaction or pyrolysis or through a synthesis route using gasification as a first step to synthesize liquid fuels. The principle routes are indicated on Figure 115.

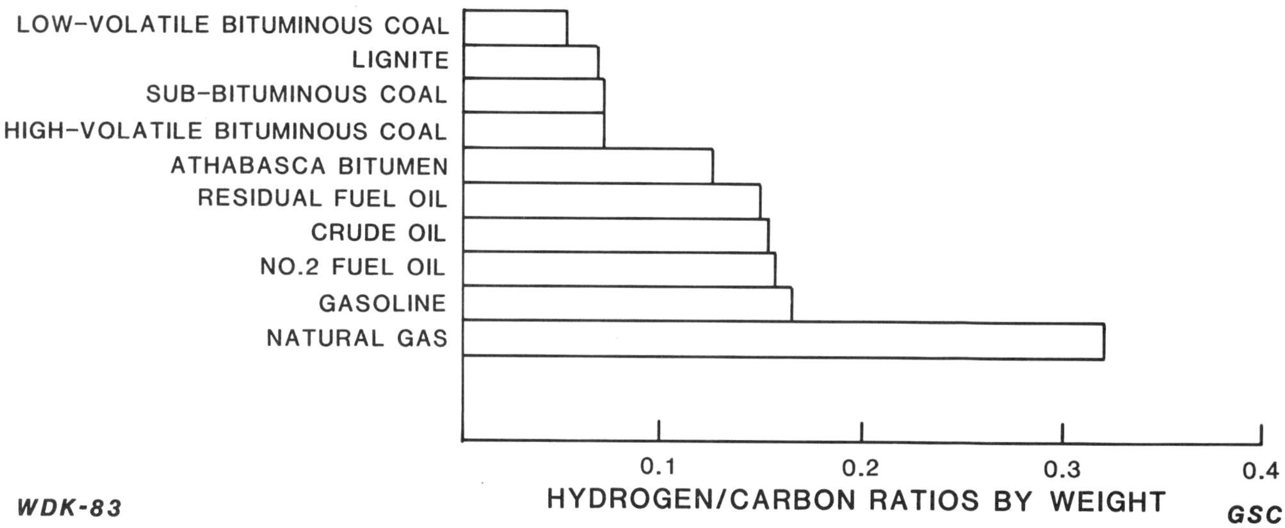

Figure 112. Typical hydrogen-to-carbon ratios by weight in Canadian fossil fuels and refined products (after Taylor, 1979).

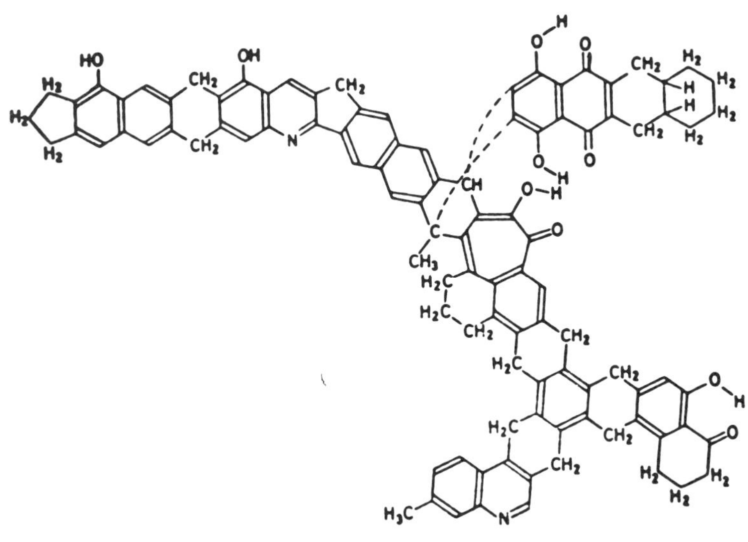

Figure 113. Proposed structural elements of coal (after Given and Peover, 1960).

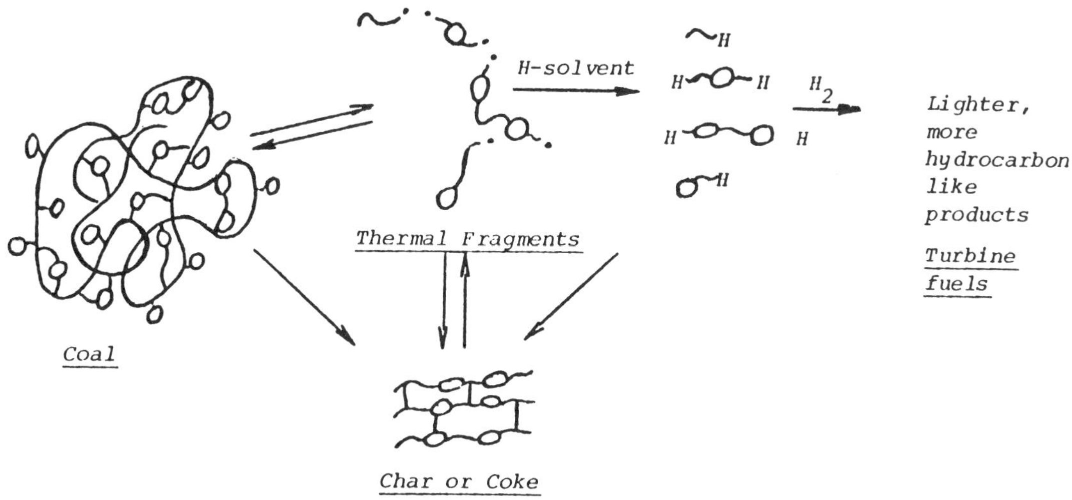

Figure 114. Schematic diagram illustrating formation of hydrogen-rich products and coke from coal (modified after Whitehurst *et al.*, 1980).

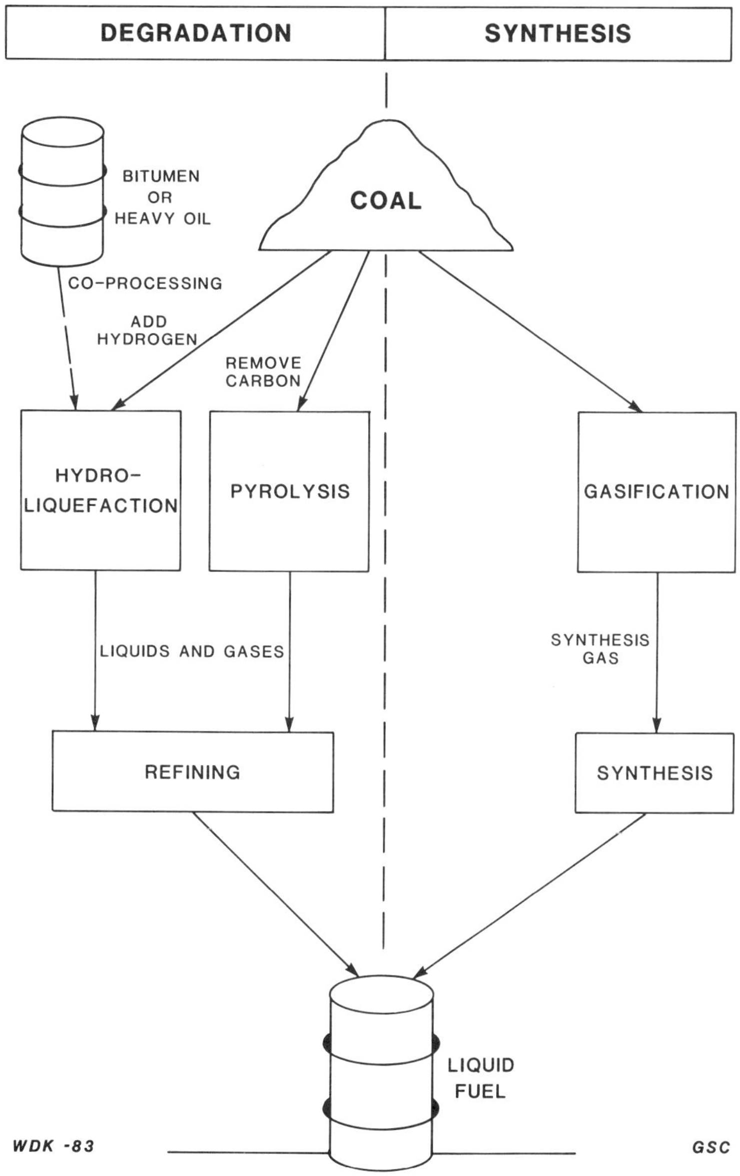

Figure 115. Principal routes to liquid fuels from coal (modified after Taylor, 1979).

In the synthesis method, also called Fischer-Tropsch synthesis, the coal is burned in the presence of oxygen and steam, generating a gas composed mainly of carbon monoxide and hydrogen. The gas is purified and then passed over a catalyst yielding liquids ranging from methanol to hydrocarbons of high molecular weight such as waxes and oils. This process is used in the Sasol I plant in South Africa, where the process has been directed primarily toward the production of motor fuels.

Pyrolysis or *carbonization*, in which coal is heated in the absence of air, is the simplest of the three processes for producing oil from coal. However, the decomposition that occurs results primarily in a carbon-rich coke or char and only to a lesser extent in hydrogen-rich tars, oils, and gases which are suitable for upgrading into liquid fuels. In addition, the coke or char can be gasified to produce hydrogen for the upgrading process. Liquid yields reported in the literature are in general rather low (maximum 35 per cent); conversion might be improved by further development of a fairly new method called flash-pyrolysis, where the coal is heated much more rapidly in the presence of hydrogen leading to a char with improved burning properties and also leading to an increase in liquid and gas yields.

In hydroliquefation the coal is fed into a reactor in the form of a slurry with a substance such as anthracene oil and reacted with hydrogen under high pressure and in the presence of catalysts. The hydrogen can either be added as a gas or donated by a hydrogen-rich solvent which can be recovered after the process. The primary liquids formed are then upgraded to form liquid fuels. The solid residues are either gasified to provide the hydrogen needed or can be converted into coke-like materials through carbonization. There are numerous modifications of this basic hydroliquefaction process under current investigation. Two of the most interesting of these at the moment are: (a) an extraction process where the coal is dissolved in a strong solvent prior to hydrogenation; and (b) a process where synthesis gas (H_2/CO mixtures) is used to increase the rate of conversion with water acting as a carrier medium and hydrogen donor.

To compare the three basic routes it is useful to compare two parameters, namely the thermal efficienty, defined as the percentage to the energy in the coal which is recovered in the final products and the liquid yields produced per ton of coal. The diagrams (Figure 116) show that by far the highest yields are obtained by using the route of hydroliquefaction, reaching values of 2.0 to 3.5 barrels of liquids per ton of coal (0.3′ - 0.5 m^3/tonne), whereas pyrolysis leads to lower amounts of liquid yields. Looking at the termal efficiencies, again hydroliquefaction is preferable over pyrolysis and synthesis, reaching thermal efficiency values about 60 per cent.

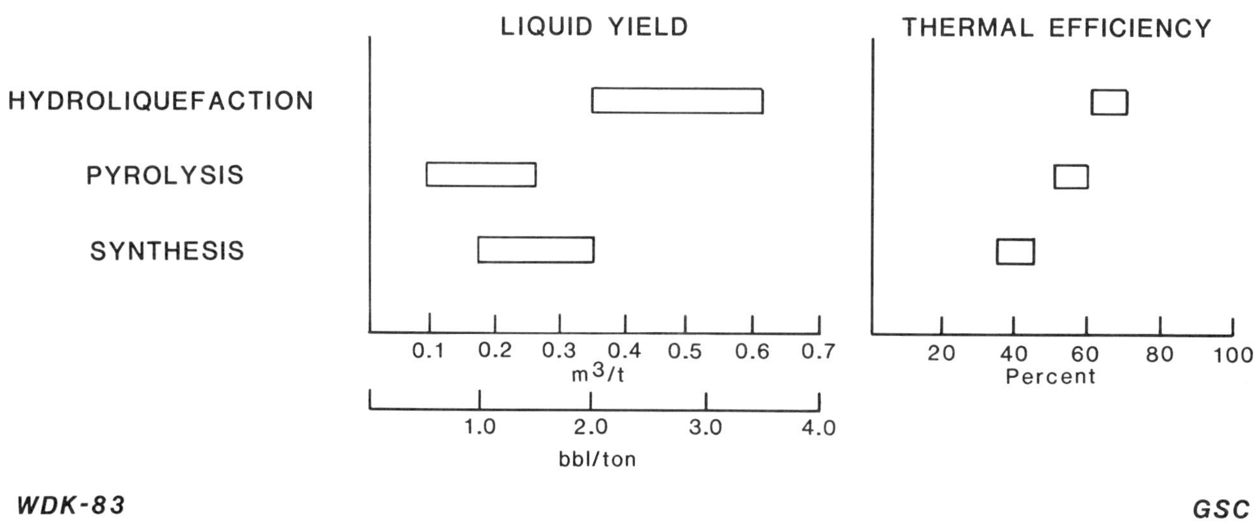

Figure 116. Typical liquid yields and thermal efficiencies in liquefaction (modified after Taylor, 1979).

10.4.2 *Influence of Rank and Petrographic Composition of Coals on Conversion*

Besides experimental conditions (amongst others, temperature, pressure and catalysts), the conversion of coal is highly dependent on rank and composition of the feed-coal, and fairly well developed relations between conversion effectiveness and rank and composition are evident. As in the case of carbonization, the macerals of the vitrinite and liptinite group are considered to be reactive, semifusinite to be partly reactive, and the other macerals of the inertinite group are considered to be non-reactive.

Rank and the ratio of reactive to inert macerals determine how a given coal sample will react in the conversion process. On Figure 117 conversion percentage is plotted against total amount of reactive macerals. In general, conversion increases with increase in reactive macerals. It appears, however, that coals of lower and higher ranks than high-volatile bituminous coals give lower yields although containing the same amount of reactive macerals. The effect of rank (as expressed by reflectance) on conversion is illustrated on Figure 118.

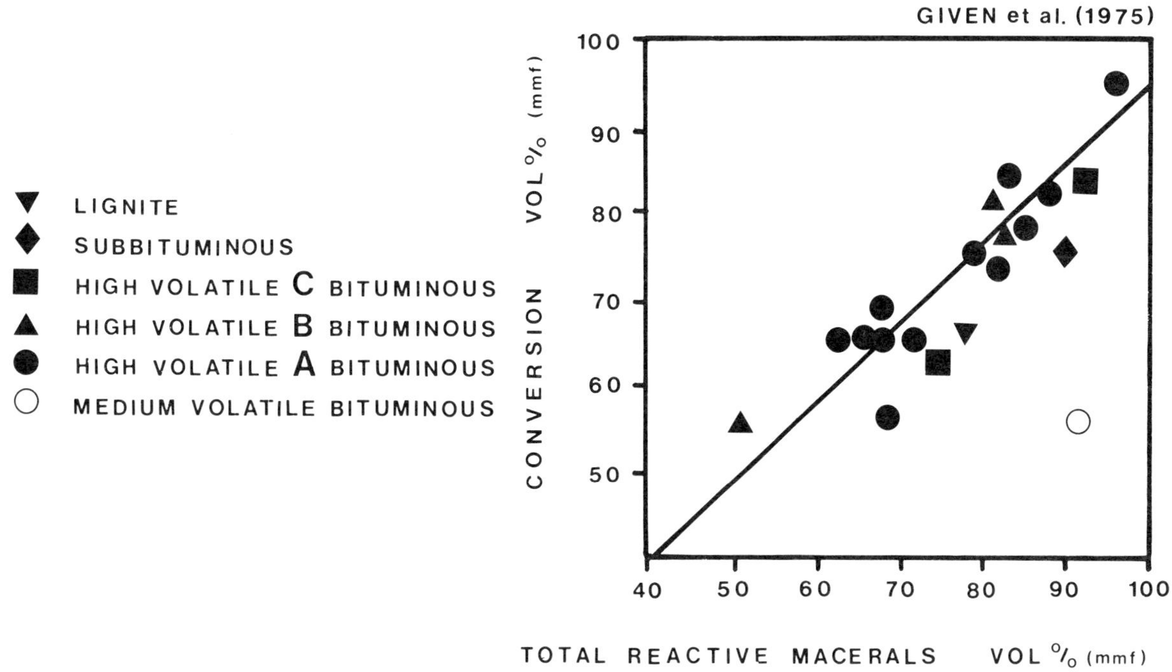

Figure 117. Dependence of conversion on reactive maceral content (after Given, 1975).

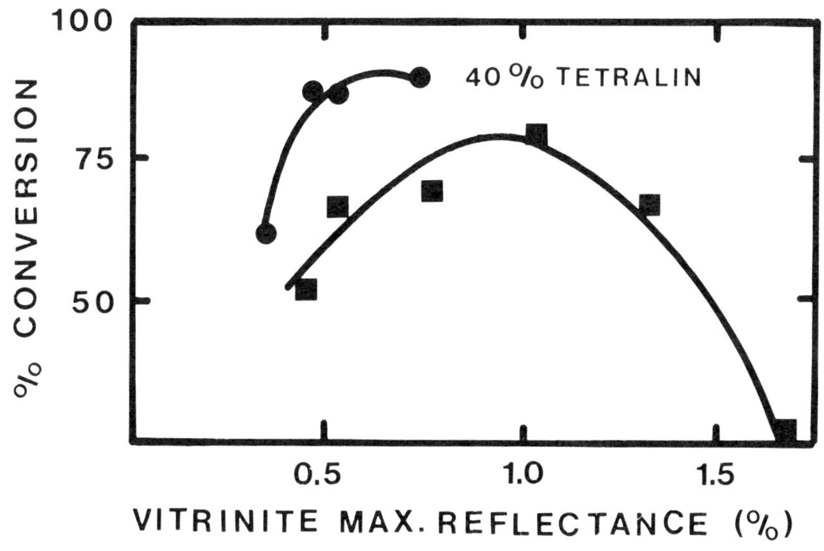

Figure 118. Relation of vitrinite reflectance on conversion of coal of similar composition (after Whitehurst *et al.*, 1980).

Based on a set of coals of similar composition, optimum conversion is achieved in the range of high-volatile bituminous coals (0.67 to 1.10 per cent reflectance range). Figure 118 also shows that conversion can be increased considerably by changing the experimental conditions, in this case by using a hydrogen-donor solvent.

Figure 119 provides a first base for the selection of a coal for the liquefaction process on the basis of rank and composition. The rectangle in the upper left corner indicates the area of predicted optimum liquefaction properties. The reflectance limits correspond to high-volatile bituminous coals. The amount of total reactive macerals should exceed 70 per cent by volume.

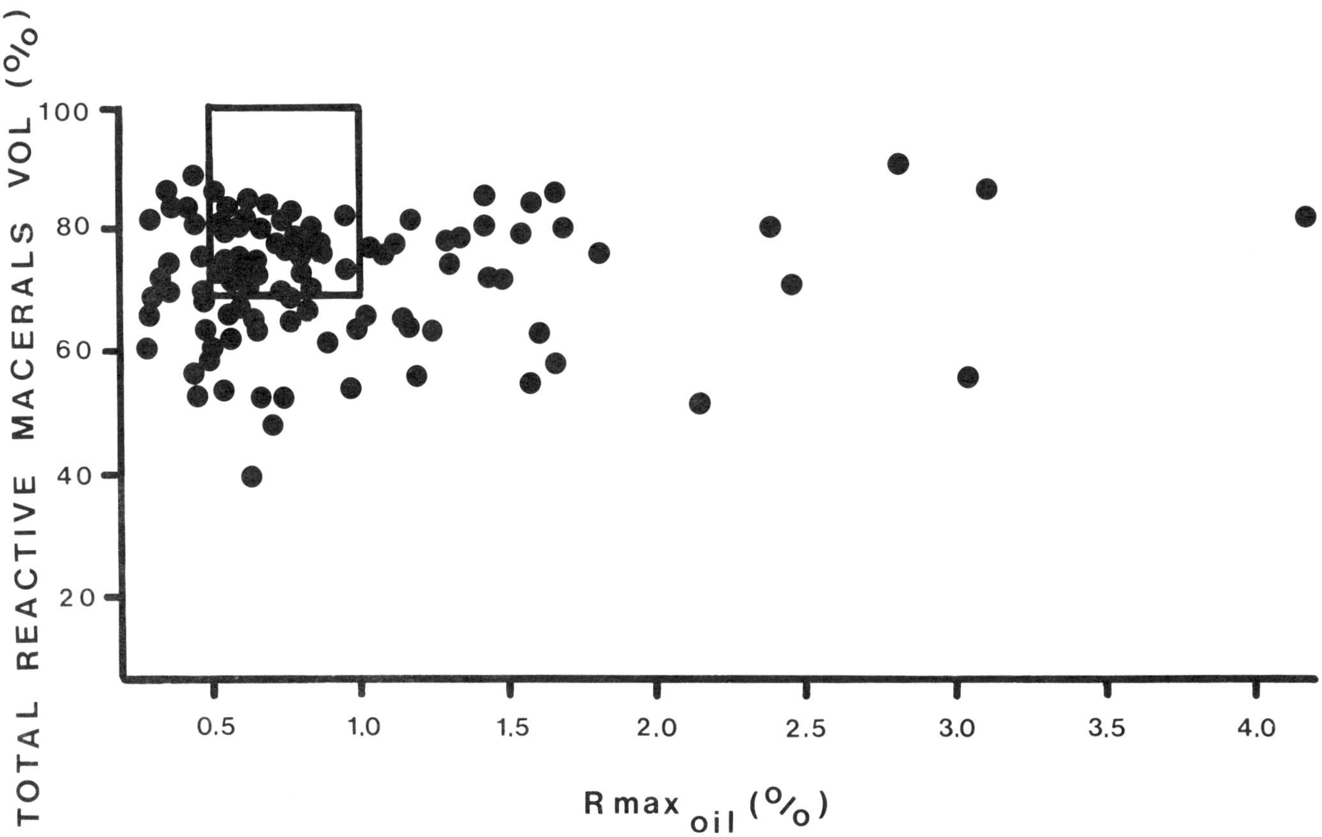

Figure 119. Petrographic control of coal utilization (after Davis *et al.*, 1976).

10.4.3 *Characterization of Residual Material*

Conversion of coals leads not only to generation of liquids and gases but also to formation of solid residual material in the autoclaves. This material may consist mainly of only slightly or nonreacted and newly formed organic material and mineral matter. Detailed study of the structure and composition of the residual material can provide valuable information about the efficiency of the conversion process, changes of the different components during conversion, and also about later use of the residual material (Kalkreuth and Chornet, 1982).

The residual material can be divided into two main categories, whose most important components are listed as follows:

(1) Unreacted or only slightly reacted components, such as minerals and various coal macerals, mainly inertinite;

(2) Materials formed during liquefaction, such as vitroplast, considered to be the once plastic degradation product of vitrinite, and mesophase, which can be formed either by prolonged thermal influence out of vitroplast (Brooks and Taylor, 1965) or by repolymerization of hydrogenated products via a unique fluid system (liquid crystals).

A general classification scheme for characterization of residual material is so far not available, although the main features of the behaviour of coal constituents during liquefaction and their resulting products have been described in detail (Mitchell *et al.*, 1977; Walker *et al.*, 1977, 1978). Table 38 shows a classification scheme of residual material derived from coal. As in the other so far published classifications, it refers to only one coal or a set of coals liquefied under certain experimental conditions. Several of the residue classes are transitional, one into another, and there are often difficulties in where to set the thresholds between the classes.

TABLE 38. QUALITATIVE BEHAVIOUR OF COAL CONSTITUENTS DURING THE LIQUEFACTION PROCESS (AFTER WALKER *ET AL.*, 1975 AND MITCHELL *ET AL* 1977).

COAL PETROGRAPHIC COMP.	DISASSOCIATION MECHANISM	INSOLUBLE FRACTION	RESIDUE COMP.
Vitrinite	→ Granular residue	granular residue	Unreacted Vit. granular res. pitch cenospheres mesophase semicoke
	→ Plastic 'solid' → volatile pitch →	cenospheres	
	→ Plastic 'solid' → non-volatile pitch →	pitch	
	→ Liquid → Liquid-crystal ⇄ mesophase →	semi-coke	
Liptinite & Resinite	Granular attritus		granular res.
Micrinite	Unknown		
Fusinite	None, mechanical fining		Fusinite
Semi-fusinite	Possible fractional contribution		Fusinite un-distinguished
Mineral Matter			
a) Carbonate	Some fining		Carbonate
b) Quartz	None		Quartz
c) Clays	Dispersion		Not recognized
d) Pyrite	Loss of sulfur		Pyrrhotite

It has been found that both composition and rank of the coal, as well as the chosen experimental conditions, are of major influence in the formation of the final products (Diessel, 1978; Diessel and Guyot, in press). The macerals of the inertinite group are considered to be 'inert' when subject to liquefaction conditions, although at least semifusinite is partly reactive (Gray, 1978). The contribution of liptinites in the formation of residual material is still uncertain, whereas the macerals of the huminite/vitrinite group seem to be the main contributors to the formation of residual material. Depending upon the nature of the coal and the experimental conditions, the macerals of this group can give rise to a variety of residue types.

Some characteristic types of liquefaction residues are illustrated on Plate 17.

Plate 17. Examples of residual materials obtained from liquefaction of peat and subbituminous coal. (Reflected light, oil immersion).

Figures 1 to 6 represent raw peat (feedstock) and peat derived residual materials from stepwise increase in temperature (200°C to 380°C). For further details the reader is referred to Kalkreuth and Chornet (1982).

Figure 1. Raw peat, *phlobaphinite* (P) in attrinitic matrix. Peat from Villeroy-Lotbiniere district, Quebec.

Figure 2. Residual material after liquefaction at 200°C. Up to this temperature level the organic material apparently has not been severely altered. This example shows a *textinite* with very distinct cell wall layering.

Figure 3. Residual material after liquefaction at 300°C. At this point the residual material consists mainly of newly formed *vitroplast* (V), characterized by vesicles of different size and inclusions of *inertinite* (I).

Figure 4. Residual material after liquefaction at 325°C. Transition of homogenized *huminitic material* (H) and *vitroplast* (V) into each other.

Figure 5. Residual material after liquefaction at 350°C. Vitroplast (V) filling cell-lumina of fusinite, confirming the once plastic stage of the vitroplast at that temperature level and the unreactive nature of the fusinite.

Figure 6. Residual material after liquefaction at 380°C and in the presence of anthracene oil. Beginning of mesophase (M) formation and granular residue (G). Crossed Nicols.

Figure 7 and 8 represent residual materials obtained from carbonization of liquefaction residues. For further details the reader is referred to Belinko *et al.* (1982).

Figure 7. *Isotropic char* produced by carbonization of liquefaction residues from 300°C liquefaction run.

Figure 8. *Semi-coke* displaying "mosaic" structure produced by carbonization of liquefaction residue from 400°C liquefaction run. Crossed Nicols.

Figures 9 and 10 represent residual materials of a subbituminous coal from Alberta. Liquefaction runs were performed at the Alberta Research Council.

Figure 9. *Mesophase-spheres* (M) of various sizes in *vitroplast* (V). Liquefaction residue from a subbituminous coal, Alberta. Crossed Nicols.

Figure 10. *Mesophase-spheres* (M) of various sizes, *vitroplast* (V) and *semi-coke* (S) formed from coalesced mesophase-spheres. Liquefaction residue from a subbituminous coal, Alberta. Crossed Nicols.

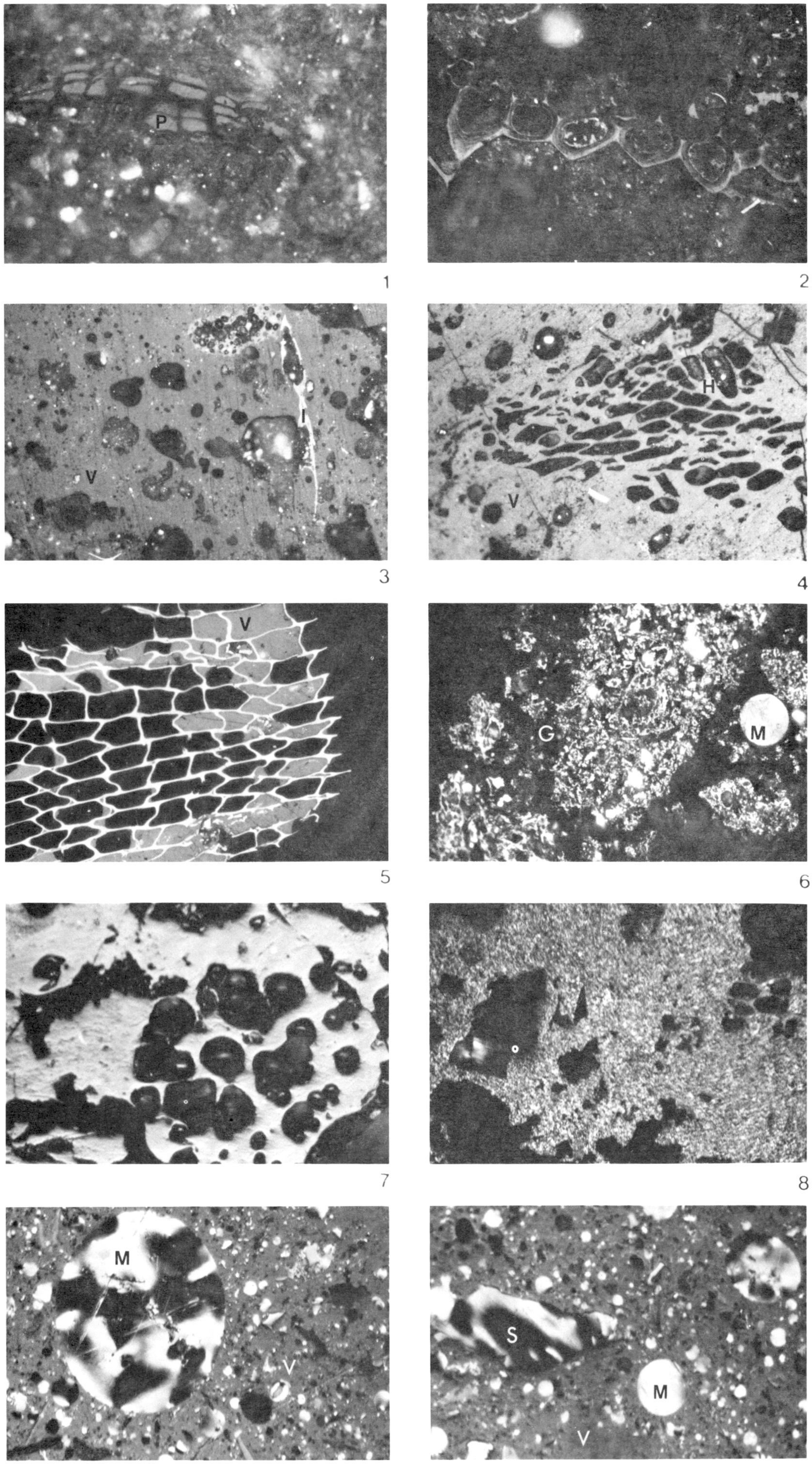
P
1
2
I
V
3
H
V
4
V
5
G
M
6
7
8
M
V
9
S
M
V
10
MICRONS
0
100

10.5 **Oil and Gas Exploration (Kerogen Analysis)**

Coal petrographic methods applied on finely dispersed organic materials have become more and more important in recent years, since many authors (amongst others, Shibaoka *et al.*, 1973; Dow, 1977; Teichmüller and Ottenjann, 1977; Bostick, 1979; Robert, 1981; Gijzel, 1981) have demonstrated that generation of oil and gas can be related to rank, amount and type of the organic material found in a given sedimentary sequence.

Finely dispersed organic matter occurs in most sedimentary rocks. The content of organic material is commonly expressed in per cent organic carbon and may vary between less than 1 per cent by weight in continental sandstones and 10 per cent by weight and more in shales of lacustrine or marine origin. There are varying opinions as to how much organic material in a sedimentary rock is necessary to generate oil. Most of the 'source' rocks which have yielded significant reserves contain several weight per cent organic carbon (Dow, 1974). The lower limit appears to be at about 0.4 weight per cent organic carbon (Hunt, 1979).

The application of coal petrographic methods is restricted mainly to kerogen, which is per definition the solid organic material insoluble in organic solvents (see Figure 120). Organic material soluble in organic solvents (0.5 to 200 mg/g total organic carbon) is obtained commonly by Soxhlet extraction and is analysed by methods of organic geochemistry.

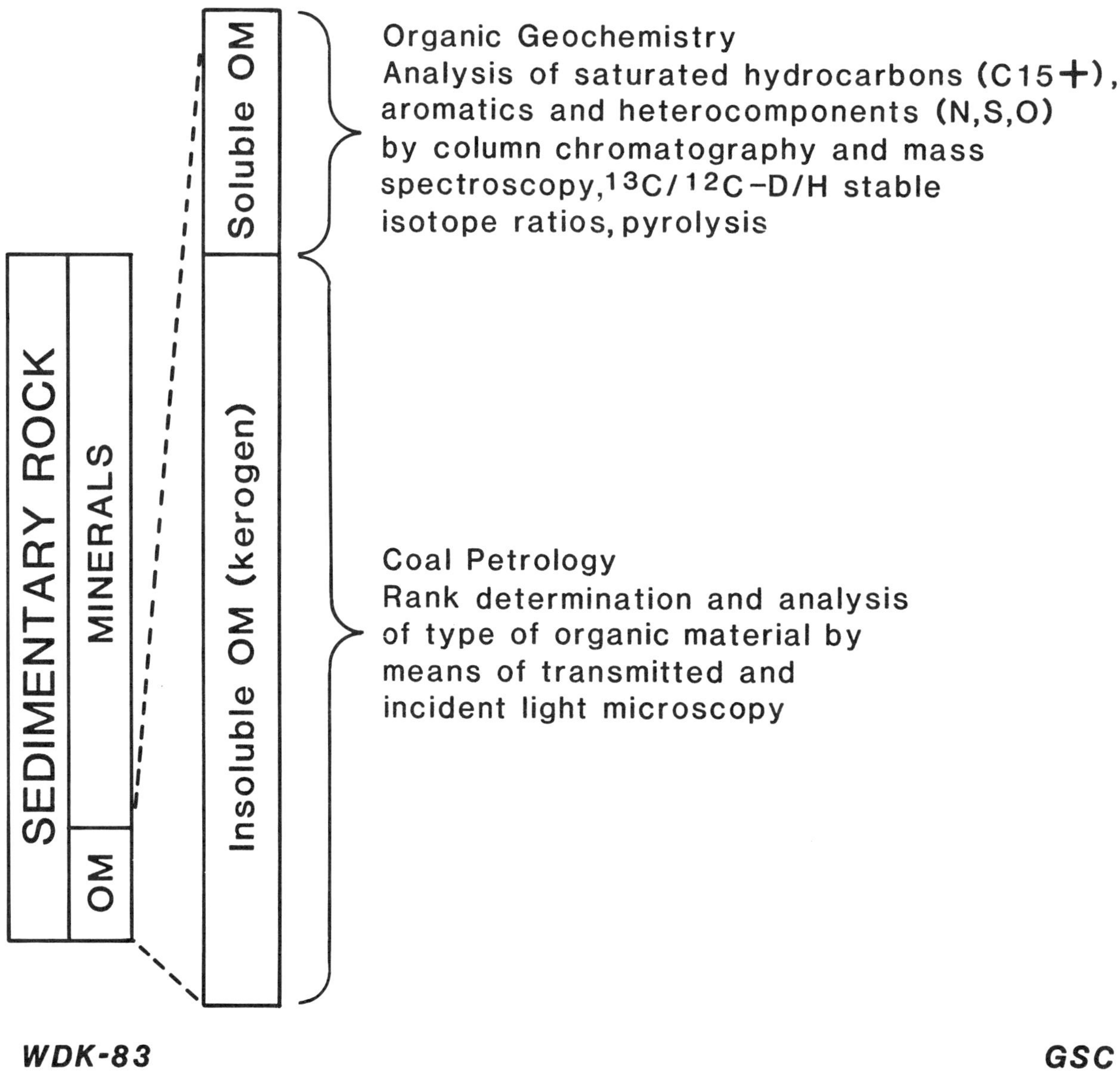

Figure 120. Generalized scheme of distribution of organic material (OM) in a sedimentary rock and methods used in coal petrology and organic geochemistry to characterize maturation level (rank) and type of organic material.

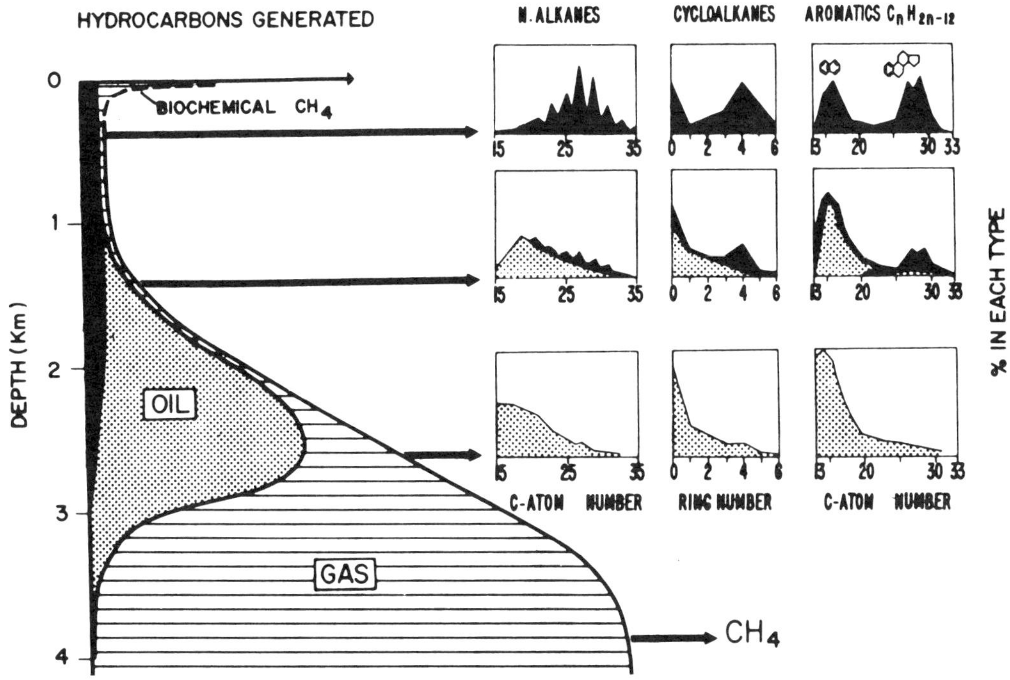

Figure 121. General scheme of hydrocarbon generation. Depth scale represented is based on examples of Mesozoic and Paleozoic source rocks. It is only indicative and may vary according to nature of organic material, burial history, and geothermal gradient (after Tissot *et al.*, 1974).

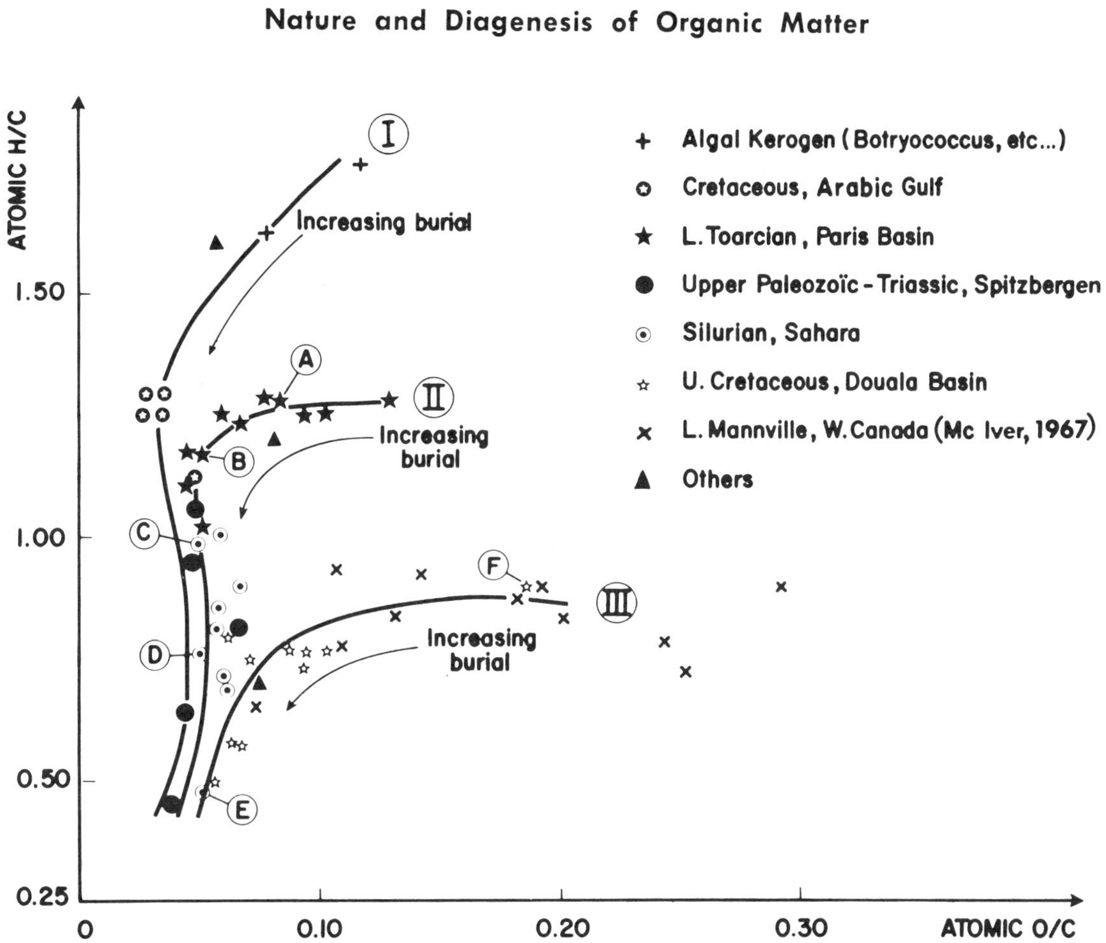

Figure 122. Examples of kerogen evolution paths. Path I includes algal kerogen and excellent source rocks from the Middle East. Path II includes good source rocks from North Africa and other basins. Path III corresponds to less oil-productive organic matter, but may include gas source rocks. Evolution of kerogen composition with depth is marked by arrow along each particular path (after Tissot *et al.*, 1974).

It is now widely accepted that oil and gas are formed by chemical breakdown from kerogen, once the organic material is buried and reaches a particular temperature (>65 degrees Celsius), which in turn is controlled by the geothermal gradient of a particular area. If the temperature exceeds a certain limit, cracking processes of carbon-carbon bonds lead to the formation of lighter hydrocarbons (natural gas) from oil and kerogen. A general scheme of hydrocarbon generation is shown on Figure 121.

The amount of oil and gas formed from a source rock is, however, not only a function of temperature (burial) and time exposed to this temperature, but is highly dependent on the type of organic matter present in the rock. On Figure 122, three main types of kerogens and their evolution paths with increasing burial are characterized by their elemental composition (H/C - O/C diagram).

Type I kerogen is rich in aliphatic structures and has an original high H/C and a relatively low O/C ratio. The organic material of this group consists mainly of alginite and bituminite (see Table 40) and represents excellent source material for the generation of oil. Kerogen Type III, characterized by high amounts of aromatic structures, is relatively rich in oxygen and consists mainly of remains of higher terrestrial plant debris. This kerogen type is much less oil-productive than Type I but may generate gas in greater amounts. Kerogen Type II represents a kerogen which is made up of large amounts of liptinitic macerals, such as sporinite and cutinite, and is considered to be a good source for oil and gas if temperature and time conditions are favourable. Most kerogens contain, however, mixtures of different types of organic materials.

To define the degree of diagenesis of the organic material at which hydrocarbon generation is initiated and the maturation level at which liquid hydrocarbons disappear, various optical methods have been developed. These include amongst others the determination of colour changes of amorphous kerogen (Peters *et al.*, 1977) and measurements of translucency of palynomorphs in samples of isolated organic material (Grayson, 1975), both of which are performed in transmitted light. In incident light measurement of vitrinite reflectance is the most widely applied method to determine rank and appears to be the most consistent and reliable technique because it is more accurate than the subjective colour estimations in transmitted light. The relationships of coal rank, vitrinite reflectance of dispersed organic material, and hydrocarbon generation and destruction is shown on Figure 123 along with some other maturation indices Dow (1977).

Oil generation from kerogen starts at a reflectance level of about 0.5 to 0.6 per cent $R_o$max (oil 'birth' line), and generation of oil ends at a reflectance level of about 1.3 per cent $R_o$max (oil 'death' line). Gaseous products from oil start to occur at a reflectance level of 0.8 per cent $R_o$max. The gas 'death' line is believed to correspond to a vitrinite reflectance of 2.5 to 3.5 per cent $R_o$max (Raynaud and Robert, 1976; Heroux *et al.*, 1979). There are indications that the onset of oil generation might be influenced by the composition of the kerogen itself. Snowdon and Powell (1982) found in parts of the Candian Frontier Basins that oil has been

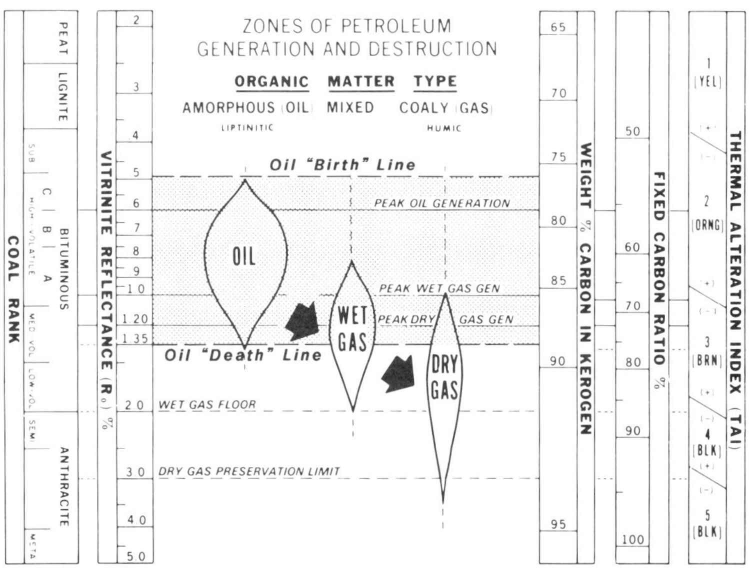

Figure 123. Correlation of the coal rank scale with various maturation indices and the zones of petroleum generation and destruction (modified after Dow, 1974).

generated at vitrinite reflectance levels as low as 0.4 per cent $R_o$max from terrestrial organic matter (kerogen Type III), which in these areas is characterized by high amounts of liptinitic (resinitic) materials. Leythaeuser *et al*. (1980) suggest that the 'birth' line for oil derived from Kerogen Type I lies at 0.5 per cent $R_o$max, for kerogen Type II at 0.55 per cent $R_o$max, and for kerogen Type III at 0.6 per cent $R_o$max. Table 39 illustrates interbasinal differences in the relationships between coal rank, rank of dispersed organic matter, and occurrence of oil and gas deposits for many parts of the world (Bostick, 1979).

10.5.1 *Determination of Vitrinite Reflectance in Kerogen*

The measurement of vitrinite reflectances in kerogen is basically done under the same standardized conditions as used for coals (see Chapter 7). The measurements can be performed either on whole rock samples or on concentrated organic material. If the analysis of whole rock yields only a few measurements, it is preferable to analyse the enriched organic concentrate obtained by one of the methods described under Section 6.2.4 in order to get statistically significant data.

Although vitrinitic material can be found in most sedimentary rocks from post-Silurian on and is considered reliable for rank determination from the stage of peat to anthracite, there are certain problems in using the vitrinite reflectance technique, some of which are listed following.

(1) The sample may contain not only one vitrinite population but two or more. Usually the population having the lowest mean represents the true maturation level. Higher reflecting vitrinite is representative of redeposited organic particles derived from erosion of older strata (recycled vitrinite on Fig. 124).

(2) In samples obtained from cuttings, caving might occur. The contaminated material might be identified when the same reflectance is repeatedly recorded over a great depth interval and the type of organic matter does not change with depth.

(3) Coaly material might have been added to the drilling mud. In particular, if these additives are of similar rank and composition as the original material, it becomes almost impossible to distinguish one from the other (an example is given on Figure 124, where lignite was added to the drilling mud).

(4) Reflectance of vitrinite in oil shales might be depressed considerably by the presence of alginite (Figure 125).

(5) The organic material in samples obtained from outcrop, in particular from porous sandstones or conglomerates, might be oxidized thus lowering the reflectance of the vitrinite considerably.

(6) Changes of vitrinite reflectance due to changes in lithology. Bostick and Foster (1975) demonstrated that vitrinite reflectances in sedimentary rocks (shales, sandstones, limestones) from the same section differed significantly from each other and differed also from associated coal seams.

10.5.2 *Examples of Application*

The primary purpose of the rank determination is to assess whether the geothermal history has been favourable for the generation of oil and gas in a given sequence, and, if they have been generated, whether they have been preserved. In boreholes this is routinely done by selecting sample material covering the complete depth interval and by measuring the rank on these samples. From the individual measurements a regression line is calculated, indicating the increase of rank with depth (coalification gradient), commonly expressed in % $R_o$max/100 metres. From the regression line the level of maturation can be obtained for any depth interval of the sedimentary sequence. Two examples are given on Figures 124 and 126.

TABLE 39. RELATIONSHIPS BETWEEN CATAGENESIS OF COAL AND DISPERSED SOLID ORGANIC MATTER AND OCCURRENCE OF OIL AND GAS DEPOSITS (FROM BOSTICK, 1979)

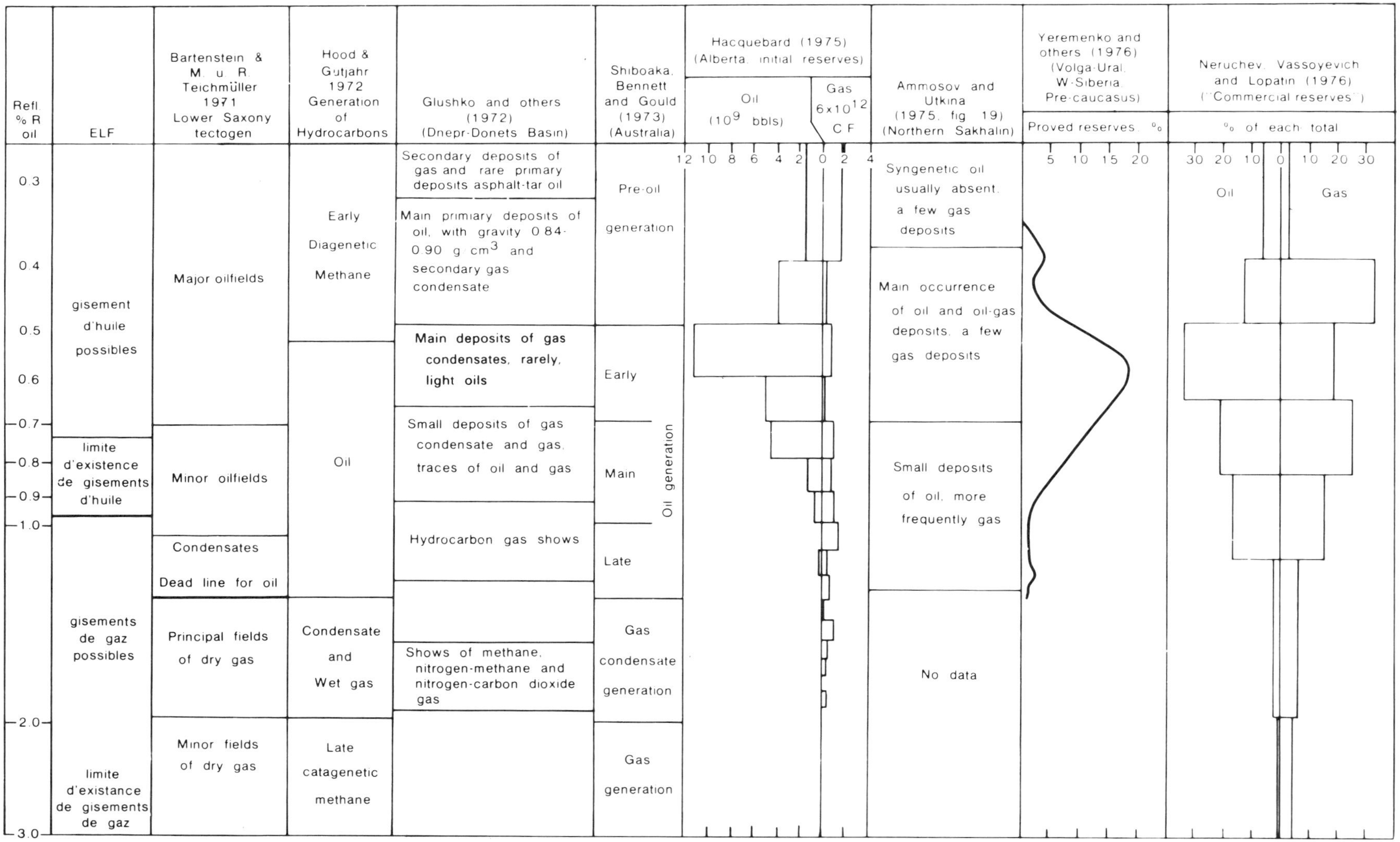

TABLE 39 CONT'D

Refl %R oil	0.3; 0.4; 0.5; 0.6; 0.7; 0.8; 0.9; 1.0; 2.0; 3.0
Volatile matter % daf	50; 45; 40; 35; 30; 20; 10
Btu lb (moist. mineral-matter free)	8,300; 9,500; 11,000; 13,000; 14,000
ASTM coal classes and groups	Lignite; Subbituminous: C, B, A; High-volatile: C, B, A; Bituminous coals: Medium-volatile, Low-volatile; Semi-anthracite; Anthracite
Internat coal classes	13; 12; 11; 9, 10; 8; 7; 6; 5; 4; 3; 2; 1
White, 1915 Eastern United States	Commercial oil fields; Oil deadline; No commercial oil fields, gas fields may occur
Fuller, 1919 North Texas, Paleozoic	Principal oil fields; Principal fields of light oil and gas; Rare high-gravity oil, common gas fields; Usually no commercial production; No oil or gas, with rare exceptions
Lilley, 1924 Eastern U.S. and Midcontinent, Pennsylvanian	Mixed base below 35°; Mixed base 35-40°; Paraffin base above 40°; Dry gas only; No hydrocarbons
Ammosov, 1961 (Intl Geol Review, 1962) Ural-Volga and North Caucasus Districts	Major oil fields; Minor oil fields, but of high quality; Rare oil occurrences
Rodionova, 1967 North Caspian Basin Paleozoic	Main formation of bituminoids
Vassoevich *et al* 1967 Review of world picture	Occasional dry gas; Naptnene-methane oil, wet gas; Methane oil increases over napthene; more solid paraffins and light aromatics, wet gas; Highly paraffinic oil, rich in normal alkanes; Gas condensates; Lowest limit of commercial oil; Gas; Dry gas; Lowest limit of commercial gas
Correia, 1967 Sahara, Paleozoic	Oil deposits; Gas and shows of oil; Gas; Hydro-carbon traces
Staplin 1969 Canadian Great Plains	Wet or dry hydrocarbons; Dry gas; Dry gas or barren
Vassoevich, *et al* 1969 Review of world picture	Methane formation in early catagenesis; Oil formation (plus gas); Wet gas and gas condensates; Late catagenetic methane
Ammosov and Gorshkov (1969, fig 14) West Siberian platform — Density of oil in near-source deposits (at 20°C, g cm^3)	0.90; 0.88; 0.86; 0.84; 0.82; 0.80
Kontorovich (1970 table 6) (Western Siberia) — Intensity of process of oil formation	Oil formation does not take place. Gas product is CO_2; Weak oil formation. Methane is main product of gas formation; Main zone of oil formation. Methane is main product of gas formation; Oil formation dies out. Gas formation continues; No oil formation. Gas formation continues, with dry methane as main product
Kontorovich (1970 table 6) (Western Siberia) — Evaluation of prospects for oil and gas	Syngenetic oil absent; **Main zone of oil and gas occurrence. Various kinds of oil**; **Subordinate oil occurrence**; Oil usually absent. Possible deposits of gas and, more rarely, gas condensate
Preobrazhenskaya, Klucheva and Ivanova (1971) (Giant oil and gas deposits of the U.S.S.R.) (% of total)	see chart below

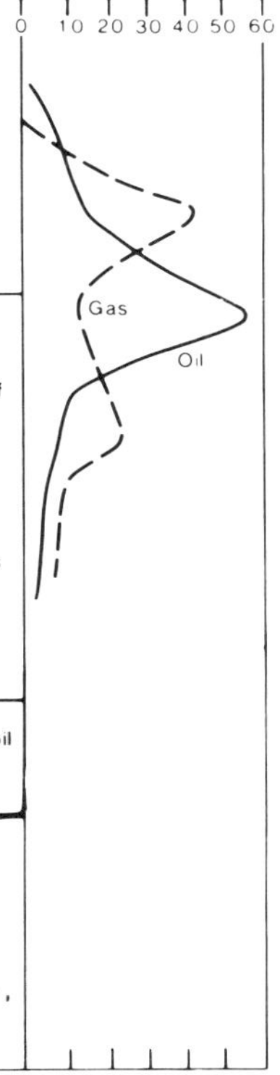

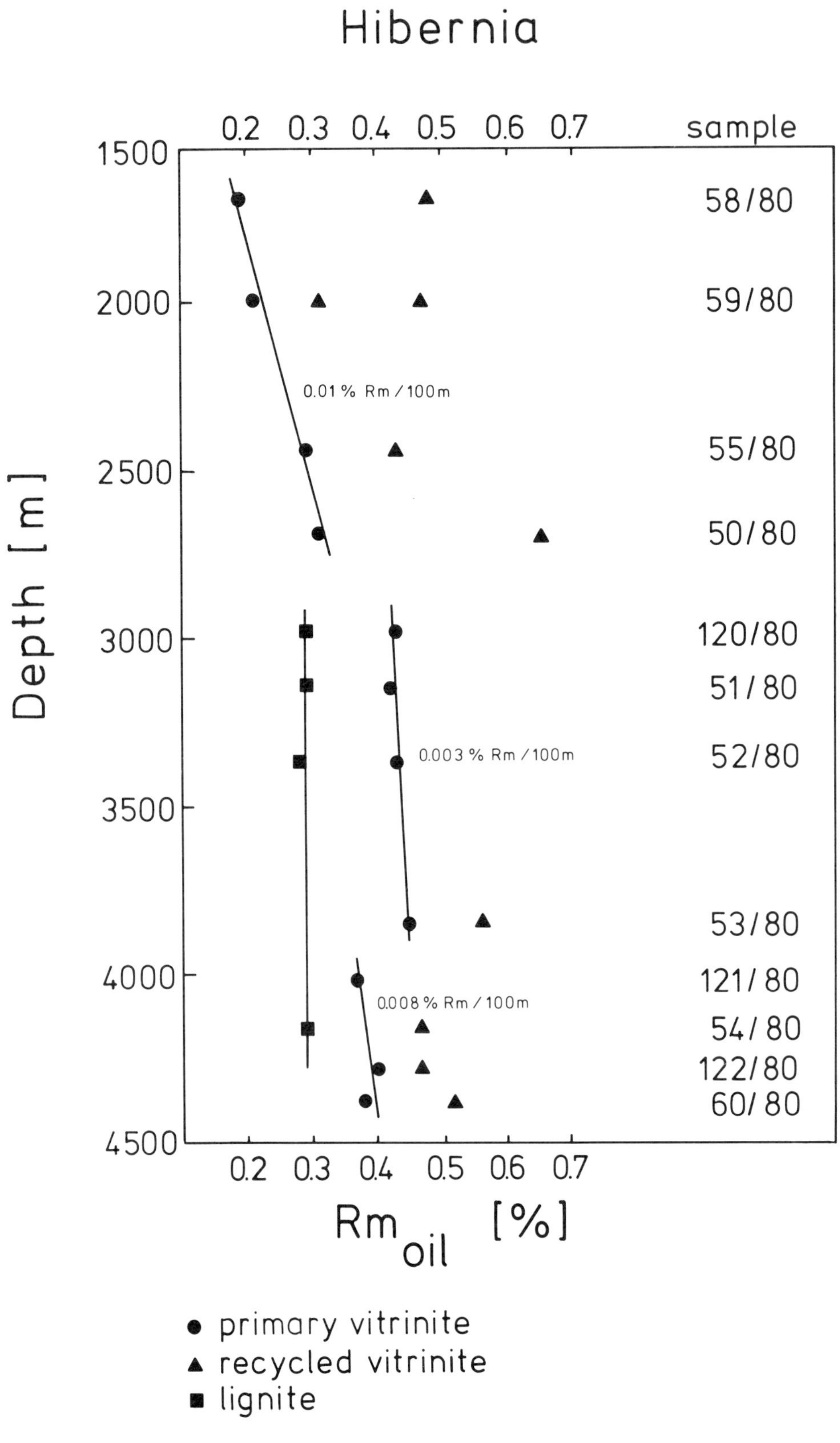

Figure 124. Rank to depth relationships for different vitrinite populations observed in borehole Hibernia P-15.

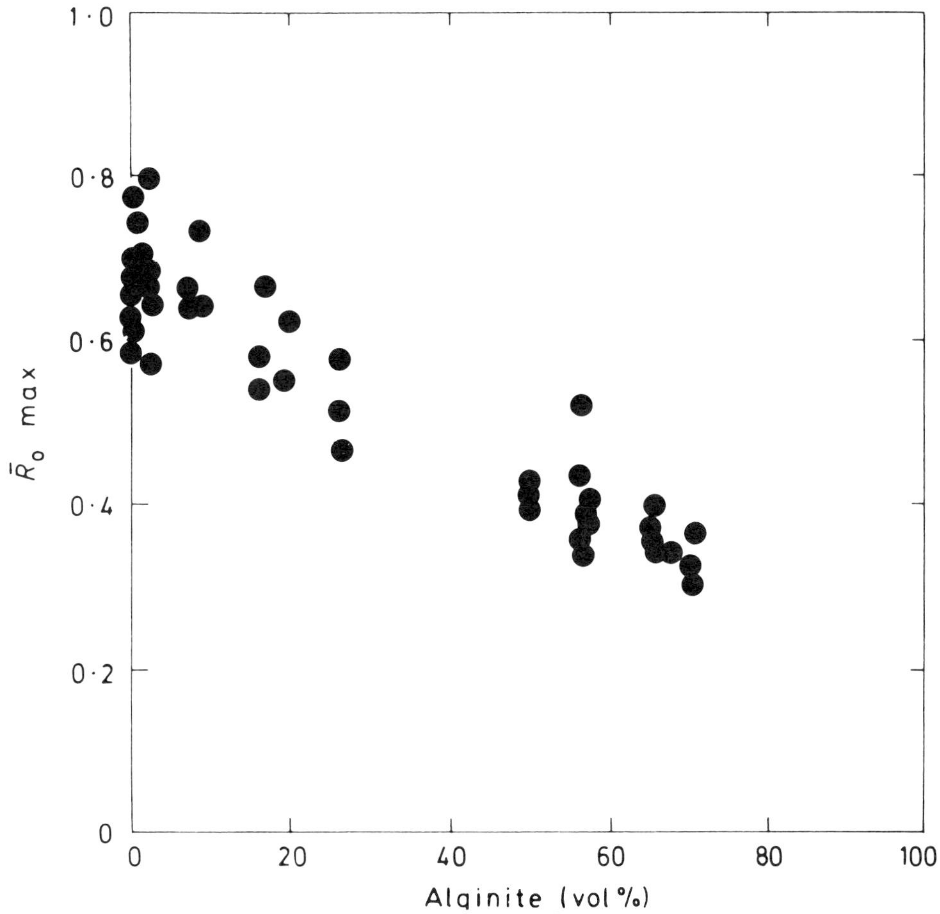

Figure 125. Relation of mean maximum reflectance of vitrinite ($\bar{R}_o$max) and the abundance of alginite (after Hutton and Cook, 1980).

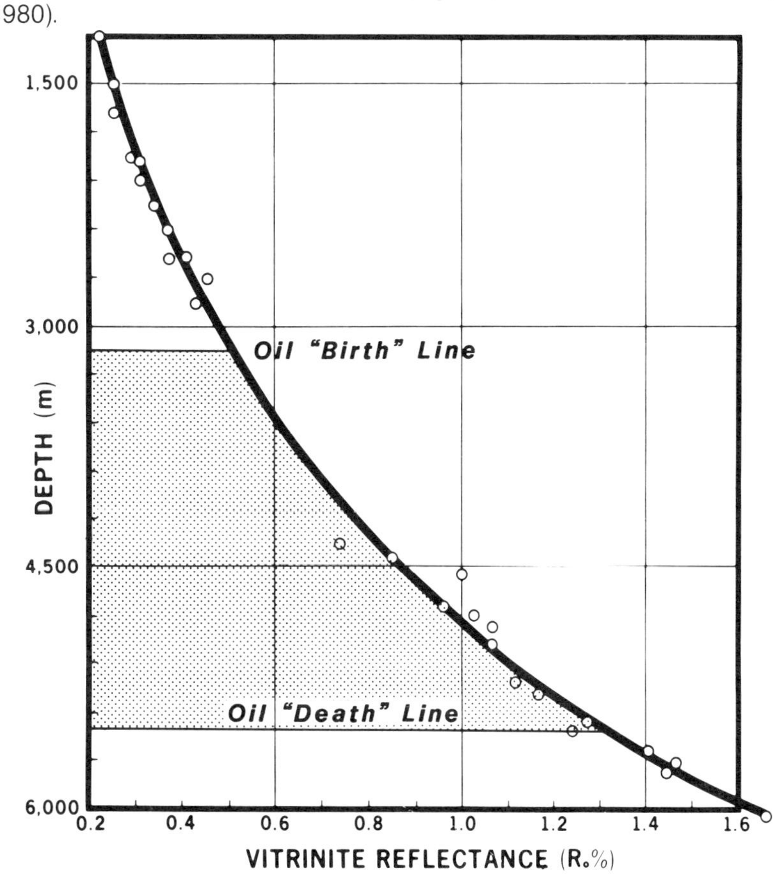

Figure 126. Kerogen maturation profile for a Louisiana Gulf Coast well (modified after Dow, 1974).

On Figure 124 a kerogen maturation profile is shown for borehole Hibernia P-15, east coast, Canada. The reflectance data of primary vitrinite, recycled vitrinite, and data obtained from added lignite are plotted versus depth. The reflectance data on the primary vitrinite at the top of the analysed sequence indicate a coalification stage of peat (0.20% $R_o$max). At greater depths the vitrinite reflectances increase slowly to values of about 0.4% $R_o$max (subbituminous rank). This indicates that even at depths close to 4 500 metres, generation of oil from kerogen should not have started. Oil located in reservoir rocks at higher levels must have been migrated from other sources, which are believed to be below 4 500 metres. The rank in Hibernia P-15 does not increase regularly with depth but shows rather three coalification tracks, the coalification gradients ranging from 0.003% $R_o$max/100 metres to 0.01% $R_o$max/100 metres. The observed irregularities could be due to changes in lithology of the associated sediments, or could be caused by minor fault systems in the regions of 2 700 to 3 000 metres and between 3 900 to 4 000 metres.

On Figure 126 a kerogen maturation profile is shown for a Louisiana Gulf Coast well, which is an example of regular increase of rank with depth. From interpolation of the reflectance data, the oil 'birth' line and the oil 'death' line can be located.

An example of regional application of rank determinations on finely dispersed organic materials is given on Figure 127. It shows coalification maps of the Lower Cretaceous and the Carboniferous of the Bramsche Massif in western Germany. Rank is expressed in the form of isoreflectances (lines of equal reflectances), and show in this area a concentric pattern which is interpreted as being caused by a large intrusive body. The isoreflectance lines on Figure 127 indicate that coalification increases from the stage of subbituminous coals at the margin of the intrusive body to the stage of anthracite right on the top of the intrusion. Main petroleum occurrences are restricted to areas where the vitrinite reflectances do not exceed 1% $R_o$max. Economic gas deposits in this region are found where vitrinite reflectances remain below 2% $R_o$max.

In sedimentary rocks where vitrinite particles are sparse or nonexistent, but liptinitic particles are abundant, as in many oil shales and carbonates, rank can be determined by measuring fluorescence properties of liptinite macerals or by measuring the fluorescence properties of the bituminous groundmass. The relationship between the fluorescence parameter and vitrinite reflectance is shown in Section 7.3.2.2.

10.5.3 *Petrographic Classification of Organic Matter in Sedimentary Rocks*

At the present time there is no generally accepted classification or nomenclature for organic constituents in sedimentary rocks. Table 40 lists proposed classifications by various authors published over the past years. Some like Teichmüller and Ottenjann (1977) follow reasonably well the maceral concept of coals, plus adding constituents commonly not observed in coals, such as faunal relicts, oil expulsions, and mineral-bituminous groundmass. Hutton *et al.* (1980) suggested much the same classification based on their studies on Australian oil shales. In addition they proposed a more detailed differentiation of alginite into alginite A and alginite B. Others like Alpern (1980) and Bostick (1979) make use only in part of the maceral concept used for coals and propose new terms in order to classify the organic material. Photomicrographs of various kerogen types are illustrated on Plate 18.

In general it is to be expected that as more research is done in this field, finer differentiation within certain groups of organic materials will be made and will be proposed. One example is the maceral bituminite. Teichmüller and Ottenjann (1977), in a study on the Jurassic Posidonian shale of western Germany, distinguish three types of bituminite based on differences of fluorescence properties and inclusions. Creaney (1980) in a study on the organic petrology of the Upper Cretaceous Boundary Creek Formation, Canada, identifies only one of these three bituminites but adds another which he calls matrix-butuminite.

In order to overcome the confusion created by the many proposed classifications, the ICCP (Internation Committee for Coal Petrology) is currently working on a generally acceptable classification of organic constituents in sedimentary rocks.

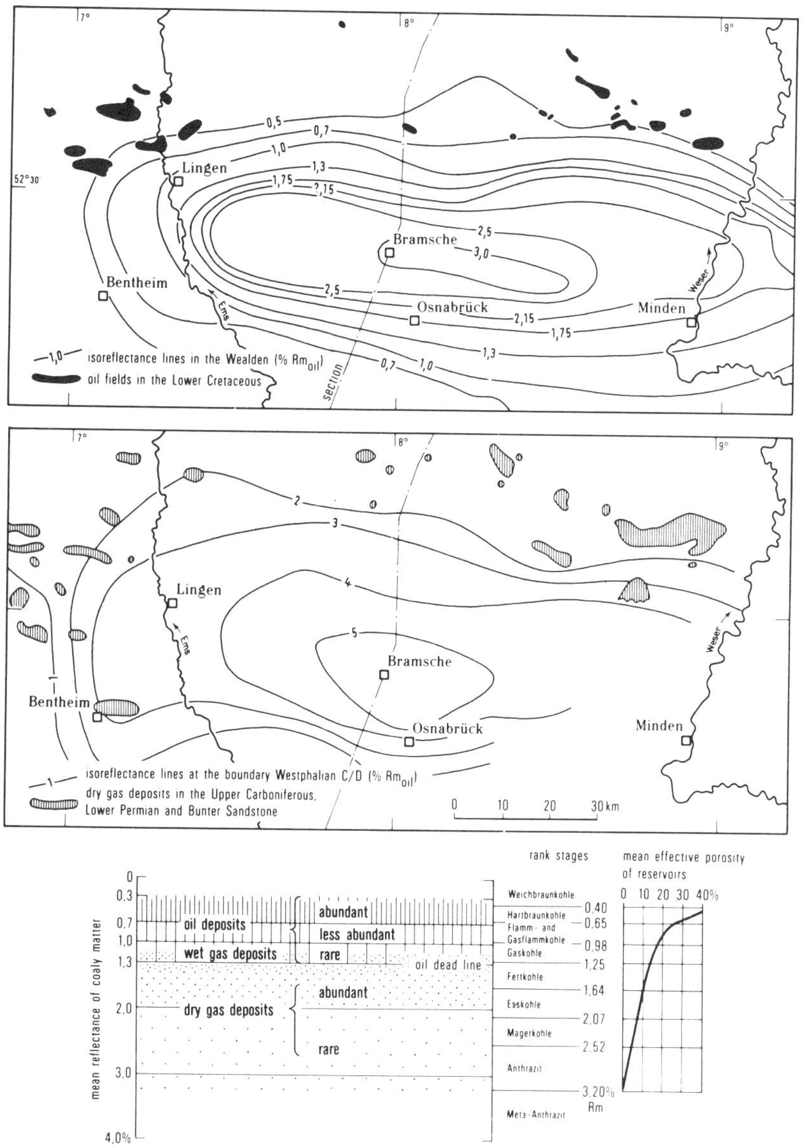

Figure 127. Coalification maps of northwest Germany for the Lower Cretaceous (Wealden) and the Carboniferous (boundary Westphalian C/D) based on vitrinite reflectances, measured mainly on finely dispersed organic material. Economic oil and gas deposits are related to particular coalification stages (from Stach, 1982).

Plate 18. Micropetrological features of kerogen. (Reflected light, oil immersion).

Figure 1a. Kerogen concentrate, normal reflected light. *Huminite* (H) and *pyrite* (P) in bituminous groundmass from an Upper Devonian black shale, Western Germany.

Figure 1b. Same section after blue-light excitation. The liptinitic component of the bituminous groundmass consists mainly of *bituminite* (B) and some *sporinite* (S). Both macerals display yellowish-orange fluorescence colours of strong intensities indicating the relatively low rank of the organic material. *Huminite* and *pyrite* do not fluoresce.

Figure 2a. Kerogen concentrate, normal reflected light. Trimaceritic coaly particle (C) associated with higher reflecting reworked materials (R). Weathering has caused the development of low reflecting oxidation rims (O). Black shale from the Upper Devonian of Western Germany.

Figure 2b. Same section after blue-light excitation. *Sporinite layers* (S) display now yellowish fluorescence colours. The huminite (H) shows weak brown fluorescence colour. Also, only visible now, becomes a liptinitic material (A) of strong fluorescence intensity. This material is probably of algal origin.

Figure 3a. Whole rock, normal reflected light. Huminite (H) in bituminous mineral matrix. Oil shale from New Brunswick, Carboniferous in age.

Figure 3b. Same section after blue-light excitation. Lamellar *alginite* B (B) and *mineral matter* (MM). Note that fluorescence accentuates the intimate association of organic material and mineral matter.

Figure 4a. Whole rock, normal reflected light. Tasmanites-type *alginite* (A) in a Devonian carbonate. Pine Point area, N.W.T.

Figure 4b. Same section after blue-light excitation. The yellowish fluorescence colour of the Tasmanites-type *alginite* (A) indicates low maturation of the organic material.

Figure 5a. Normal reflected light, porous *asphalt* associated with *quartz grains* (Q) in a bituminous sandstone, Wadi-Girr/Djebel, Syria.

Figure 5b. Same section after blue-light excitation, *asphalt* displays greenish to yellowish fluorescence colour.

Note the different scales on this plate.

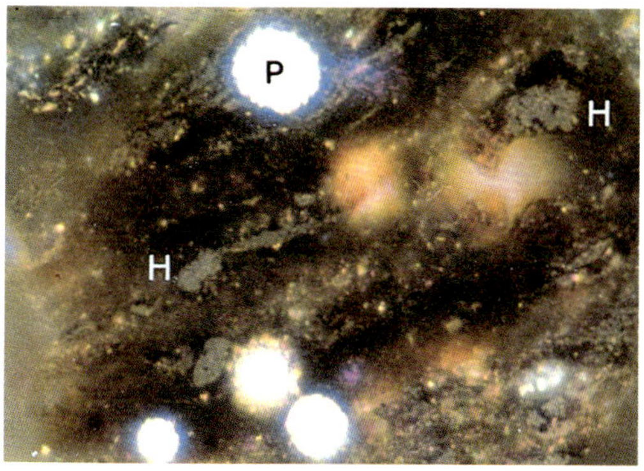

1a

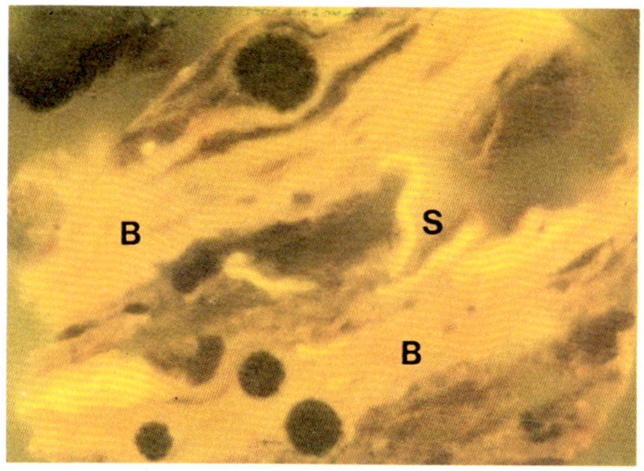

1b

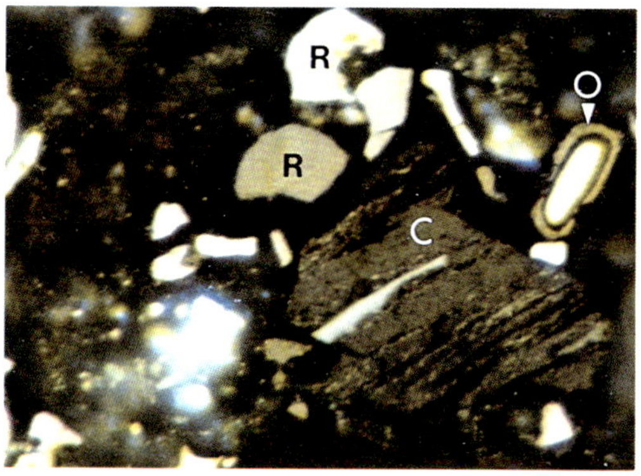

2a

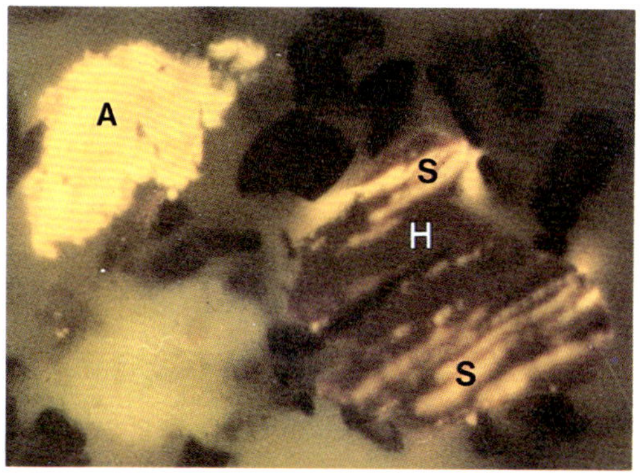

2b

MICRONS

0 100

3a

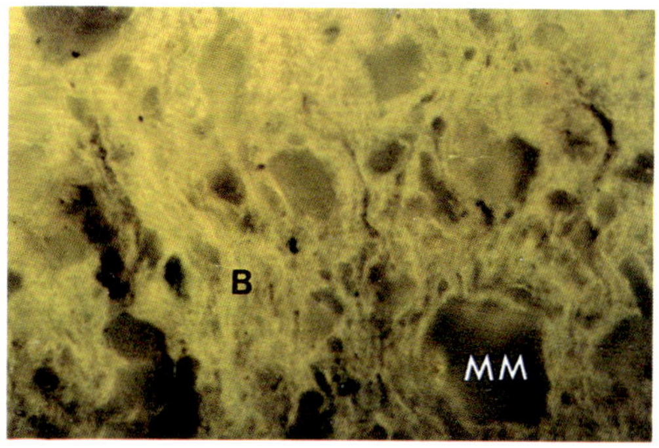

3b

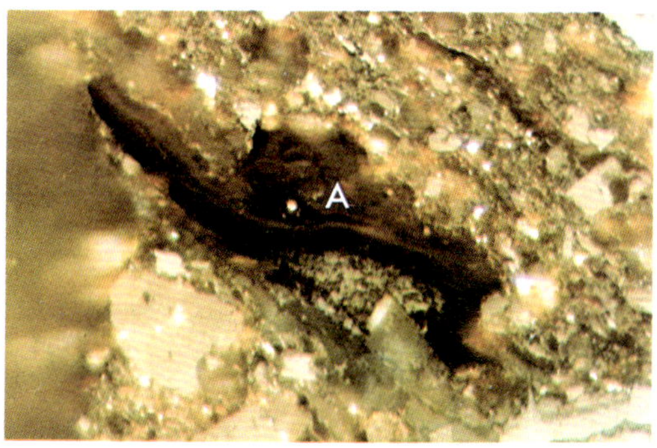

4a

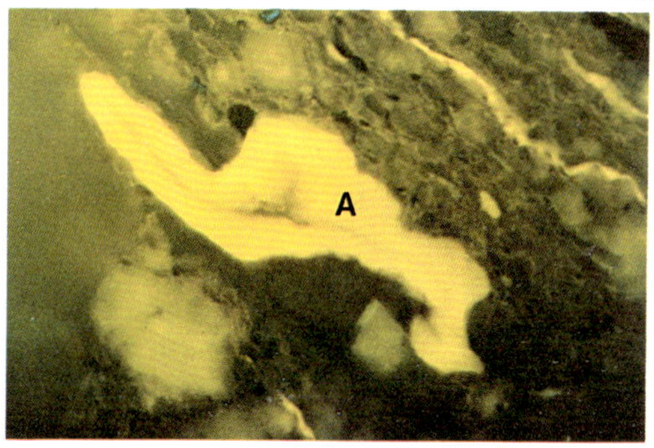

4b

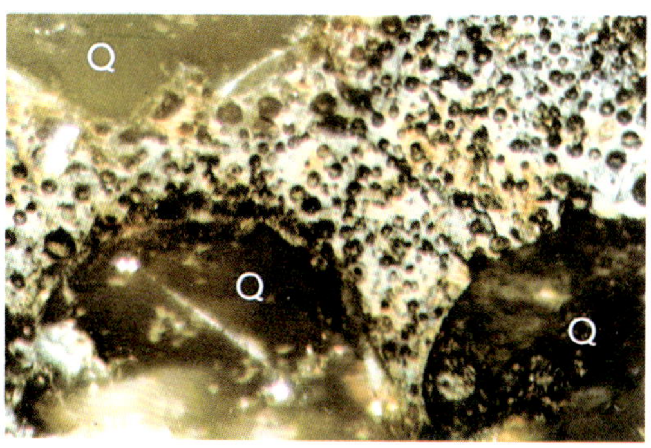

5a

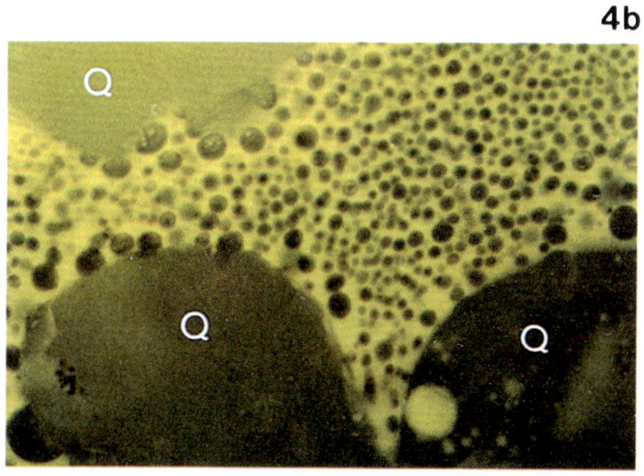

5b

MICRONS

0 100

TABLE 40. PETROGRAPHIC CLASSIFICATIONS OF ORGANIC CONSTITUENTS IN ROCKS INCLUDING OIL-SOURCE ROCKS (FROM STACH, 1982)

Method	Reflected light incl. fluorescence (coal petrology)								Transmitted light (palynology)	
Author	M. TEICHMÜLLER & OTTENJANN (1977)	ALPERN & CHEYMOL (1978)	ROBERT (1979)		BOSTIK (1979)	HUTTON *et al.* (1980)	ALPERN (1980)		palynologists, after HUNT (1979)	MANUM & THRONDSEN (1978 b)
Principal term	maceral	matière organique	matière organique	ensemble*	phytoclast	maceral	organoclast	class**	kerogen type	morphological category of kerogen
Humic	huminite/vitrinite telinite collinite corpocollinite vitrodetrinite	vitrinite, gels tissues gelifiés gels cellulosiques	vitrinite telinite collinite	I	low-gray vitrinite	vitrinite corpohuminite desmocollinite	gelitextite gelite detrite	I	woody	hylogen
Inertinitic	inertinite fusinite semifusinite sclerotinite macrinite inertodetrinite micrinite	inertinite tissues fusinisés	inertinite fusinite sclérotinite	I	high-gray v. and fusinite	inertinite inertodetrinite micrinite	fusitexite sclérotes	I II	coaly	melanogen
Liptinitic	liptinite sporinite cutinite suberinite fluorinite resinite chlorophyllinite liptodetrinite	exinite: spores, pollen cuticules subérines microlamelles	exinite sporinite cutinite fluorinite resinite, terpenite	I	liptinite (= exinite) spores, pollen cuticles resin	exinite sporinite cutinite liptodetrinite	sporites epidermes inclusions	III I	herbaceous	phyrogen
	alginite	algues	algues	I	algal bodies	alginite A	algues planctoniques	IV	algal	
	bituminite	lamelles noir-rougeâtres or: bituminite noir-rouge	alginite, bituminite	II	floccules and groundmass (="colloalginite", "sorbomixtinite"	alginite B, bituminite or: fine lamellar alginite	gels algaires et bactériens (sapropelite)	V	amorphous	amorphogen
	exsudatinite oil expulsions		bitumes, exsudatinite huile (micrinite)	III	solid bitumen	bitumen oil stains	migrabitumen	VI	amorphous	amorphogen
Faunal	faunal relics						zooplankton, zoobenthos	IV		
Mineral matrix	mineral-bituminous groundmass (= matrix bituminite of CREANEY, 1980 b)		matrice fluorescent (humique) (sapropelique)	II					(amorphous)	(amorphogen)

* ensemble (ROBERT): I constituants primaires, biologiquement "identifiables"
II matières primaires non figurées
III produits "secondaires"

** class (ALPERN): I phytoclasts, II fungi, III spores, IV organisms autonomes, V bacteries, VI bitumes

REFERENCES

Abramski, C. and Mackowsky, M. Th., 1952, Methoden und Ergebnisse der angewandten Koksmikroskopie; in Freund, H., editor, *Handbuch der Mikroskopie in der Technik*; Band II, Teil I, Umschau Verlag- Frankfurt-am-Main, pp. 311-410.

Albers, G., Lenart, L., and Oeler, H-H, 1974, Spectroscopic studies on chemically related (oxidized) coal; Fuel, vol. 53, pp. 47-53.

Allan, J. A. and Douglas, A. G., 1977, Variations in the content and distribution of n-alkanes in a series of Carboniferous vitrinites and sporinites of bituminous rank; Geochimica Cosmochimica, Acta vol. 41, pp. 1223-1230.

Allardice, D. J., George, A. M., Hauser, D., Neubert, K. H., and Smith, G. C., 1977, The variation of Latrobe Valley brown coal properties and utilization parameters with lithotype; Rept. no. 342, State Electricity Commission of Victoria, Australia.

Allen, E.A.D., 1978, Petrology and stratigraphy of Holocene coastal marsh deposits along the western shore of Delaware Bay; unpublished Ph.D. Thesis, University of Delaware, 284 pp.

Allen, J.R.L., 1970, Sediments of the modern Niger Delta, a summary and review; *in* Morgan, J. P. and Shaver, R. H., editors, *Deltaic Sedimentation*; Society of Economic Paleonotolgists and Mineralogists Special Publication 15, pp. 138-151.

Alpern, B., 1980, Petrographie du kerogene; *in* Durand, B., editor, *Kerogen;* Technip. Paris, pp. 339-371.

Alpern, B., 1981, Pour une classification synthetique universelle des combustibles solides; Bull. Centres Rech. Explor.-Prod. Elf-Aquitaine, 5,2, pp. 271-290.

Alpern, B. and Cheymol, D., 1978, Reflectance et fluorescence des organoclastes du Toarcien du Bassin de Paris en fonction de la profondeur et de la temperature; Rev. Inst. Franc. Petrole, 33, 515-535, Paris.

Alpern, B. and Lemos de Sousa, M., 1970, Sur le pouvoir reflecteur de la vitrinite et de la fusinite des houilles; Centre Rech. Acad. Sc. Paris, 271, pp. 956-959.

American Society of Testing Materials (ASTM), D-2797 (Reapproved, 1980), Preparing coal samples for microscopical analysis by reflected light; American Society of Testing Materials, Philadelphia, Pennsylvania.

American Society of Testing Materials (ASTM), D-2799 (Reapproved, 1980), Microscopic determination of volume percent of physical components of coal; (Part 26), American Society of Testing Materials, Philadelphia, Pennsylvania.

American Society of Testing Materials (ASTM), D-388 (Reapproved, 1977), Standard classification of coals by rank; American Society of Testing Materials, Philadelphia, Pennsylvania, 5 pp.

American Society of Testing Materials (ASTM), 1981, Annual book of ASTM standards; (Part 26), American Society of Testing Materials, Philadelphia, Pennsylvania.

Ammosov, I. I., 1961, Stadii izmeneniya osadochnykh porod i parageneticheskie otnosheniya goryuchikh iskopaemykh (Stages in the alteration of sedimentary rocks and paragenetic relations of combustible materials); Sov. Geol., 1961, no. 4, pp. 7-24 (English transl., International Geology Review, v. 4, pp. 1105-1117).

Ammosov, I. I., Eremin, I. V., Sukhenko, S. F., and Oshurkova, L. S., 1957, Calculation of coking charges on the basis of petrographic characteristics of coals; Koks i Khim. 12, pp. 9-12.

Ammosov, I. I. and Gorshkov, V. I., 1969, Vzaimosvyaz' katageneza i neftegazonosnosti otlozhenii Zapadno-Sibirskoy Nizmennosti (The relationship between catagenesis and petroleum content of strata in the West-Siberian lowland); *in* Rasseyannye vklyucheniya uglya v osadochnykh porodakh; Nauka Press, Moscow, pp. 5-80.

Ammosov, I. I. and Utkina, A. I., 1975, Paleotemperatury, litifikatsiya i neftegazonosnost' neogenovykh otlozheniy severnogo Sakhalina (Paleotemperatures, lithification and oil and gas occurrence in Neogene deposits of northern Sakhalin); *in* Yeremin, I. V., editor, *Paleotemperatury zon nefteobrazovaniya*; Nauka Press, Moscow, pp. 70-93.

Anderson, J.A.R., 1964, The structure and development of peat swamps of Sarawak and Brunei; Journal Tropical Geography, vol. 18, pp. 7-16.

Balsley, J. K., 1980, Cretaceous wave dominated delta systems, east central Utah: a field guide; Amoco Production Company, Denver, 163 pp.

Barker, C. E., 1979, Vitrinite reflectance geothermetry in the Cerro Prieto geothermal system, Baja California, Mexico; Unpublished Ph.D. Thesis, Dept. of Earth Sciences, University of California, Riverside, California, 127 pp.

Bartenstein, H., Teichmüller, M., and Teichmüller, R., 1971, Die umwandlung der organischen Substanz im Dach des Bramscher Massivs; Fortschr. Geol. Rheinld. u. Westf., 18, 501-538.

Bartenstein, H. and Teichmüller, R., 1974, Inkohlungsuntersuchungen, ein Schlussel zur Prospektierung von palaozoischen Kohlenwasserstofflagerstatten; Fortschr. Geol. Rheinld. u. Westf., 24, 129-160.
Bayer, J. L. and Denton, G. H., 1966, Application of coke microscopy to plant problems; Blast Furnace and Steel Plant, vol. 54, no. 12, pp. 1133-1142.
Belinko, K., Kalkreuth, W., and Chornet, E., 1982, Anisotropic coke-like structures produced by carbonization of residues derived from aqueous-phase hydrogenolysis of peat with syngas; Fuel Processing Technology, vol. 6, pp. 75-81.
Benedict, L. G. and Berry, W. F., 1964, Recognition and measurement of coal oxidation; Bituminous Coal Research Report, Monroeville, Pennsylvania, 41 pp.
Benedict, L. G., Thompson, R. R., Shigo, J., and Aikman, R., 1968, Pseudovitrinite in Appalachian coking coals; Fuel, vol. 47, pp. 125-143.
Benedict, L. G., Thompson, R. R., and Wenger, R. O., 1968, Relationship between coal petrographic composition and coke stability; Blast Furnace and Steel Plant, vol. 56, no. 3, pp. 217-224.
Benedict, L. G. and Thompson, R. R., 1973, Coal and coke petrography and coke reactivity; Paper presented at Northeast Section Meeting, Geological Society of America, Allentown, Pennsylvania, March 22, 1973.
Benedict, L. G. and Thompson, R. R., 1980, Coke/carbon reactions in the study of factors affecting coke quality; International Journal of Coal Geology, vol. 1, no. 1, pp. 19-34.
Bennett, A.J.R. and Taylor, G. H., 1970, A petrographic basis for classifying Australian coals; Australian Institute of Mineral and Metallurgy Proceedings, 233, pp. 1-6.
Berkowitz, W., 1979, *An introduction to coal technology*; Academic Press, New York, pp. 95-108.
Berner, R. A., 1971, *Principles of chemical sedimentology*; McGraw Hill, Toronto, 240 pp.
Berry, W. F., Cameron, A. R., and Nandi, B. C., 1967, The development of coal petrology in North America; Proc. of Symposium on the Science and Technology of Coal, Ottawa, 1967, CANMET, Ottawa, pp. 6-29.
Bostick, N. and Damberger, H., 1971, The carbon ratio rule and petroleum potential in NPC region 9; Illinois State Geological Survey, Illinois Petroleum, vol. 95, pp. 142-151.
Bostick, N. and Foster, N., 1975, Comparison of vitrinite reflectance in coal seams and in kerogen of sandstones, shales, and limestones in the same part of a sedimentary section; Colloque Int. Petr. Org. et Pot. Petrol., pp. 13-25.
Bostick, N. and Nicksic, C. P., 1975, Bibliography and index of coal and dispersed organic matter in sedimentary rocks: petrography, catagenesis, relation to petroleum and natural gas, and geochemistry; Illinois State Geological Survey, Illinois Petroleum, vol. 108, 92 pp.
Bostick, N. and Alpern, B., 1977, Principles of sampling, preparation and constituent selection for microphotometry in measurement of maturation of sedimentary organic matter; Journal of Microscopy, vol. 109, pt. 1, pp. 41-47.
Bostick, N. H., 1979, Microscopic measurement of the level of catagenesis of solid-organic matter in sedimentary rocks to aid exploration for petroleum and to determine former burial temperatures - A review; in Scholle, P. and Schluger, P. R., editors, *Aspects of Diagenesis*; Society of Economic Paleontologists and Mineralogists Special Publication 25, pp. 17-43.
Brown, H. R., Cook, A. C., and Taylor, G. H., 1964a, Variations in the properties of vitrinite in isometamorphic coal; Fuel, vol. 43, pp. 111-124.
Brown, H. R., Taylor, G. H., and Cook, A. C., 1964b, Prediction of coke strength from the rank and petrographic composition of Australian coals; Fuel, vol. 43, no. 1, pp. 43-54.
Brooks, J. and Taylor, G., 1965, The formation of graphitizing carbons from the liquid phase; Carbon, vol. 3, pp. 185-193.
Buntebarth, G., 1978, The degree of metamorphism of organic matter in sedimentary rocks as a paleogeothermometer, applied to the Upper Rhine Graben; Paleogeophysics, vol. 117, pp. 83-91.
Burstlein, E., 1957, Progres et developements recents en metiere de preparation des charbons en vue de leur cokefaction d'apres la procede Burstlein-Longwy; l'Echo des Mines et de la Metallurgie. Nr. 3506.
Bustin, R. M., Hills, L. V., and Gunther, P. R., 1977, Implications of coalification levels, Eureka Sound Formation, northeastern Arctic Canada; Canadian Journal of Earth Sciences, vol. 14, pp. 1588-1597.
Bustin, R. M., 1980, Oxidation characteristics of some sheared coal seams of the Mist Mountain Formation, southeastern Canadian Cordillera; Geological Survey of Canada, Paper 80-18, pp. 249-254.
Bustin, R. M., 1982, The effect of shearing on the quality of some coals in the southeastern Canadian Cordillera; Canadian Institute of Mining and Metallurgy Bulletin, vol. 75, pp. 76-83.
Bustin, R. M., 1983, Heating during thrust faulting: friction or fiction?; Tectonophysics, vol. 95, pp. 305-328.

Cady, G. H., 1942, Modern concepts of the physical constitution of coal; Journal of Geology, vol. 50, pp. 337-356.

Cameron, A. R., 1971, Some petrological aspects of the Harbour coal seam, Sydney Coalfield, Nova Scotia; Geological Survey of Canada Bulletin 175, 74 pp.

Cameron, A. R., 1972, Petrography of Kootenay coals in the upper Elk River and Crowsnest areas, British Columbia and Alberta; Research Council of Alberta, Information Series no. 60, pp. 31-45. Cameron, A. R., 1975, Principles of coal petrography; Research Council of Alberta, Information Series no. 76, Symposium on Coal Evaluation, pp. 51-67.

Cameron, A. R., 1978, Megascopic description of coal with particular reference to seams in southern Illinois; *in* Dutcher, R. R., editor, *Field Description of Coal*; ASTM Special Technical Publication 661, Philadelphia, Pennsylvania, pp. 9-32.

Cameron, A. R. and Babu, S. K., 1968, The petrology of the No. 10 (Balmer) coal seam in the Natal area of the Fernie Basin, British Columbia; Geological Survey of Canada Paper 68-35, 57 pp.

Cameron, A. R. and Kalkreuth, W. D., 1982, Petrological characteristics of Jurassic-Cretaceous coals in the Foothills and Rocky Mountains in western Canada; Proceedings Fifth Symposium on the Geology of Rocky Mountain Coal, Utah Geological and Mineral Survey Bull. 118, pp. 163-167.

Campbell, D. D., Jory, L. T., and Saunders, C. R., 1977, Geology of Hat Creek coal deposits; Canadian Institute of Mining and Metallurgy Bull., vol. 70, pp. 99-108.

Carmichael, S.M.M., 1982, Depositional environments of coal development in the Gates Member, northeastern British Columbia; 11th International Congress on Sedimentology, Hamilton, Ontario, Abstracts of Papers, p. 57.

Casagrande, D. J., 1971, Geochemistry of amino acids in selected Florida peats; Diss. Abstracts 32, No. 1.

Casagrande, D. J., Siefert, K., Berschinsky, C., Sutton, N., 1977, Sulphur in peat forming systems of the Okefenokee Swamp and Florida Everglades: origins of sulphur in coal; Geochemica et Cosmochimica Acta, vol. 41, pp. 161-167.

Castano, J. R. and Sparks, D. M., 1974, Interpretation of vitrinite reflectance measurements in sedimentary rocks and determination of burial history using vitrinite reflectance and authigenic minerals; *in* Dutcher, R. R., Hacquebard, P. A., Schopf, J. M., and Simon, J. A., editors, *Carbonaceous Materials as Indicators of Metamorphism*; Geological Society of America Special Paper 163, pp. 31-52.

Cecil, C. B., Renton, J. J., Stanton, R. W., Finkelman, R. B., 1979, Mineral matter in coals of the central Appalachian Basin; 9th International Congress, Carboniferous Stratigraphy and Geology Abstracts, p. 33.

Chandra, D., 1958, Reflectance of oxidized coals; Economic Geology, vol. 53, pp. 102-108.

Chandra, D., 1962, Reflectance and microstructure of weathered coals; Fuel, vol. 41, pp. 185-193.

Chandra, D., 1982, Oxidized coals; *in* Stach, E., Mackowsky, M-Th, Teichmüller, M., Taylor, G. H., Chandra, D., and Teichmüller, R., editors, *Stach's Textbook of Coal Petrology*; Gebruder Borntraeger Berlin, pp. 198-206.

Cobb, J. C., Steele, J. D., Treworgy, C. G., and Ashby, J. F., 1980, The abundance of zinc and cadmium in sphalerite-bearing coals in Illinois; Illinois State Geological Survey Mineral Note 74, 28 pp.

Cohen, A. D. and Spackman, W., 1972, Methods of peat petrology and their application to reconstruction of paleoenvironments; Geological Society of American Bulletin, vol. 83, pp. 129-142.

Cohen, A. D. and Spackman, W., 1977, Phytogenic organic sediments and sedimentary environments in the Everglades-mangrove complex; Palaeontographica, Abt. B, 162, 144 pp.

Cohen, A. D. and Spackman, W., 1980, Phytogenic organic sediments and sedimentary environments in the Everglades-mangrove complex of Florida - Part III, The alteration of plant material in peat and origin of coal macerals; Palaeontographica, Abt. B, 172, pp. 125-149.

Coleman, J. M., Sherwood, M., Gagliano, A., Smith, W. G., 1970, Sedimentation in a high tide tropical delta; *in* Morgan, J. P. and Shaver, R. H., editors, *Deltaic Sedimentation*; Society of Economic Paleontologists and Mineralogists Special Publication 15, pp. 185-197.

Cook, A. C., 1980, Optical techniques for the examination of organic matter in oil shales; *in* Cook, A. and Kantsler, A., editors, *Oil Shale Petrology Workshop*; Wollongong, 1980, Keiraville Kopiers, pp. 1-15.

Cook, A. C. and Edwards, G. E., 1971, Vitrinite content and coke strength; Fuel, vol. 50, pp. 41-52.

Cook, A. C., Hutton, A. C., and Sherwood, N. R., 1981, Classification of oil shales; Bull. Centres Rech. Explor. - Prod. Elf-Aquitaine, 5, pp. 353-381.

Cooper, B. S. and Murchison, D. G., 1969, Organic geochemistry of coal; *in* Eglinton, G. and Murphy, M.T.J., editors, *Organic Geochemistry*; Springer-Verlag, New York, pp. 699-726.

Correia, M., 1967, Relations possibles entre l'etat de conservation des elements figures de la matiere organique (microfossils, palynoplanctoligiques) et l'existence de gisements d'hydrocarbures; Inst. Francais Petrole Rev., vol. 22, pp. 1285-1306.
Cotter, E., 1982, Upper Cretaceous coals of the Sabinas Basin, northeastern Mexico: Products of back-barrier deposition and changing rates of progradation; 11th International Congress on Sedimentology, Hamilton, Ontario, Abstracts of Papers, p. 57.
Creaney, S., 1979, Organic material in the Tranquille Beds of the (Tertiary) Kamloops Volcanic Group near Afton Mine, Kamloops, British Columbia; Geological Survey of Canada Paper 79-1A, pp. 381, 382.
Creaney, S., 1980a, Petrographic texture and vitrinite reflectance variation on the Alston Block, northeast England; Proceedings of the Yorkshire Geological Society, vol. 42, pp. 553-580.
Creaney, S., 1980b, The organic petrology of the Upper Cretaceous Boundary Creek Formation, Beaufort-Mackenzie Basin; Bulletin of Canadian Petroleum Geology, vol. 28, pp. 112-119.
Crelling, J. C., Schrader, R. H., and Benedict, L. G., 1979, Effects of weathered coal on coking properties and coke quality; Fuel, vol. 58, pp. 542-546.
Davis, A., 1978a, The reflectance of coal; *in* Karr, editor, *Analytical Methods for Coal and Coal Products*; vol. 1, Academic Press, New York, pp. 27-81.
Davis, A., 1978b, Compromise in coal seam description; *in* Dutcher, R. R., editor, *Field Description of Coal*; ASTM Special Technical Publication 661, American Society of Testing Materials, Philadelphia, Pennsylvania, pp. 33-40.
Davis, A., Spackman, W., and Given, P., 1976, The influence of the properties of coals and their conversion into clean fuels; Energy Sources, vol. 3, no. 1, pp. 55-81.
Davis, J. C., 1973, *Statistics and Data Analysis in Geology*; Wiley and Sons, Toronto, 550 pp.
De Boosere, O., 1926, The macroscopic constituents of Campine coals; Fuel, vol. 5, pp. 522-527.
Denton, G. H., Bayer, J. L., and Hassel, R. E., 1967, Progress and problems in routine petrographic evaluation of coals for coke plant use; Journal of Metals, vol. 19, no. 5, pp. 88-92.
Diessel, C.F.K., 1965, Correlation of macro- and micropetrography of some New South Wales coals; Proceedings 8th Commonwealth Mining and Metallurgical Congress, vol. 6, pp. 669-677.
Diessel, C.F.K., and Guyot, R.E., (in press), Petrographic studies on solid residues derived from the hydrogenation of some Australian coals; Proceedings 9th International Congress, Stratigraphy and Geology of the Carboniferous, Urbana, Illinois.
Donaldson, A. C., 1974, Pennsylvanian sedimentation of central Appalachians; *in* Briggs, G., editor, *Carboniferous of the Southeast United States Geological Society of America Special Paper 148, pp. 47-78.*
Donaldson, A. C., Martin, R. H., and Kanes, W. H., 1970, Holocene Guadelupe Delta of Texas Gulf Coast; in Morgan, J. P., editor, *Deltaic Sedimentation*; Society of Economic Paleontologists and Mineralogists Special Publication 15, pp. 107-137.
Dow, W., 1977, Kerogen studies and geological interpretations; Journal of Geochemical Exploration, vol. 7, pp. 79-99.
Duparque, A., 1934, Structure microscopique des charbons du bassin houiller du Nord et du Pas-de-Calais; Mem. Soc. Geol. Nord. 11 (2 vols.), 756 pp.
Elliot, M. A., 1981, *Chemistry of Coal Utilization, Second Supplementary Volume*; Wiley Interscience, New York, 2374 pp.
Elliot, M. A. and Yohe, G. R., 1981, The coal industry and coal research and development in perspective - Chapter 1; *in* Elliot, M. A., editor, *Chemistry of Coal Utilization, Second Supplementary Volume*; Wiley Interscience, New York, pp. 1-54.
Eremin, I. V., Ruban, V. A., Suprunenko, O. I., and Tsikarev, 1976, A comparison of the classification of coals used in individual countries with the International Classification; Khimiya Tverdogo Topliva, 10, pp. 9-16.
Ethridge, F. G., Jackson, T. J., and Youngberg, A. D., 1981, Floodbasin sequence of a fine-grained meander belt subsystem: The coal bearing Lowe Wasatch and Upper Fort Union Formations, southern Powder River Basin, Wyoming; *in* Ethridge, F. G. and Flores, R. M., editors, *Recent and Ancient Nonmarine Depositional Environments: Models for Exploration*; Society of Economic Paleotologists and Mineralogists Special Publications 31, pp. 191-212.
Facer, R. A., Cook, A. C., and Beck, A. E., 1980, Thermal properties and coal rank in rocks and coal seams of the southern Sydney Basin, New South Wales: A paleogeothermal explanation of coalification; Internation Journal of Coal Geology, vol. 1, pp. 1-17.
Falini, F., 1965, On the formation of coal deposits of lacustrine origin; Geological Society of America Bulletin, vol. 76, pp. 1317-1346.

Fayol, H., 1887, Terrain houiller de commentry, Saint Etienne, Livre premier; Lithologie et Stratigraphie.

Ferm, J. C., 1974, Carboniferous environmental methods in eastern United States and their significance; *in* Briggs, G., editor, *Carboniferous of the Southeastern United States*; Geological Society of America Special Paper 148, pp. 79-95.

Finkleman, R. B., 1980, Modes of occurrence of trace elements in coal; Unpublished Ph.D. Thesis, University of Maryland, 301 pp.

Fisher, W. L. and McGowen, J. H., 1969, Depositional systems in the Wilcox Group (Eocene) of Texas and their relationship to occurrence of oil and gas; American Association of Petroleum Geologists Bulletin, vol. 53, pp. 30-54.

Fisk, H. N., 1960, Recent Mississippi River sedimentation and peat accumulation; Compte Rendu 4, Congress International Stratigraphy Geology Carboniferous, 1, pp. 187-199.

Francis, W., 1961, Coal, its formation and composition; Edward Arnold, Ltd., London, 806 pp.

Frazier, D. E. and Osanik, A., 1969, Recent peat deposits - Louisiana coastal plain; *in* Dapples, E. C. and Hopkins, M. E., editors, *Environments of Coal Deposition*; Geological Society of America Special Paper 114, pp. 63-85.

Fuller, M. L., 1919, Relation of oil to carbon ratios of Pennsylvanian coals in north Texas; Economic Geology, vol. 14, pp. 536-542.

Gibson, D. W., 1979, Sedimentology facies in the Jura-Cretaceous Kootenay Formation, Crowsnest Pass area, southwestern Alberta and southeastern British Columbia; Canadian Society of Petroleum Geologists Bulletin, vol. 25, pp. 767-791.

Gijzel, P., 1963, Autofluorescence of fossil pllen and spores with special reference to age determination and coalification; Leidse Geol. Mededel., vol. 40, pp. 263-317.

Gijzel, P., 1981, Applications of the geomicrophotometry of kerogen, solid hydrocarbons and crude oils to petroleum exploration; *in* Brooks, J., editor, Organic Maturation Studies and Fossil Fuel Exploration; Academic Press, London, pp. 351-377.

Gijzel, P. van, 1975, Polychromatic UV-Fluorescence microphotometry of fresh and fossil plant substances, with special references to the location and identification of dispersed organic matter in rocks; *in* Alpern, B., editor, *Petrographie de la Matiere Organique des Sediments, Relations Avec la Paleotemperature et le Potential Petrolieu*; Centre Nationale Researche Scientifique, Paris, pp. 67-91.

Given, P. H. and Peover, M., 1960, Polarographic and electrolytic reduction of the aromatic systems in solvent extracts of coals; Fuel, vol. 39, pp. 463-473.

Given, P. H., Binder, C. R., and Hill, L. W., 1966, Electron spin resonance study of pure macerals; Coal Science, Advances in Chemistry Series 55, American Chemical Society, Washington, D.C., pp. 344-362.

Given, P. H., Cronauer, D. C., Spackman, W., Davis, A., and Biswas, B., 1975, Coal liquefaction behaviour; Fuel, vol. 54, pp. 34-39.

Given, P. H., Cronauer, D. C., Spackman, W., Davis, A., and Biswas, B., 1975, Dependence of coal liquefaction behaviour on coal characteristics, 2. Role of petrographic composition; Fuel, vol. 54, pp. 40-49.

Gluskoter, H. J., Rich, R. R., Miller, W. G., Cahill, R. A., Dreher, G. B., and Kuhn, J. K., 1977, Trace elements in coal: Occurrence and distribution; Illinois State Geological Circular 499, 154 pp.

Goodarzi, F. and Murchison, D. G., 1972, Optical properties of carbonized vitrinites; Fuel, vol. 51, pp. 322-328.

Goodarzi, F. and Murchison, D. G., 1978, Influence of heating-rate variation on the anisotropy of carbonized vitrinites; Fuel, vol. 57, pp. 273-284.

Goscinski, J. S., Robinson, J. W., and Chun, D., 1978, Megascopic description of coal drill cores; *in* Dutcher, R. R., editor, *Field Description of Coal*; ASTM Special Technical Publication 661, American Society of Testing Materials, Philadelphia, Pennsylvania, pp. 50-57.

Gould, K. W. and Smith, J. W., 1979, The genesis and isotopic composition of carbonates associated with some Permian Australian coals, Chemical Geological, 24, pp. 137-150.

Grand'eury, F. C., 1882, Memoire sur la formation de la houille; Ann. Mines, 1, ser. 8, 106 pp.

Gray, D., 1978, Inherent mineral matter in coal and its effect upon hydrogenation; Fuel, vol. 57, pp. 213-216.

Gray, R. J., Schapiro, N., and Coe, G.D., 1963, Distribution and forms of sulfur in a high-volatile Pittsburgh seam coal; Trans. AIME, vol. 226, pp. 113-121.

Gray, R. J., Rhoades, A. H., and King, D. T., 1976, Detection of oxidized coals and the effect of oxidation on technological properties; Transaction Society Mineral Engineers, vol. 260, pp. 334-340.

Gray, R. J., Goscinski, J. S., and Shoeberger, R. W., 1978, Selection of coals for coke making; Paper presented a joint conference of Iron and Steel Society (AIME) and Society of Mining Engineers (AIME), Pittsburgh, Pennsylvania, October 3, 1978.

Grayson, J., 1975, Relationship of palymorph translucency to carbon and hydrocarbons in clastic rocks; *in* Alpern, B., editor, *Petrographie Organique et Potential Petrolier*; Centre Nationale Researche Scientifique, Paris, pp. 261-273.

Hacquebard, P. A., 1952, A petrographic evaluation of the Tracy seam of the Sydney Coalfield, Nova Scotia; Proceedings 2nd Conference on Origin and Constitution of Coal, Crystal Cliffs, Nova Scotia, pp. 293-318.

Hacquebard, P. A., 1975a, Correlation between coal rank, paleotemperature and petroleum occurrences in Alberta; Geological Survey of Canada Paper 75-1, Part B, pp. 5-8.

Hacquebard, P. A., 1975b, Pre- and post-deformational coalification and its significance for oil and gas exploration; *in* Alpern, B., editor, *Petrographic Organique et Potential Petrolier*; CNRS, Paris, pp. 225-241.

Hacquebard, P. A., Birmingham, T. F., and Donaldson, J. R., 1967, Petrography of Canadian coals in relation to environment of deposition; Symposium in the Science and Technology of Coal, Ottawa, March, 1967, Energy, Mines, and Resources, Ottawa, pp. 84-97.

Hacquebard, P. A. and Donaldson, J. R., 1969, Carboniferous coal deposition associated with flood-plain and limnic environments in Nova Scotia; *in* Dapples, E. C. and Hopkins, M. E., editors, *Environments of Coal Deposition*; Geological Society of America Special Paper 114, pp. 143-191.

Hacquebard, P. A. and Donaldson, J. R., 1970, Coal metamorphism and hydrocarbon potential in the Upper Paleozoic of the Atlantic provinces, Canada; Candian Journal of Earth Sciences, vol. 7, pp. 1139-1163.

Hacquebard, P. A. and Donaldson, J. R., 1974, Rank studies of coals in the Rocky Mountains and inner Foothills Belt, Canada; Geological Society of America Special Paper 153, pp. 75-94.

Hagemann, H. W., 1978, Macropetrographic classification of brown coal; Proposal Presented to Members of the ICCP at the Essen Meeting, April, 1978.

Hagemann, H. W., 1980, Part I. Identification of lithotypes in lignites of southern Saskatchewan, Part II Macropetrographic examination and collecting samples from cores stored at the Hat Creek Mine site, British Columbia; Geological Survey of Canada Open File Report, 67 pp.

Hagemann, H. W. and Hollerbach, A., 1979, Relationship between the macropetrographic and organic geochemical composition of lignites; *in* Douglas, A. G. and Maxwell, R., editors, *Advances in Organic Geochemistry 1979*; Pergamon Press, pp. 631-638.

Hagemann, H. W. and Hollerbach, A., 1980, Spektralfuorimetrische Analysen von Sedimentextrakten. Reifegradbestimmung von Kohlen und Kohlenwasserstoffmuttergesteinen; Erdol und Kohle, Bd. 33, Heft 12, 577.

Hagemann, H. W. and Hollerbach, A., 1981, Spectral fluorimetric analysis of extracts - a new method for determining the degree of maturity of organic matter in sedimentary rocks; Bull. Centr. Rech., Explor.-Prod., Elf-Aquitane, 5, 2, pp. 635-650.

Hagemann, H. W. and Hollerbach, A., 1983, The spectral fluorometric analysis of the soluble organic matter applied to hydrocarbon source rock evaluation, *in* M. Bjoroy and others ed. Advances in organic geochemistry 1981; Chichester, Wiley, pp. 72-75.

Hampson, A. J., Becken, W. H., Schat, R. R., and Whyte, P. A., 1980, Coke quality control by on-line coal quality monitoring at Dofasco; Report by Dominion Foundries and Steel, Limited (DOFASCO), Hamilton, Ontario, 20 pp.

Harrison, J. A., Jackman, H. W., and Simon, J. A., 1964, Predicting coke stability from petrographic analysis of Illinois coals; Illinois State Geological Survey, Circular 366, 20 pp.

Hebda, R. J., 1977, The paleoecology of a raised bog and associated deltaic sediments of the Fraser River Delta, British Columbia; Unpublished Ph.D. Thesis, Department of Botany, University of British Columbia, Vancouver, 200 pp.

Heroux, Y., Chagnon, A., and Bertrand, R., 1979, Compilation and correlation of major thermal maturation indicators; American Association of Petroleum Geology Bulletin, vol. 63, pp. 2128-2144.

Hickling, H.G.A. and Marshall, C. E., 1932, The microstructure of the coal in certain fossil trees, Transaction Institute of Mineral Engineers, vol. 84, pp. 13-23.

Hilt, C., 1873, Die Beiziehungen zwischen der Zusammensetzung und den technischen Eigenschaften der Steinkohle; Sitzber. Aachener Bezirksvereinigung VDI, H. 4.

Hoffmann, E., 1933, Neue Erkenntnisse uber die Vorgange der Flozbildung; Bergbau 7, pp. 89-94.

Hoffmann, E. and Jenkner, A., 1932, Die Inkohlung und ihre Erkennung im Mikrobild; Gluckauf 68, pp. 81-88.

Hood, A. and Castano, J. R., 1974, Organic metamorphism: Its relationship to petroleum generation and application to studies of authigenic minerals; United Nations ESCAP, CCOP, Technical Bulletin 8, pp. 87-118.

Hood, A. and Gutjahr, C.C.M., 1972, Organic metamorphism and the generation of petroleum; Geological Society of America Abstract with Programs, vol. 4, pp. 542, 543.

Horne, J. C., Fern, J. C., Carvecio, T., and Baganz, B. P., 1978, Depositional models in coal exploration and mine planning in Appalachian region; American Association of Petroleum Geologists Bulletin, vol. 62, pp. 2379-2409.

Horne, J. C., McKenna, L. L., Levey, R. A., and Petranoff, T. V., 1982, Wave dominated deltas: An important economic depositional model for Upper Cretaceous of southwestern Wyoming; American Association of Petroleum Geologists Short Course, Coal Bearing Sequences.

Hornsby, D. T. and Leja, J., (in press), Effect of weathering, petrographic composition and rank on floatability; Journal of Chemical Engineering.

Huck, G. and Patteisky, K., 1964, Inkohlungsreaktionen unter Druck; Fortschr. Geol. Rheinld. U. Westf., vol. 12, pp. 551-558.

Hunt, J., 1979, Petroleum geochemistry and geology; Freeman & Company, San Fransico, 617 pp.

Hutton, A., Kantsler, A., Cook, A., and McKirdy, D., 1980, Organic matter in oil shales; Australian Petroleum Exploration Association J., vol. 20, pp. 44-68.

Hutton, A. and Cook, A., 1980, Influence of alginite in the reflectance of vitrinite from Joadja, NSW, and some other coals and oil shales containing alginite; Fuel, vol. 59, no. 10, pp. 711-716.

Inculet, I. I., Bergougnow, M. A., and Brown, J. D., 1982, Electrostatic benification of coal; *in* Liu, Y. A., editor, *Physical Cleaning of Coal*; Energy, Power and Environment no. 15, Marcek Derkfer Inc., New York, pp. 87-132.

International Committee for Coal Petrology (1957), International handbook of coal petrography; Centre National de la Recherche Scientifique, Paris, France.

International Committee for Coal Petrology (1963), International handbook of coal petrography - 2nd edition; Centre National de la Recherche Scientifique, Paris, France.

International Committee for Coal Petrology (1971), International handbook of coal petrography, 1st supplement to 2nd edition; Centre National de la Recherche Scientifique, Paris, France.

International Committee for Coal Petrology (1975), Analysis subcommission, fluorescence microscopy and fluorescence photometry; in Internation Handbook of Coal Petrography, 2nd supplement to 2nd edition; Centre National de la Recherche Scientifique, Paris, France.

Irvine, J. A., Whitaker, S. H., Broughton, P. L., and Tibbets, T. E., 1974, Evaluation of borehole sampling methods applicable to lignite resource exploration; Proceedings of Symposium - Fuels: A Geological Appraisal, Saskatchewan Geological Survey Special Publication no. 2, pp. 203-227.

Jacob, H., 1964, Neue Erkenntnisse auf dem Gebiet der Lumineszenzmikroskopie fossiler Brennstoffe; Fortschr. Geol. Rheinld. u. Westf., vol. 12, pp. 569-588.

Jacob, H., 1969, Luminiszenzmikroskopie der organopetrographischen Bestandteile von Sedimentgesteinen; Leitz-Mitt., Wiss und Technik, Wetzlar, vol. 4, pp. 250-254.

Jacob, H., 1973, Kombination von Fluoreszenz- und Reflexions- Mikroskopphotometrie der organischen Stoffe von Sedimenten und Boden; Leitz-Mitt., Wiss. and Technik, Westzlar, vol. 6, pp. 21-27.

Jacob, H., 1974, Fluoreszenz-Mikroskopie und Photometrie der Organischen Substanz von Sedimenten und Boden; *in* Freund, H., editor, *Handbuch der Mikroskopie in der Technik 4, Part 2*; Frankfurt, Umschau Verlag., pp. 369-391.

Jacob, H., 1976, Petrologie, Nomenklatur und Genesis naturlicher fester Erdolbitumina; Comp. 76/77, Erganzungsband Erdol u. Kohle, pp. 36-49.

Jongmans, W. J., Koopmans, R. G., and Roose, G., 1936, Nomenclature of coal petrography; Fuel, vol. 15, pp. 14, 15.

Juckes, L. M. and Pitt, G. J., 1977, The standardization of petrographic analysis of coal; Journal of Microscopy, vol. 109, pp. 13-21.

Jurasky, K. A., 1929, Neue Untersuchungen und Gedanken zur Entstehung fossiler Holzkohle; Schr. Geb. Brennstoff-Geologie, vol. 2, pp. 23-41.

Jurasky, K. A., 1936, Deutschlands Braunkohlen und ihre Entstehung; Btsh. Boden 2, Gebr. Bortraeger, 165 pp.

Kalkreuth, W., 1979, Das Inkohlungsbild des Ostsauerlander Hauptsattels im Rhenoherzynikum mit besonderer Berucksichtigung der Trendflachenanalysen; Fortschr. Geol. Rheinld. u. Westf., vol. 27, pp. 277-321.

Kalkreuth, W., 1982, Rank and petrographic composition of selected Jurassic-Lower Cretaceous coals of British Columbia, Canada; Bulletin Canadian Petroleum Geology, vol. 30, no. 2, pp. 112-139.

Kalkreuth, W. and Chornet, E., 1982, Peat hydrogenolysis using H_2/CO mixtures: Micropetrological and chemical studies of original material and reaction residues; Fuel Processing Technology, vol. 6, pp. 93-122.

Kalkreuth, W. and McMechan, M., 1984, Regional patterns of thermal maturation as determined from coal rank studies, Rocky Mountain Foothills and Front Ranges North of Grande Cache, Alberta—Implications for Petroleum exploration; Bulletin of Canadian Petroleum Geology, vol. 32, pp. 271-279.

Karmasin, K. von, 1952, Deutung des Fazieswechsel in der Frozen Erda und Aginr auf Grund mikropetrographischer Schlitzprobenuntersuchungen; Bergbau Archiv 13, Heft 1-2, pp. 74-100.

Karweil, J., 1956, Die Metamorphose der Kohlen vom Standpunkt der physikalischen Chemie; Z. Deutsch. Geol. Ges., vol. 107, pp. 132-139.

Kim, H., 1979, Depositional environment and stratigraphic subdivision, Hat Creek No. 1 deposit, British Columbia, Canada; Program with Abstracts, 4th Annual Canadian Institute of Mining and Metallurgy Meeting, Vancouver, British Columbia, October 25-27, 1979.

Koensler, W., 1980, Das Verhalten des Inertinits westkanadischer Kreidekohlen bei der Verkokung; Diss. Techn. Hochsch. Aachen, 118 pp.

Kontorovich, A. E., 1970, Teoreticheskiyeosnovy ob'yemno-geneticheskogo metoda otsenkipotensial'-nykh resuorv nefti i gaza (The theoretical foundations of a volume-genetic method for evaluating oil and gas resources); Sibir. Nauchno-Issled. Inst. Geologii, Geofizikii Minerl'nogo Syr'ya Trudy, vol. 95, pp. 4-51.

Kotter, K., 1960, Die mikroskopische Reflexionsmessung mit dem Photomultiplier und ihre Anwendung auf die Kohlenuntersuchung; Brennst. Chemie, vol. 41, pp. 263-272.

Krevelen, D. W. van, 1961, Coal; Elsevier, Amsterdam, London, New York, Princeton, 514 pp.

Krevelen, D. W. van and Schuyer, J., 1957, Coal Science; Elsevier Publishing Co., New York, 352 pp.

Kroger, C., Pohl, A., Horestadt, H., Kuthe, F., and Burger, H., 1957a, Die physikalischen und chemischen Eigenschaften der Steinkohlegefugebestandteile (Macerale), I. Chemische Eigenschaften und Inkohlungsverhalten der Steinkohlengefugebestandteile; Brennst. Chemie, vol. 38, pp. 33-42.

Kroger, C. and Pohl, A., 1957b, Die physikalischen und chemischen Eigenschaften der Steinkohlengefugebestandteile (Macerale), III. Das Entgasungsverhalten; Brennst. Chemie, vol. 38, pp. 102-107.

Kroger, C., Kuthe, F., and Gondermann, H., 1957c, Die physikalischen und chemischen Eigenschaften der Steinkohlenmacerale, IV. Swelling-Index, Dilatometertest; Brennst. Chemie, vol. 38, pp. 147-151.

Kroger, C., Pohl, A., and Kuthe, F., 1957d, Uber die Isolierung der Steinkohlengefugebestandteile aus Glanz-und Mattkohlen von Ruhrflozen; Gluckauf, vol. 93, pp. 122-135.

Kuhlwein, F. and Hoffmann, E., 1952, Petrographie und Mikroskopie der Steinkohle in Wissenschaft und Praxis; *in* Freund, H., editor, *Handbuch der Mikroskopie in der Technik*; Band II, Teil I, Umschau Verlag - Frankfurt-am-Main, pp. 65-234.

Leeder, M. R., 1982, Sedimentology; George Allen and Unwin Publishers, London, 344 pp.

Leythaeuser, D., 1973, Effects of weathering on organic matter in shales; Geochimica et Cosmochimica Acta, vol. 37, pp. 113-120.

Leythaeuser, D., Hagemann, H. W., Hollerbach, A., and Schaefer, R., 1980, Hydrocarbon generation in source beds as a function of type and maturation of their organic matter: A mass balance approach; Proceedings World Petroleum Congress, Heyden: London-Philadelphia- Rheine, vol. 2, pp. 31-41.

Lilley, E. R., 1924, Coal as an aid in oil exploration; Engineering and Mining Journal, vol. 117, pp. 1009-1012.

Lopatin, N. V., 1971, Temperature and geologic time as factors in coalification; Akad. Nauk SSSR, Ser. Geol. Izvestiya, no. 3, pp. 95-106. (Russian original - translated by N. Bostick).

McCartney, J. T. and Teichmüller, M., 1972, Classification of coals according to degree of coalification by reflectance of the vitrinite component; Fuel, vol. 51, pp. 64-68.

McIver, R., 1967, Composition of kerogen - clue to its role in the origin of petroleum; World Petroleum Congress 7th, vol. 2, pp. 25-36.

McKenzie, D. P., 1981, The variation of temperature with time and hydrocarbon maturation in sedimentary basins formed by extension; Earth and Planetary Science Letters, vol. 55, pp. 87-98.

Mackowsky, M.-Th., 1968a, Mineral matter in coal; *in* Murchison, D. G. and Westoll, T. S., editors, *Coal and Coal-bearing Strata*; Oliver and Boyd, Edinburgh, pp. 309-321.

Mackowsky, M.-Th., 1968b, European Carboniferous coalfields and Permian Gondwana coalfields; *in* Murchison, D. G. and Westoll, T. S., editors, *Coal and Coal-bearing Strata*; Oliver and Boyd, Edinburgh and London, pp. 325-345.

Mackowsky, M.-Th., 1974, Das Atzen der Steinkohle, ein Aufschlussreiches Hilfsmittel bei der Kohlenmikroskopie; Compte Rendu Septieme Congres International de Stratigraphie et de Geologie du Carbonifere, Krefeld 23, 28.8.1971, Band 3, pp. 375-383.

Mackowsky, M.-Th., 1975, Minerals and trace elements occurring in coal; *in* Stach's Textbook of Coal Petrology, 2nd Edition; pp. 121-132.

Mackowsky, M.-Th., 1977, Prediction methods in coal and coke microscopy; Journal of Microscopy, vol. 109, part 1, pp. 119-137.

Mackowsky, M.-Th., 1982a, Minerals and trace elements occurring in coal; *in* Stach, E., Mackowsky, M.-Th., Teichmüller, M., Taylor, G. H., Chandra, D., and Teichmüller, R., editors, *Coal Petrology, 3rd Edition*; Gebruder Borntraeger, Berlin-Stuttgart, pp. 153-171.

Mackowsky, M.-Th., 1982b, Methods of coal examination; *in* Stach, E., Mackowsky, M.-Th., Teichmüller, M., Taylor, G. H., Chandra, D., and Teichmüller, R., editors, *Coal Petrology, 3rd Edition*; Gebruder Borntraeger, Berlin-Stuttgart, pp. 300-348.

Mackowsky, M.-Th., 1982c, The application of coal petrography in technical process; *in* Stach, E., Mackowsky, M.-Th., Teichmüller, M., Taylor, G. H., Chandra, D., and Teichmüller, R., editors, *Coal Petrology, 3rd Edition*; Gebruder Borntraeger, Berlin-Stuttgart, pp. 413-474.

Mackowsky, M.-Th. and Hoffmann, E., 1960, Die rohstofflichen Eigenschaften der deutschen Steinkohlen in ihrer Bedeutung fur die Aufbereitung; *in* Der deutsche Steinkohlenbergbau, Verlag Gluckauf Essen, vol. 4, pp. 32-52.

Mackowsky, M.-Th. and Simonis, W., 1969, Die Kennzeichnung von Kokskohlen fur die mathematische Beschreibung der Hochtemperaturverkokung im Horizontalkammerofen bei Schuttbetrieb durch Ergebnisse mikroskopischer Analysen; Gluckauf-Forschungsheft, vol. 30, pp. 25-37.

Malan, O., 1971, Zur Definition des Begriffes Lithotyp und seiner Anwendung in der petrographischen Klassifikation der Braunkohlen; Neue Bergbautechnik, vol. 8, pp. 619-623.

Manum, S. and Throndsen, T., 1978, Dispersed organic matter (kerogen) in the Spitzbergen Tertiary; Norsk Polarinst. Arb., 1977, pp. 179-187.

Marchioni, D. L., 1978, The effect on vitrinite reflectance of elevated temperatures during sample preparation; Geological Survey of Canada Paper 78-1C, pp. 125, 126.

Marchioni, D. L., 1980, Petrography and depositional environment of the Liddell seam, upper Hunter Valley, New South Wales; International Journal of Coal Geology, vol. 1, pp. 35-61.

Marchioni, D. L., 1981, Influence of petrographic composition on the free-swelling index of coal and on its use as a weathering parameter; Coal Miner, May 1981, pp. 22-24.

Marchioni, D. L., 1983, The detection of weathering in coal by petrographic, rheologic and chemical methods; International Journal of Coal Geology, vol. 2, pp. 231-259.

Marchioni, D. L., (in press), Reflectance studies of brown coals - an example from the Hat Creek deposit of British Columbia; Compte Rendu 9th Congress of Carboniferous Stratigraphy and Geology.

Marsh, H. and Smith, J., 1978, The formation and properties of anisotropic cokes from coals and coal derivatives studied by optical and scanning electron microscopy; *in* Karr, C. Jr., editor, *Analytical Methods for Coal and Coal Products, Chapter 30*; Academic Press, New York, San Francisco, London, vol. II, pp. 371-414.

Marshall, C. E., 1954, Introduction to a study of the nature and origin of fusain (fusinite); Fuel, vol. 33, pp. 134-144.

Marshall, R. J., 1976, Classification of coking coals; Fuel, vol. 55, pp. 346-348.

Matsuoka, H., 1975, Requirements for coals in Japanese coking blends; *in* Cook, A. C., editor, *Australian Black Coal - Its Occurrence, Mining Preparation and Use*; Australian Institute of Mining and Metellurgy, Illawarra Branch, pp. 251-263.

Meriaux, E., 1972, Les tonsteins de la veine de charbon No. 10 (Balmer) a Sparwood Ridge dans le bassin de Fernie (Columbia Britannique); Geological Survey of Canada Paper 72-1, Part B, pp. 11-22.

Middleton, M. F., 1982, Tectonic history from vitrinite reflectance; Geophysical Journal Royal Astronomical Society, vol. 68, pp. 121-132.

Mitchell, T., Davis, A., and Spackman, W., 1977, A petrographic classification of solid residues derived from the hydrogenation of bituminous coals; *in* Eglington, G., editor, *Liquid Fuels from the Hydorgenation of Bituminous Coals*; Academic Press, New York, pp. 255-271.

Montgomery, W. J., 1978, Standard laboratory test methods for coal and coke; *in* Karr, C. Jr., *Analytical Methods for Coal and Coal Products*; Academic Press, New York, pp. 191-246.

Moore, L. R., 1968a, Cannel coals, bogheads and oil shales; *in* Murchison, D. G. and Westoll, T. S., editors, *Coal and Coal-bearing Strata*; Oliver and Boyd, Edinburgh, pp. 19-29.

Moore, L. R., 1968b, Some sediments closely associated with coal seams; *in* Murchison, D. G. and Westoll, T. S., editors, *Coal and Coal-bearing Strata*; Oliver and Boyd, Edinburgh.

Moses, R., 1976, Use of petrographic analysis for the prediction of coke stability; Report form Bituminous Coal Research, Inc., Monroeville, Pennsylvania, 19 pp.

Murchison, D. G., 1978a, Optical properties of carbonized vitrinites; *in* Karr, C. Jr., editor, *Analytical Methods for Coal and Coke Products*; Academic Press, New York, San Francisco, London, vol. II, pp. 415-464.

Murchison, D. G., 1978b, Butcher, baker, candlestickmaker, but organic petrologists all; Presidential address to the Royal Microscopical Society of London.

Nandi, B. N., Brown, T. D., and Lee, G. K., 1977, Inert coal macerals in combustion; Fuel, vol. 56, pp. 125-130.

Nandi, B. N., Ghosh, A., and Ciavaglia, L. A., 1978, Anomalous microhardness impressions of some weathered coals; Fuel, vol. 57, pp. 317-319.

Neavel, R. C., 1981, Origin, petrography and classification of coal; *in* Elliott, M. A., editor, *Chemistry of Coal Utilization - 2nd Supplementary Volume*; John Wiley & Sons Inc., New York, Toronto, pp. 91-158.

Okuyama, Y., Miyazu, T., Sugimura, H., and Kumagai, M., 1970, Prediction of the coking property of coal by microscopic analysis; Journal of the Fuel Society of Japan, vol. 49, (522), pp. 736-743.

Ottenjann, K., 1980, Spektrale Fluoreszenz-Mikrophotometrie von Kohlen und Olschiefern; Leitz-Mitt., Wiss. und Technik, Bd VII, Nr 8, pp. 262-272.

Ottenjann, K., 1981-1982, Verbesserungen bei der mikroskop- photometrischen Fluoreszenzmessung an Kohlenmazeralen; Zeiss- Inform., Oberkochem, 26, pp. 40-46, Heft 93, 1981-1982.

Ottenjann, K., Teichmüller, M., and Wolf, M., 1974, Spektrale Fluoreszenzmessungen an Sporiniten mit Auflicht-Anregung, eine mikroskopische Methode zur Bestimmung des Inkohlungsgrades gering inkohlter Kohlen; Fortschr. Geol. Rheinld. u. Westf., vol. 24, pp. 1-36.

Ottenjann, K., Teichmüller, M., and Wolf, M., 1975, Spectral fluorescence measurements of sporinites in reflected light and their applicability for coalification studies; *in* Alpern, B., editor, *Petrographie Organique et Potentiel Petrolier*; Centre Nationale Recherche Scientifique, Paris, France, pp. 67-91.

Ottenjann, K. and Teichmüller, M., 1977, Liptinite und lipoide Stoffe in einem Erdolmuttergestein; Erdol und Kohle, vol. 30, pp. 387-398.

Ottenjann, K., Wolf, M., and Wolff-Fischer, E., 1981, Beziehungen zwischen der Fluoreszenz von Vitriniten und den technologischen Eigenschaften von Kohlen; Proceedings International Conference Coal Science, Dusseldorf, 7.9.9 1981, pp. 86-91.

Ottenjann, K., Wolf, M., and Wolff-Fischer, E., 1982, Das Fluoreszenzverhalten der Vitrinite zur Kennzeichnung der Kokungseigenschaften von Steinkohlen; Gluckauf-Forschungshefte, 43, 4, pp. 173-179.

Parks, B. C. and O'Donnel, H. J., 1956, Petrography of American coals; United States Bureau of Mines Bulletin 550, 193 pp.

Patrick, J. W., Reynolds, M. J., and Shaw, F. H., 1973, Development of optical anisotropy in vitrains during carbonisation; Fuel, vol. 52, pp. 198-204.

Paulencu, H. N. and Readyhough, P. J., 1974, Interpreting coal properties for utilization in commercial coke making; Proceedings of Symposium on Coal Evaluation, Alberta Research Council, Information Series 76, pp. 108-133.

Pearson, D. E., 1980, The quality of western Canadian coking coal; Canadian Institute of Mining and Metallurgy Bulletin, January 1980, vol. 73, pp. 1-15.

Pearson, D. E., Gigliotti, F. B., Grieve, D. A., and Ollerenshaw, N. C., 1977, Geology of Crowsnest Coalfield, northwest part; B.C. Ministry of Energy, Mines and Petroleum Resources, Preliminary Map 24.

Pearson, D. E. and Kwong, J., 1979, Mineral matter as a measure of oxidation of a coking coal; Fuel, vol. 58, pp. 63-66.

Pearson, D. E. and Grieve, D. A., (in press), Rank distribution and coalification pattern, Crowsnest Coalfield, British Columbia, Canada; Compte Rendu, 9th International Congress of Carboniferous Stratigraphy and Geology.

Pearson, D. E. and Grieve, D. A., (in preparation), Petrographic evaluation of Crowsnest Coalfield, southeastern British Columbia.

Peters, K., Ishiwarari, R., and Kaplan, I., 1977, Color of kerogen as index of organic maturity; Bulletin of American Association of Petroleum Geologists, vol. 61, no. 4, pp. 504-510.

Potonie, H., 1920, Die Entstehung der Steinkohle und der Kaustobiolithe uberhaupt; 6. Aufl., Borntraeger, Berlin, 233 pp.
Potonie, R., 1924, Einfuhrung in die allgemeine Kohlenpetrographie; Brontraeger, Berlin, 285 pp.
Preobrazhenskaya, G. S., Klucheva, N. Yu, and Ivanova, K. P., 1971, Geologicheskiye zakonomemosti rasprostraneniya krupnykh mestorozhdeniy nefti i gaza v SSR (Distribution features of giant oil and gas deposits in the USSR); Vses. Nauchno-Issled. Geol. Inst. Trudy, vol. 286, 175 pp.
Radke, M., Schaeffer, R., Leythaeuser, D., and Teichmüller, M., 1980, Composition of soluble organic matter in coals: Relation to rank and liptinite fluorescence; Geochimica et Cosmochimica Acta, vol. 44, pp. 1787-1800.
Rahmani, R. A., 1981, Facies relationships and paleoenvironments of a Late Cretaceous tide-dominated delta, Drumheller, Alberta; *in* Field Guides to Geology and Mineral Deposits; Calgary 1981, Geological Association of Canada, Mineralogical Association of Canada Joint Meeting, 1981, pp. 159-176.
Rao, C. P. and Gluskoter, H. J., 1973, Occurrence and distribution of minerals in Illinois coals; Illinois State Geological Survey Circular 476, 56 pp.
Raymond, R., 1978, Relative abundance of common elements in coal macerals; Los Alsmos Scientific Laboratory Preprint, LA-VR 78-3334, 6 pp.
Raynaud, J. and Robert, P., 1976, Les methodes d'etude optique de la matiere organique; Bull. Centr. Rech. Pau-SNPA 10, 1, pp. 109-127.
Reinson, G. E., 1979, Facies models 6, Barrier Island systems; *in* Walker, R. G., editor, *Facies Models*; Geoscience Canada Reprint, Series 1, pp. 57-74.
Robert, P., 1971, Etude petrographique des matieres organiques insolubles par la mesure de leur pouvoir reflecteur: Contribution a l'exploration petroliere et a la connaissance des bassins sedimentaires; Rev. Inst. Franc. Petrole, Ann. Comb. Liquide, vol. 26, pp. 105-135.
Robert, P., 1979, Classification des matieres organiques en fluorescence. Application aux roches-meres petrolieres; Bull. Centr. Rech. Explor.-Prod. Elf-Aquitane, vol. 3, 1, pp. 223-263, Pau.
Robert, P., 1981, Classification of organic matter by means of fluorescene: Application to hydrocarbon source rocks; International Coal Geology, vol. 1, pp. 101-137.
Rodionova, K. F., 1967, Organicheskoe veshchestovo i neftematerinskie porody devona Volgo-Uralskoi neftegazonosnoi oblasti (Organic matter and oil source rocks of the Devonian in the Volga-Ural petroleum region); Nauka Press, Moscow, 358 pp.
Schapiro, N., Gray, R. J., and Euser, G. R., 1961, Recent developments in coal petrography; Proceedings of Blast Furnace, Coke Oven, and Raw Materials Committee, AIME, vol. 20, pp. 89-112.
Schapiro, N. and Gray, R. J., 1963, Relation of coke structure to reactivity; Blast Furnace and Steel Plant, vol. 51, pp. 273-280.
Schapiro, N. and Gray, R. J., 1964, The use of coal petrography in coke making; Journal of the Institute of Fuel, vol. XI, no. 30, pp. 234-242.
Schochardt, M., 1936, Zur Praxis der Luminiszenz-Mikroskopie mit dem Epi-Kondensor W.; Zeiss-Nachr. 2. 16.
Schopf, J. M., 1956, A definition of coal; Economic Geology, vol. 51, pp. 521-527.
Schopf, J. M., 1960, Field description and sampling of coal beds; United States Geological Survey Bulletin 1111B, pp. 25-67.
Seyler, C. A., 1929, The microscopical examination of coal; D.S.I.R., Fuel Research, Physics and Chemical Survey, National Coal Resources, no. 16, H.M.S.O., 67 pp.
Seyler, C. A., 1943, Recent progress in the petrology of coal; Journal Institute Fuel, vol. 16, pp. 134-141.
Shibaoka, M., 1969, Combustion of coal in thin sections; Fuel, vol. 48, pp. 285-295.
Shibaoka, M., Bennett, A.J.R., and Gould, K. W., 1973, Diagenesis of organic matter and occurrences of hydrocarbons in some Australian sedimentary basins; Australian Petroleum Exploration Association Journal, vol. 13, pp. 73-80.
Shibaoka, M. and Bennett, A.J.R., 1976, Use of swelling-index profiles of the Bulli seam in New South Wales, and their general application in coalfield geology; Fuel, vol. 55, pp. 99-104.
Shibaoka, M. and Smyth, M., 1975, Coal petrology and the formation of coal seams in some Australian sedimentary basins; Economic Geology, vol. 70, pp. 1463-1473.
Shibaoka, M. and Ueda, S., 1978, Formation and stability of mesophase during coal hydrogenation. 1. Formation of mesophase; Fuel, vol. 57, pp. 667-672.
Smith, A.H.V., 1968, Seam profiles and seam characters; *in* Murchison, D. G. and Westoll, T. S., editors, *Coal and Coal-bearing Strata*; Oliver and Boyd, Edinburgh, pp. 30-42.
Smith, A.H.V., 1973, Calculation of Micum 40 from petrographic data based on 250 kg test oven results; Communication to the International Committee for Coal Petrology.

Smith, W. H., Roux, H. J., and Steyn, J.G.D., (in press), The classification of coal macerals and their relation to certain chemical and physical parameters of coal; Proceedings of International Conference on Applied Mineralogy in the Mineral Industry, Johannesburg, South Africa, June 1981, sponsored by South African National Institute for Metallurgy and Geological Society of South Africa.

Smyth, M., 1966, A siderite-pyrite association in Australian coals; Fuel, vol. 45, pp. 221-231.

Snowdon, L. and Powell, T., 1982, Immature oil and condensate - modification of hydrocarbon generation model for terrestrial organic matter; Bulletin of American Association of Petroleum Geologists, vol. 66, pp. 775-788.

Solomon, M., 1963, Counting and sampling errors in model analysis; Journal of Petrology, vol. 4, pp. 367-382.

Sontag, E. and Suss, M., 1976, Zur makropetrographischen Analyse von Weichbraunkohlen Teil I: Erfahrungen und Probleme der praktischen Anwendung; Teil II Empfehlung eines Internationalen Klassifikation systems fur Weichbraunkohlen, Z Angew. Geol. 12, pp. 222-226, 313-319.

Spackman, W., 1958, The maceral concept and the study of modern environments as a means of understanding the nature of coal; Transactions, New York Academic Scientific Series II, vol. 20, no. 5, pp. 411-423.

Spackman, W., Cohen, A. D., Given, P. H., and Casagrande, D. J., 1974, The comparative study of the Okefenokee Swamp and the Everglades-mangrove-swamp-marsh complex of southern Florida; Geological Society of America Field Guidebook, Trip No. 16, November 1974, 265 pp.

Stach, E., 1927, Der Kohlenreliefschliff, ein neues Hilfsmittel fur die angewandte Kohlenpetrographie; Mitt. Abt. Gestein-Erz-, Kohle- und Salzuntersuchungen, vol. 2, pp. 75-94.

Stach, E., 1935, Lehrbuch der Kohlenpetrographie; Borntraeger, Berlin, 293 pp.

Stach, E., 1982a, The microscopically recognizable constituents of coal; *in* Stach, E., Mackowsky, M.-Th., Teichmüller, M., Taylor, G. H., Chandra, D., and Teichmüller, E., editors, *Coal Petrology*; Gebruder Borntraeger, Berlin-Stuttgart, pp. 87-139.

Stach, E., 1982b, The lithotypes of humic and sapropelic coals; *in* Stach, E., Mackowsky, M.-Th., Teichmüller, M., Taylor, G. H., Chandra, D., and Teichmüller, E., editors, *Coal Petrology*; Gebruder Borntraeger, Berlin-Stuttgart, pp. 171-177.

Stach, E., Mackowsky, M.-Th., Teichmüller, M., Taylor, G. H., Chandra, D., and Teichmüller, R., 1975, Coal petrology, 2nd edition, 428 pp.; 1982, 3rd edition, 535 pp., Gebruder Borntraeger, Berlin-Stuttgart.

Staplin, F. L., 1969, Sedimentary organic matter, organic metamorphism, and oil and gas occurrences; Canadian Society of Petroleum Geologists Bulletin, vol. 17, pp. 47-56.

Staub, J. R. and Cohen, A. D., 1979, The Snuggedy Swamp of South Carolina: A back barrier estuarine coal forming environment; Journal of Sedimentary Petrology, vol. 49, pp. 133-144.

Steyn, J.G.D. and Smith, W. H., 1977, Coal petrography in the evaluation of South African coals; Reprint from Coal, Gold and Base Metals, pp. 107-117.

Stone, I. J. and Cook, A. C., 1979, The influence of some tectonic structures upon vitrinite reflectance; Journal of Geology, vol. 87, pp. 497-508.

Stopes, M. C., 1919, On the four visible ingredients in banded bituminous coals; Proceedings of the Royal Society, B, vol. 90, pp. 470-487.

Stopes, M. C., 1935, On the petrology of banded bituminous coals; Fuel, vol. 14, pp. 4-13.

Stopes, M. C. and Wheeler, R. V., 1923, The spontaneous combustion of coal in relation to its composition and structure; Fuel, vol. 2, pp. 29-41, 83-92, 122-132.

Strauss, P. G., Russell, N. J., Bennett, A.J.R., and Atkinston, C. M., 1976, Coal petrography as an exploration aid in the west cirum-Pacific; *in* Muir, W.L.G., editor, *Coal Exploration*; Miller Freeman Publications, pp. 401-443.

Styan, W. B., 1981, The sedimentology, petrography and geochemistry of some Fraser Delta peat deposits; Unpublished M.Sc. Thesis, University of British Columbia, 188 pp.

Styan, W. B. and Bustin, R. M., (1983a), Petrography of some Fraser Delta peat deposits: Coal maceral and microlithotype precursors in temperate-climate peats; International Journal of Coal Geology, vol. 2, pp. 321-370.

Styan, W. B. and Bustin, R. M., (1983b), Sedimentology of Fraser River Delta peat: A modern analogue for some ancient deltaic coals; International Journal of Coal Geology, vol. 3, pp. 101-143.

Tasch, K. H., 1960, Die Moglichkeiten der Flozgleichstellung unter Zuhilfenahme von Flozbildungsdiagrammen; Bergbau-Rdsch., vol. 12, pp. 153-157.

Taylor, G. H., 1979, Liquid fuels from Canadian coals; CANMET Report 79-13, 43 pp.

Taylor, G. H., Mackowsky, M-Th., and Alpern, B., 1967, The behaviour of inertinite during carbonization; Fuel, vol. 46, pp. 431-440.

Taylor, G. H. and Warne, S. St. J., 1960, Some Australian coal petrological studies and their geological implications; Proceedings of International Commission of Coal Petrology, no. 3, pp. 75-83.

Teichmüller, M., 1962, Die Genese der Kohle; Compte Rendu 4th Congress Stratigraphy Geology Carboniferous, Heerlen, 1958, vol. 3, pp. 699-722.

Teichmüller, M., 1966, Geological causes of coalification; Coal Science, Advances in Chemistry Series 55, pp. 133-155.

Teichmüller, M., 1968, Zur Petrographie und Diagenese eines fast 200 m machtigen Torfprofils (mit Ubergangen zur Weichbraunkohle?) im Quartar von Philippi (Mazedonien); Geol. Mitt., vol. 8, pp. 65-110.

Teichmüller, M., 1971, Anwendung kohlenpetrographischer Methoden bei der Erdol - und Erdgasprospektion; Erdol and Kohle, Bd. 24, pp. 69-76.

Teichmüller, M., 1974a, Entstehung und Veranderung bituminoser Substanzen in Kohlen in Beziehung zur Entstehung und Umwandlung des Erdols; Fortschr. Geol. Rheinld. und Westf., vol. 24, pp. 65-112.

Teichmüller, M., 1974b, Uber neue Macerale der Liptinit-Gruppe und die Entstehung des Micrinits; Fortschr. Geol. Rheinld. und Westf., vol. 24, pp. 37-64.

Teichmüller, M., 1979, Die Diagenese der kohligen Substanzen in den Gesteinen des Tertiars und Mesozoikums des mittleren Oberrhein-Grabens; Fortschr. Geol. Rheinld. und Westf., vol. 27, pp. 19-49.

Teichmüller, M., 1982a, Application of coal petrological methods in geology including oil and natural gas prospecting; *in* Stach, E., Mackowsky, M.-Th., Teichmüller, M., Taylor, G. H., Chandra, D., and Teichmüller, R., editors, *Coal Petrology, 3rd Edition*; Gebruder Borntraeger, Berlin-Stuttgart, pp. 381-413.

Teichmüller, M., 1982b, Origin of the petrographic constituents of coal; *in* Stach, E., Mackowsky, M.-Th., Teichmüller, M., Taylor, G. H., Chandra, D., and Teichmüller, R., editors, *Coal Petrology, 3rd Edition*; Gebruder Borntraeger, Berlin-Stuttgart, pp. 219-294.

Teichmüller, M. and Ottenjann, K., 1977, Art und Diagenese von Liptiniten und lipoiden Stoffen in einem Erdolmuttergestein auf Grund fluoreszenzmikroskopischer Untersuchungen; Erdol und Kohle, Bd. 30, pp. 387-398.

Teichmüller, M. and Teichmüller, R., 1982, The geological basis of coal formation; *in* Stach, E., Mackowsky, M.-Th., Teichmüller, M., Taylor, G. H., Chandra, D., and Teichmüller, R., editors, *Coal Petrology, 3rd Edition*; Gebruder Borntraeger, Berlin-Stuttgart, pp. 5-86.

Teichmüller, M. and Thomson, P. W., 1958, Vergleichende mikroskopische und chemische Untersuchungen der wichtigsten Fazies-Typen im Hauptfloz der niederrheinischen Braunkohle; Fortschr. Geol. Rheinld. und Westf. vol. 2, pp. 573-598.

Teichmüller, M. and Wolf, M., 1976, Application of fluorescence microscopy in coal petrology and oil exploration; Journal of Microscopy, vol. 109, Part 1, pp. 49-73.

Thiessen, R., 1920, Structure in Paleozoic bituminous coals; United States Bureau of Mines Bulletin 117, 296 pp.

Thiessen, R., 1925, Origin of the boghead coals; United States Geological Survey Professional Paper 132/1, pp. 121-135.

Thiessen, R. and Sprunk, G. C., 1934, The effect of heat on coal as revealed by the microscope; Fuel, vol. 13, pp. 116-125.

Thiessen, R. and Sprunk, G. C., 1936, The origin of the finely divided or granular opaque matter in splint coals; Fuel, vol. 15, pp. 304-315.

Thompson, R. R. and Benedict, L. G., 1974, Vitrinite reflectance as an indicator of coal metamorphism for coke making; *in* Dutcher, R. R., Hacquebard, P. A., Schopf, J. M., and Simon, J. A., editors, *Carbonaceous Materials as Indicators of Metamorphism*; Geological Society of America Special Paper 153, pp. 95-108.

Thompson, R. R. and Benedict, L. G., 1978, Coal composition and its influence on coke making; Paper given at 25th Anniversary Meeting of ICCP, Essen, Germany.

Thompson, R. R., Shigo, J. J. III, Benedict, L. G., and Aikman, R. P., 1965, The use of coal petrography at Bethlehem Steel Corporation; Paper presented at joint meeting Eastern and Western States Blast Furnace and Coke Oven Association, Pittsburgh, Pennsylvania, no. 1965.

Ting, F.T.C., 1977, Microscopical investigation of the transformation (diagenesis) from peat to lignite; Journal of Microscopy, vol. 109, pp. 75-83.

Ting, F.T.C., 1978, Petrographic techniques in coal analysis; *in* Karr, C. Jr., editor, *Analytical Methods for Coal and Coal Products*; Academic Press, vol. 1, pp. 3-26.

Tissot, B., Durand, B., Espitalie, J., and Combaz, A., 1974, Influence of nature and diagenesis of organic matter in formation of petroleum; Bulletin of American Association of Petroleum Geologists, vol. 58, no. 3, pp. 499-506.

Tissot, B. and Welte, D. H., 1978, Petroleum formation and occurrence; Springer-Verlag, New York, 538 pp.

Tompos, E., 1982, An application of coal petrography to coking coal; *in* Taweel, A. M., editor, *Coal Phoenix of the 80's*; Proceedings 64th CIC Coal Symposium, Halifax, Nova Scotia, pp. 9-13.

Vassoyevich, N. B., Korchagina, Yu. I., Lopatin, N. V., and Chernyshev, V. V., 1969, Glavnaya faza nefteobrazovaniya (The main phase of oil formation); Moskov. Univ. Vestnik, Ser. Geol., vol. 24, no. 6, pp. 3-27.

Vassoyevich, N. B., 1973, Osnovnyye zakonomernosti, kharakterizuyuschiye organicheskoye veshchestvo sovremennykh i iskopayemykh osadkov (Characterization of organic matter in recent and fossil sediments); *in* Prioroda organicheskogo veschestva sovremennykh i iskopayemykh osadkov; Akad. Nauk SSSR Otdeleniye Geologii, Geofiziki, Geikhimi, Kom. Osad. Porodam, Nauka Press, Moscow, pp. 11-59.

Wachowska, H. M., Nandi, B. N., and Montgomery, D. S., 1974, Oxidation studies on coking coal related to weathering. 4. Oxygen linkages influencing dilatometric properties and the effect of cleavage of ether linkages; Fuel, vol. 53, pp. 212-219.

Walker, P., Spackman, W., Given, P., White, E., Davis, A., and Jenkins, R., 1977, Characterization of mineral matter in coals and coal liquefaction residues; Pennsylvania State University, Reports EPRI 366-1 (1975), EPRI AF-417 Project 366-1. Wandless, A. M. and Macrae, J. C., 1934, The banded constituents of coals; Fuel, vol. 13, pp. 4-15.

Wanless, H. R., Baroffio, J. R., Trescott, P. C., 1969, Conditions of deposition of Pennsylvanian coal beds; *in* Dapples, C. and Hopkins, M. E., editors, *Environments of Coal Deposition*; Geological Society of America Special Paper 114, pp. 105-142.

Waples, D. W., 1980, Time and temperature in petroleum formation: Application of Lopatin's method to petroleum exploration; American Association of Petroleum Geologists Bulletin, vol. 64, pp. 916-929.

Waters, J. G., Ortuglio, C., and Wolfson, D. E., 1971, Prediction of coke strength of American coals by the Ruhr dilatometer method; United States Bureau of Mines, RI 7564, 17 pp.

Weimer, R. J., 1977, Stratigraphy and tectonics of western coals; *in* Murray, D. K., editor, *Geology of Rocky Mountain Coals*; Colorado Geological Survey Series 1, pp. 9-27.

Welte, D. H., Yukler, A., 1980, Evolution of sedimentary basins from the standpoint of petroleum origin and accumulation — an approach for a quantitative basin study; Organic Geochemistry, vol. 2, pp. 1-8.

White, C. D., 1915, Some relations in origin between coal and petroleum; Washington Academic Scientific Journal, vol. 5, pp. 189-212.

Whitehurst, D., Mitchell, T., and Farcasiu, M., 1980, Coal liquefaction; Academic Press, New York, 378 pp.

Wollrab, V. and Streibl, M., 1969, Earth waxes, peat, montan wax, and other organic brown coal constituents; *in* Eglinton, G. and Murphy, M.T.J., editors, *Organic Geochemistry*; Springer-Verlag, New York, pp. 576-598.

Yorath, C. J. and Hyndman, R. D., 1983, Subsidence and thermal history of the Queen Charlotte Basin; Canadian Journal of Earth Sciences, vol. 20, pp. 135-159.

Zimmerman, R. E., 1979, Evaluating and testing the coking properties of coal; Miller Freeman Publications, San Francisco, 114 pp.